Manipulation and Expression
of Genes in Eukaryotes

Manipulation and Expression
of Genes in Eukaryotes

Proceedings of an International Conference,
held in conjunction with the 12th International
Congress of Biochemistry, at Monash University,
9–13 August 1982

Edited by

Phillip Nagley
Department of Biochemistry
Monash University

Anthony W. Linnane
Department of Biochemistry
Monash University

W. J. Peacock
Division of Plant Industry
Commonwealth Scientific and Industrial Research Organization

J. A. Pateman
Department of Genetics
Research School of Biological Sciences
Australian National University

ACADEMIC PRESS
A Subsidiary of Harcourt Brace Jovanovich, Publishers
Sydney   New York   London
Paris   San Diego   San Francisco   São Paulo   Tokyo   Toronto
1983

ACADEMIC PRESS AUSTRALIA
Centrecourt, 25–27 Paul Street North
North Ryde, N.S.W. 2113

United States Edition published by
ACADEMIC PRESS INC.
111 Fifth Avenue
New York, New York 10003

United Kingdom Edition published by
ACADEMIC PRESS, INC. (LONDON) LTD.
24/28 Oval Road, London NW1 7DX

Printed in Australia

National Library of Australia Cataloguing-in-Publication Data

Manipulation and expression of genes in eukaryotes.

   Bibliography.
   Includes index.

   International Conference on Manipulation and Expression of
Genes in Eukaryotes (1982: Monash University).
ISBN 0 12 513780 x.

   1. Genetic transcription—Congresses.
   2. Eukaryotic cells—Congresses. I. Nagley,
Phillip. II. Title.

574.8'7322

Library of Congress Catalog Card Number: 82-73671

Academic Press Rapid Manuscript Reproduction

Contents

Part III   Simple Eukaryotes

Contributors

Numbers in parentheses indicate the pages on which the authors' contributions begin.

S. M. Abmayr (107), Department of Biology, Washington University, St. Louis, Missouri, USA

Jan Abramczuk[1] (355), The Wistar Institute, Philadelphia, Pennsylvania, USA

Jerry M. Adams (29, 33, 41), The Walter and Eliza Hall Institute of Medical Research, Melbourne, Victoria, Australia

Michael J. Adang (203), Agrigenetics Research Park, 5649 East Buckeye Road, Madison, Wisconsin, USA

Etienne Agsteribbe (313), Laboratory of Physiological Chemistry, State University Medical School, Groningen, The Netherlands

W. Michael Ainley (269), Division of Molecular Biology, Department of Biochemistry, The University of Texas Health Science Center at Dallas, Dallas, Texas, USA

S. M. Archer (55), Department of Microbiology, University of Toronto, Toronto, Ontario, Canada

A. R. Baker (11), Department of Physiology, The University of Sydney, NSW, Australia

Giuseppe Baldacci (279), Laboratoire de Génétique Moléculaire, Institut de Recherche en Biologie Moléculaire, Paris, France

John Bedbrook[2] (229), Division of Plant Industry, CSIRO, Canberra, ACT, Australia

A. R. Bellamy (371), Department of Cell Biology, University of Auckland, Auckland, New Zealand

Alan J. D. Bellett (365), John Curtin School of Medical Research, Australian National University, Canberra, ACT, Australia

Ora Bernard (29, 33, 41), The Walter and Eliza Hall Institute of Medical Research, Melbourne, Victoria, Australia

Giorgio Bernardi (279), Laboratoire de Génétique Moléculaire, Institut de Recherche en Biologie Moléculaire, Paris, France

Christopher G. Bingham (291), Department of Biochemistry, Monash University, Clayton, Victoria, Australia

G. W. Both (357, 371), Molecular and Cellular Biology Unit, CSIRO, Delhi Rd, North Ryde, NSW, Australia

Warwick Bottomley (247), Division of Plant Industry, CSIRO, Canberra, ACT, Australia

Marc Boutry (151), Department of Biochemistry, University of Texas Health Science Center, San Antonio, Texas, USA

A. Boyd (29), The Walter and Eliza Hall Institute of Medical Research, Melbourne, Victoria, Australia

Antony W. Braithwaite[3] (365), John Curtin School of Medical Research, Australian National University, Canberra, ACT, Australia

Leslie Burnett (347), Department of Clinical Biochemistry, Royal Prince Alfred Hospital, Camperdown, NSW, Australia

Ronald A. Butow (269), Division of Molecular Biology, Department of Biochemistry, The University of Texas Health Science Center at Dallas, Dallas, Texas, USA

G. R. Cam (11), Department of Physiology, The University of Sydney, NSW, Australia

P. Cantatore (325), Dipartimento di Biologia Cellulare, Università della Calabria, Cosenza, Italy

A. Caputo (25), School of Biochemistry, University of New South Wales, Kensington, NSW, Australia

D. F. Catanzaro (11), Department of Physiology, The University of Sydney, NSW, Australia

A. K. Chakravorty (237), Department of Biochemistry, University of Queensland, St. Lucia, Queensland, Australia

V. L. Chan (55), Department of Microbiology, University of Toronto, Toronto, Ontario, Canada

Brian F. Cheetham[4] (365), John Curtin School of Medical Research, Australian National University, Canberra, ACT, Australia

G. D. Clark-Walker (159, 303), Department of Genetics, Research School of Biological Sciences, Australian National University, Canberra, ACT, Australia

John Coghlan (21), Howard Florey Institute of Experimental Physiology and Medicine, University of Melbourne, Parkville, Victoria, Australia

Leeanne S. Coles (73), Department of Biochemistry, University of Adelaide, South Australia, Australia

Yves Colin (279), Laboratoire de Génétique Moléculaire, Institut de Recherche en Biologie Moléculaire, Paris, France

Catherine M. Corrick (167), Department of Genetics, University of Melbourne, Parkville, Victoria, Australia

Suzanne Cory (29, 33, 41), The Walter and Eliza Hall Institute of Medical Research, Melbourne, Victoria, Australia

R. G. H. Cotton (27), Birth Defects Research Institute, Royal Children's Hospital, Parkville, Victoria, Australia

Ulrike Courage-Tebbe (221), Institut für Genetik, Universität zu Köln, Köln, FRG

Alan F. Cowman (33, 185), The Walter and Eliza Hall Institute of Medical Research, Melbourne, Victoria, Australia

Robert Crawford[5] (95), Department of Biochemistry, University of California, San Francisco, California, USA

Martha L. Crouch[6] (193), Department of Biology, University of California, Los Angeles, California, USA

Bob Dalgleish (185), Tick Fever Research Centre, Brisbane, Queensland, Australia

Richard D'Andrea (73), Department of Biochemistry, University of Adelaide, South Australia, Australia

Earl W. Davie (13), Department of Biochemistry, University of Washington, Seattle, Washington, USA

Ronald W. Davis (123), Department of Biochemistry, Stanford University School of Medicine, Stanford, California, USA

C. De Benedetto (325), Istituto di Chimica Biologica, Università di Bari e Centro Studio Mitocondri e Metabolismo Energetico CNR, Bari, Italy

Diane de Cicco (99), Department of Embryology, Carnegie Institution of Washington, Baltimore, Maryland, USA

Jenny C. de Jonge (313), Laboratory of Physiological Chemistry, State University Medical School, Groningen, The Netherlands

E. S. Dennis (213), Division of Plant Industry, CSIRO, Canberra, ACT, Australia

Hans de Vries (313), Laboratory of Physiological Chemistry, State University Medical School, Groningen, The Netherlands

Miklos de Zamaroczy (279), Laboratoire de Génétique Moléculaire, Institut de Recherche en Biologie Moléculaire, Paris, France

Carol L. Dieckmann (141), Department of Biological Sciences, Columbia University, New York, N.Y., USA

Hans-Peter Döring (221), Institut für Genetik, Universität zu Köln, Köln, FRG

Michael G. Douglas (151), Department of Biochemistry, University of Texas Health Science Center, San Antonio, Texas, USA

C. H. Doy (171), Department of Genetics, Research School of Biological Sciences, Australian National University, Canberra, ACT, Australia

Pamela Dunsmuir (229), Division of Plant Industry, CSIRO, Canberra, ACT, Australia

S. C. R. Elgin (107), Department of Biology, Washington University, St. Louis, Missouri, USA

Duncan R. Ersland (203), Agrigenetics Research Park, 5649 East Buckeye Road, Madison, Wisconsin, USA

Bronwyn A. Evans (3), Centre for Recombinant DNA Research, and Department of Genetics, Australian National University, Canberra, ACT, Australia

Dale Fahey (59), Children's Medical Research Foundation, The University of Sydney, Camperdown, NSW, Australia

Shu-Chen Fang (369), Department of Biochemistry, National Yang-Ming Medical College, Taipei, Taiwan, China

Godeleine Faugeron-Fonty (279), Laboratoire de Génétique Moléculaire, Institut de Recherche en Biologie Moléculaire, Paris, France

Barbara K. Felber (335), Laboratory of Biochemistry, National Cancer Institute, National Institutes of Health, Bethesda, Maryland, USA

Miriam Fischer[7] (241), Department of Agriculture, New South Wales, Australia

Sandra J. Friezner Degen (13), Department of Biochemistry, University of Washington, Seattle, Washington, USA

G. Gadaleta (325), Istituto di Chimica Biologica, Università di Bari e Centro Studio Mitocondri e Metabolismo Energetico CNR, Bari, Italy

Cesira L. Galeotti[8] (159), Department of Genetics, Research School of Biological Sciences, Australian National University, Canberra, ACT, Australia

R. Gallerani (325), Dipartimento di Biologia Cellulare, Università della Calabria, Cosenza, Italy

Ann K. Ganesan (45), Department of Biological Sciences, Stanford University, Stanford, California, USA

Martin Geiser[9] (221), Institut für Genetik, Universität zu Köln, Köln, FRG

W. L. Gerlach (213), Division of Plant Industry, CSIRO, Canberra, ACT, Australia

Steven D. Gerondakis (29, 33, 41), The Walter and Eliza Hall Institute of Medical Research, Melbourne, Victoria, Australia

Dalip S. Gill (373), Adelaide University Centre for Gene Technology, Department of Biochemistry, University of Adelaide, South Australia, Australia

Robert B. Goldberg (193), Department of Biology, University of California, Los Angeles, California, USA

Karl H. J. Gordon (373), Adelaide University Centre for Gene Technology, Department of Biochemistry, University of Adelaide, South Australia, Australia

Allan R. Gould (373), Adelaide University Centre for Gene Technology, Department of Biochemistry, University of Adelaide, South Australia, Australia

Regina Goursot (279), Laboratoire de Génétique Moléculaire, Institut de Recherche en Biologie Moléculaire, Paris, France

René Goursot (279), Laboratoire de Génétique Moléculaire, Institut de Recherche en Biologie Moléculaire, Paris, France

Keith Gregg (65), Department of Biochemistry, University of Adelaide, South Australia, Australia

A. Grimes (27), Birth Defects Research Institute, Royal Children's Hospital, Parkville, Victoria, Australia

Lawrence I. Grossman (269), Department of Cellular and Molecular Biology, Division of Biological Sciences, The University of Michigan, Ann Arbor, Michigan, USA

John B. Gurdon (83), MRC Laboratory of Molecular Biology, University Medical School, Hills Road, Cambridge, England

S. Guttman (55), Department of Microbiology, University of Toronto, Toronto, Ontario, Canada

Timothy C. Hall (203), Agrigenetics Research Park, 5649 East Buckeye Road, Madison, Wisconsin, USA

Dean H. Hamer (335), Laboratory of Biochemistry, National Cancer Institute, National Institutes of Health, Bethesda, Maryland, USA

Philip C. Hanawalt (45), Department of Biological Sciences, Stanford University, Stanford, California, USA

Alan W. Harris (29, 33), The Walter and Eliza Hall Institute of Medical Research, Melbourne, Victoria, Australia

Richard P. Harvey (73), Department of Biochemistry, University of Adelaide, South Australia, Australia

R. J. Hill (107), Molecular and Cellular Biology Unit, CSIRO, Delhi Rd, North Ryde, NSW, Australia

Peter Hobart (95), Department of Biochemistry, University of California, San Francisco, California, USA

Leslie M. Hoffman (203), Agrigenetics Research Park, 5649 East Buckeye Road, Madison, Wisconsin, USA

M. Holtrop (313, 325), Laboratory of Physiological Chemistry, State University Medical School, Groningen, The Netherlands

M. J. Howell (189), Department of Zoology, Australian National University, Canberra, ACT, Australia

Peter Hudson (21), Howard Florey Institute of Experimental Physiology and Medicine, University of Melbourne, Parkville, Victoria, Australia

Michael E. Hudspeth (269), Department of Cellular and Molecular Biology, Division of Biological Sciences, The University of Michigan, Ann Arbor, Michigan, USA

S. M. Hunt (27), Birth Defects Research Institute, Royal Children's Hospital, Parkville, Victoria, Australia

Alain Huyard (279), Laboratoire de Génétique Moléculaire, Institut de Recherche en Biologie Moléculaire, Paris, France

Michael J. Hynes (167), Department of Genetics, University of Melbourne, Parkville, Victoria, Australia

D. O. Irving (189), Department of Zoology, Australian National University, Canberra, ACT, Australia

Maliyakal E. John (91), Institut für Molekularbiologie und Biochemie, Freie Universität Berlin, Berlin, FRG

Mark Johnston (123), Department of Biochemistry, Stanford University School of Medicine, Stanford, California, USA

P. Juranka (55), Department of Microbiology, University of Toronto, Toronto, Ontario, Canada

Laura Kalfayan (99), Department of Embryology, Carnegie Institution of Washington, Baltimore, Maryland, USA

Heather Kane (171), Department of Genetics, Research School of Biological Sciences, Australian National University, Canberra, ACT, Australia

David J. Kemp (33, 185), The Walter and Eliza Hall Institute of Medical Research, Melbourne, Victoria, Australia

Caroline Le Van Kim (279), Laboratoire de Génétique Moléculaire, Institut de Recherche en Biologie Moléculaire, Paris, France

Julie A. King (167), Department of Genetics, University of Melbourne, Parkville, Victoria, Australia

Walter Knöchel (91), Institut für Molekularbiologie und Biochemie, Freie Universität Berlin, Berlin, FRG

Laurence Jay Korn (83), Department of Genetics, Stanford University School of Medicine, Stanford, California, USA

Paul A. Krieg (73), Department of Biochemistry, University of Adelaide, South Australia, Australia

A. M. Kroon (313, 325), Laboratory of Physiological Chemistry, State University Medical School, Groningen, The Netherlands

Satya Kunapuli (151), Department of Biochemistry, University of Texas Health Science Center, San Antonio, Texas, USA

Michael A. Kuziora (131), Department of Biochemistry, Baylor College of Medicine, Houston, Texas, USA

Eileen M. Lafer (107), Department of Biochemistry and Pharmacology, Tufts University, Boston, Massachusetts, USA

C. Lanave (325), Istituto di Chimica Biologica, Università di Bari e Centro Studio Mitocondri e Metabolismo Energetico CNR, Bari, Italy

Francisco J. S. Lara (113), Department of Biochemistry, Institute of Chemistry, University of São Paulo, São Paulo, Brazil

Philip J. Larkin (241), Division of Plant Industry, CSIRO, Canberra, ACT, Australia

John Langridge (241), Division of Plant Industry, CSIRO, Canberra, ACT, Australia

Yan-Hwa Wu Lee (369), Department of Biochemistry, National Yang-Ming Medical College, Taipei, Taiwan, China

Joseph Levine (99), Department of Embryology, Carnegie Institution of Washington, Baltimore, Maryland, USA

Tse-Jia Lieu (369), Department of Obstetrics and Gynecology, Department of Surgery, Veterans General Hospital, Taipei, Taiwan, China

Anthony W. Linnane (17, 257, 293), Department of Biochemistry, Monash University, Clayton, Victoria, Australia

D. Llewellyn (213), Division of Plant Industry, CSIRO, Canberra, ACT, Australia

H. Lörz (213, 241), Division of Plant Industry, CSIRO, Canberra, ACT, Australia

Ky Lowenhaupt (107), Department of Biology, Washington University, St. Louis, Missouri, USA

H. B. Lukins (257), Department of Biochemistry, Monash University, Clayton, Victoria, Australia

Eric McCairns (59), Children's Medical Research Foundation, University of Sydney, Camperdown, NSW, Australia

Ross T. A. MacGillivray[10] (13), Department of Biochemistry, University of Washington, Seattle, Washington, USA

Patricia McGraw (141), Department of Biological Sciences, Columbia University, New York, N.Y., USA

Ian F. C. McKenzie (61), Research Centre for Cancer and Transplantation, Department of Pathology, University of Melbourne, Parkville, Victoria, Australia

A. G. Mackinlay (25), School of Biochemistry, University of New South Wales, Kensington, NSW, Australia

Ian G. Macreadie (257), Department of Biochemistry, Monash University, Clayton, Victoria, Australia

Gabrielle L. McMullen (17), Department of Biochemistry, Monash University, Clayton, Victoria, Australia

Marguerite Mangin (279), Laboratoire de Génétique Moléculaire, Institut de Recherche en Biologie Moléculaire, Paris, France

Carl Mann (123), Department of Biochemistry, Stanford University School of Medicine, Stanford, California, USA

J. M. Manners (237), Department of Biochemistry, University of Queensland, St. Lucia, Queensland, Australia

Renzo Marotta (279), Laboratoire de Génétique Moléculaire, Institut de Recherche en Biologie Moléculaire, Paris, France

Anthony J. Mason (3), Centre for Recombinant DNA Research, and Department of Genetics, Australian National University, Canberra, ACT, Australia

Ronald J. Maxwell (257), Department of Biochemistry, Monash University, Clayton, Victoria, Australia

F. Meffe (55), Department of Microbiology, University of Toronto, Toronto, Ontario, Canada

J. F. B. Mercer (27), Birth Defects Research Institute, Royal Children's Hospital, Parkville, Victoria, Australia

Birgit A. Metz (179), Max-Planck-Institut für Biochemie, D–8033 Martinsried bei München, FRG

Wolfgang Meyerhof (91), Institut für Molekularbiologie und Biochemie, Freie Universität Berlin, Berlin, Germany

Claudia A. Mickelson (61), Research Centre for Cancer and Transplantation, Department of Pathology, University of Melbourne, Parkville, Victoria, Australia

Alex A. Moen (313), Laboratory of Physiological Chemistry, State University Medical School, Groningen, The Netherlands

Peter L. Molloy (65), Molecular and Cellular Biology Unit, CSIRO, Delhi Rd, North Ryde, NSW, Australia

G. Morahan (33), The Walter and Eliza Hall Institute of Medical Research, Melbourne, Victoria, Australia

B. J. Morris (11), Department of Physiology, The University of Sydney, NSW, Australia

M. R. Mott (107), Molecular and Cellular Biology Unit, CSIRO, Delhi Rd, North Ryde, NSW, Australia

Mark Murphy (17), Department of Biochemistry, Monash University, Clayton, Victoria, Australia

Michael J. Murray (203), Agrigenetics Research Park, 5649 East Buckeye Road, Madison, Wisconsin, USA

George E. O. Muscat (59), Children's Medical Research Foundation, The University of Sydney, Camperdown, NSW, Australia

Phillip Nagley (257, 291, 293), Department of Biochemistry, Monash University, Clayton, Victoria, Australia

Heung-Tat Ng (369), Department of Obstetrics and Gynecology, Department of Surgery, Veterans General Hospital, Taipei, Taiwan, China

Hugh Niall (21), Howard Florey Institute of Experimental Physiology and Medicine, University of Melbourne, Parkville, Victoria, Australia

Charles E. Novitski (257), Department of Biochemistry, Monash University, Clayton, Victoria, Australia

Jane Olsen (171), Department of Genetics, Research School of Biological Sciences, Australian National University, Canberra, ACT, Australia

Stuart H. Orkin (335), Division of Hematology and Oncology, Children's Hospital Medical Center and The Sidney Farber Cancer Institute, Department of Pediatrics, Harvard Medical School, Boston, Massachusetts, USA

Suki Parks (99), Department of Embryology, Carnegie Institution of Washington, Baltimore, Maryland, USA

J. A. Pateman (171), Department of Genetics, Research School of Biological Sciences, Australian National University, Canberra, ACT, Australia

W. J. Peacock (213), Division of Plant Industry, CSIRO, Canberra, ACT, Australia

Li Peng (365), John Curtin School of Medical Research, Australian National University, Canberra, ACT, Australia

Jenny Penschow (21), Howard Florey Institute of Experimental Physiology and Medicine, University of Melbourne, Parkville, Victoria, Australia

G. Pepe (325), Istituto di Chimica Biologica, Università di Bari e Centro Studio Mitocondri e Metabolismo Energetico CNR, Bari, Italy

J. V. Possingham (255), Division of Horticultural Research, CSIRO, Adelaide, South Australia, Australia

Jennifer Price (83), Department of Genetics, Stanford University School of Medicine, Stanford, California, USA

A. J. Pryor (213), Division of Plant Industry, CSIRO, Canberra, ACT, Australia

C. Quagliariello (325), Dipartimento di Biologia Cellulare, Università della Calabria, Cosenza, Italy

K. C. Reed (189), Department of Biochemistry, Australian National University, Canberra, ACT, Australia

Robert I. Richards (3), Centre for Recombinant DNA Research, and Department of Genetics, Australian National University, Canberra, ACT, Australia

Allan Robins (73), Department of Biochemistry, University of Adelaide, South Australia, Australia

George E. Rogers (65), Department of Biochemistry, University of Adelaide, South Australia, Australia

Peter B. Rowe (59), Children's Medical Research Foundation, The University of Sydney, Camperdown, NSW, Australia

W. J. Rutter (95), Department of Biochemistry, University of California, San Francisco, California, USA

C. Saccone (325), Istituto di Chimica Biologica, Universitá di Bari e Centro Studio Mitocondri e Metabolismo Energetico CNR, Bari, Italy

M. M. Sachs (213), Division of Plant Industry, CSIRO, Canberra, ACT, Australia

Jo Saltzgaber (151), Department of Biochemistry, University of Texas Health Science Center, San Antonio, Texas, USA

John Samallo (313), Laboratory of Physiological Chemistry, State University Medical School, Groningen, The Netherlands

E. Sbisà (325), Istituto di Chimica Biologica, Università di Bari e Centro Studio Mitocondri e Metabolismo Energetico CNR, Bari, Italy

Stewart Scherer[11] (123), Department of Biochemistry, Stanford University School of Medicine, Stanford, California, USA

K. J. Scott (237), Department of Biochemistry, University of Queensland, St. Lucia, Queensland, Australia

N. Steele Scott (255), Division of Horticultural Research, CSIRO, Adelaide, South Australia, Australia

William R. Scowcroft (241), Division of Plant Industry, CSIRO, Canberra, ACT, Australia

John Shine (3), Centre for Recombinant DNA Research, and Department of Genetics, Australian National University, Canberra, ACT, Australia

Agda M. Simpson (307), Biology Department and Molecular Biology Institute, University of California, Los Angeles, California, USA

Larry Simpson (307), Biology Department and Molecular Biology Institute, University of California, Los Angeles, California, USA

M. J. Sleigh (357), Molecular and Cellular Biology Unit, CSIRO, Delhi Rd, North Ryde, NSW, Australia

Jerry L. Slightom (203), Agrigenetics Research Park, 5649 East Buckeye Road, Madison, Wisconsin, USA

D. R. Smyth (239), Department of Genetics, Monash University, Clayton, Victoria, Australia

Terence W. Spithill[12] (307), Biology Department and Molecular Biology Institute, University of California, Los Angeles, California, USA

Graciela Spivak (45), Department of Biological Sciences, Stanford University, Stanford, California, USA

Allan C. Spradling (99), Department of Embryology, Carnegie Institution of Washington, Baltimore, Maryland, USA

K. S. Sriprakash (303), Department of Genetics, Research School of Biological Sciences, Australian National University, Canberra, ACT, Australia

Peter Starlinger (221), Institut für Genetik Universität zu Köln, Köln, FRG

Fred Stauder (21), Howard Florey Institute of Experimental Physiology and Medicine, University of Melbourne, Parkville, Victoria, Australia

A. F. Stewart (25), School of Biochemistry, University of New South Wales, Kensington, NSW, Australia

B. David Stollar (107), Department of Biochemistry and Pharmacology, Tufts University, Boston, Massachusetts, USA

J. E. Street (371), Department of Cell Biology, University of Auckland, Auckland, New Zealand

Vivien R. Sutton (61), Research Centre for Cancer and Transplantation, Department of Pathology, University of Melbourne, Parkville, Victoria, Australia

Robert H. Symons (373), Adelaide University Centre for Gene Technology, Department of Biochemistry, University of Adelaide, South Australia, Australia

Marjorie Thomas (123), Department of Biochemistry, Stanford University School of Medicine, Stanford, California, USA

A. R. Thompson (25), School of Biochemistry, University of New South Wales, Kensington, NSW, Australia

Edith Tillmann (221), Institut für Genetik, Universität zu Köln, Köln, FRG

Peter Timms (185), Tick Fever Research Centre, Brisbane, Queensland, Australia

Glen H. Tobias (61), Research Centre for Cancer and Transplantation, Department of Pathology, University of Melbourne, Parkville, Victoria, Australia

Brett M. Tyler (33), The Walter and Eliza Hall Institute of Medical Research, Melbourne, Victoria, Australia

Alexander Tzagoloff (141), Department of Biological Sciences, Columbia University, New York, N.Y., USA

A. Underwood (107), Molecular and Cellular Biology Unit, CSIRO, Delhi Rd, North Ryde, NSW, Australia

Peter Upcroft (351), Recombinant DNA Unit, Queensland Institute of Medical Research, Brisbane, Queensland, Australia

Peter van 't Sant (313), Laboratory of Physiological Chemistry, State University Medical School, Groningen, The Netherlands

Paul R. Vaughan (293), Department of Biochemistry, Monash University, Clayton, Victoria, Australia

Laurie von Kalm (239), Department of Genetics, Monash University, Clayton, Victoria, Australia

Salih J. Wakil (131), Department of Biochemistry, Baylor College of Medicine, Houston, Texas, USA

Barbara Wakimoto (99), Department of Embryology, Carnegie Institution of Washington, Baltimore, Maryland, USA

Linda Walling (193), Department of Biology, University of California, Los Angeles, California, USA

Thomas E. Ward[13] (179), Max-Planck-Institut fur Biochemie, D-8033 Martinsreid bei Munchen, FRG

F. Watt (107), Molecular and Cellular Biology Unit, CSIRO, Delhi Rd, North Ryde, NSW, Australia

Elizabeth A. Webb (29, 33, 41), The Walter and Eliza Hall Institute of Medical Research, Melbourne, Victoria, Australia

Ed Weck (221), Institut für Genetik, Universität zu Köln, Köln, FRG

Dennis L. Welker (179), Max-Planck-Institut für Biochemie, D-8033 Martinsried bei München, FRG

Julian R. E. Wells (73), Department of Biochemistry, University of Adelaide, South Australia, Australia

Wolfgang Werr (221), Institut für Genetik, Universität zu Köln, Köln, FRG

Paul R. Whitfeld (247), Division of Plant Industry, CSIRO, Canberra, ACT, Australia

Jennifer Whiting (73), Department of Biochemistry, University of Adelaide, South Australia, Australia

Keith L. Williams (179), Max-Planck-Institut für Biochemie, D-8033 Martinsried bei München, FRG

I. M. Willis (25), School of Biochemistry, University of New South Wales, Kensington, NSW, Australia

Stephen D. Wilton (65), Department of Biochemistry, University of Adelaide, South Australia, Australia

Sam W. Woo (293), Department of Biochemistry, Monash University, Clayton, Victoria, Australia

Graeme C. Woodrow (17), Department of Biochemistry, Monash University, Clayton, Victoria, Australia

H. Peter Zassenhaus (269), Division of Molecular Biology, Department of Biochemistry, The University of Texas Health Science Center at Dallas, Dallas, Texas, USA

Gerard Zurawski[14] (247), Division of Plant Industry, CSIRO, Canberra, ACT, Australia

Present Addresses:

1. John Curtin School of Medical Research, Australian National University, Canberra, ACT, Australia
2. Advanced Genetic Sciences Inc., P. O. Box 3266, Berkeley, California, USA

3. Research School of Biological Sciences, Australian National University, Canberra, ACT, Australia
4. University of California, Santa Barbara, California, USA
5. Howard Florey Institute, University of Melbourne, Parkville, Victoria, Australia
6. Indiana University, Bloomington, Indiana, USA
7. Division of Plant Industry, CSIRO, Canberra, ACT, Australia
8. Department de Biologie Moléculaire, Université de Genève, Genève, Switzerland
9. Friedrich Miescher Institut, Basel, Switzerland
10. Department of Biochemistry, University of British Columbia, Vancouver, B.C., Canada
11. California Institute of Technology, Pasadena, California, USA
12. The Walter and Eliza Hall Institute of Medical Research, Parkville, Victoria, Australia
13. University of Massachusetts, Worcester, Massachusetts, USA
14. DNAX Research Institute, Palo Alto, California, USA

Preface

The presence in Australia of many prominent scientists for the 12th International Congress of Biochemistry in August 1982 provided an opportunity to hold a satellite meeting on eukaryote molecular biology. The manipulation, structure and expression of genes was particularly emphasized at the satellite meeting. More than 200 participants attended this meeting, which was held at Monash University on 9–13 August 1982.

This international conference was neither too wide-ranging nor too highly specialized: the participants had a shared interest in eukaryote molecular biology and represented many different areas of biological research.

This book includes most of the papers presented at the conference. The chapters are arranged in six parts, each based on a particular group of organisms or genetic systems. Each part has a brief introduction that summarizes the main advances in, and problems associated with, research into the particular group.

The diversity of topics addressed in the sixty chapters—covering basic research into animals, plants, simple eukaryotes, viruses and organelles—emphasizes how modern molecular biology and recombinant DNA research can yield valuable information. This has led to the current world-wide interest in developing new biotechnologies which are based on molecular biology and which can be applied to significant industrial, medical and agricultural problems.

The organizers and editors of this volume gratefully acknowledge the sponsorship of the conference by the International Union of Biochemistry, the Australian Biochemical Society, the Commonwealth of Australia, Monash University, CSIRO, the Australian National University, and a number of commercial organizations, from whom financial support was received. Appreciation is expressed to the scientific and technical staff of the Department of Biochemistry at Monash University for their willing assistance, and also to Monash University for making its extensive facilities available.

Part I
Mammals and Birds

Introduction

Current research on the genes of higher eukaryotes, especially mammalian genes, is focussed on two aspects—the sequence, structure and arrangement of genes in the genome, and the expression of genes during development.

The isolation of desired genes from higher eukaryotes frequently involves the following procedures. Initially, cDNA clones are prepared from specialized tissues that contain relatively abundant mRNA species for particular proteins. Following characterization of the cDNA clones, the chromosomal genes corresponding to particular mRNAs are identified by screening genomic DNA libraries prepared in lambda phage vectors. Chapter 4 describes the use of chemically synthesized oligonucleotide probes to screen directly a human genomic DNA library.

Several chapters illustrate that a high level of genotypic complexity often underlies metabolic function. The kallikrein gene family, which codes for a series of proteases that process individual polypeptide hormones (Chapter 1) and the avian genes that determine feather and scale keratins (Chapter 15) provide examples of gene evolution by duplication and subsequent divergence. The chicken histone gene family (Chapter 16) should be useful in future studies of gene expression in development and for furthering our understanding of chromatin structure. This analysis of histone genes shows that the conservation of histones and of chromatin structure in a wide variety of organisms is not paralleled by a common genomic topography of the histone genes themselves.

The immunoglobulin genes represent a unique system in which the expression of the genetic information is manifested through physical rearrangements of the chromosomal DNA. Some chapters deal with aspects of gene splicing and gene expression in this and related systems. Chapter 9 provides evidence for a *trans*-acting factor with a specific cleavage function in the processing of transcripts derived from genes coding for immunoglobulin heavy chains, and Chapter 8 describes work on pre-B lymphoma cells which reveals some early events in immunoglobulin gene expression.

There are chapters on a diversity of other topics. Chapter 5 reports a method for analysing gene expression in endocrine tissues, where ^{125}I cDNA probes were used for high-resolution hybridization histochemistry. Chapter 13 suggests that post-transcriptional modification is of major importance in altered gene expression during transformation of human lymphocytes induced by phytohaemagglutinin. Chapter 11 illustrates the usefulness of gene transfer techniques in dealing with problems of molecular genetics in mammalian cells.

1

1

The Kallikrein Multi-Gene Family: A General Role in Prohormone Processing

John Shine, Anthony J. Mason, Bronwyn A. Evans
and Robert I. Richards

Centre for Recombinant DNA Research, and
Department of Genetics, Australian National
University, Canberra City, ACT.

Peptide hormones fall into two general categories according to their biosynthetic pathways. The prolactin, placental lactogen, growth hormone family (1,2,3) and the glycoprotein hormone family (4,5) are examples of those synthesized as prepeptides with a putative signal sequence which is cleaved off during transport across the membrane. This is usually the only peptide cleavage involved and the remaining sequence is the complete active peptide. There may be exceptions to this in particular cases, e.g. growth hormone where several proteolytically modified forms have been detected (6).

The second group of peptide hormones include examples such as ACTH, β-endorphin, several growth factors, insulin, relaxin, somatostatin, vasopressin, oxytocin, and calcitonin. Although these hormones are also translated with a pre-peptide which is lost during membrane transport, they contain peptide sequences additional to those needed for biological function. These are often referred to as "pro" sequences. Hormones in this pro-form are usually inactive (7). Further proteolytic cleavage is required to release the active hormone peptide and this is an essential process in the biosynthetic pathway of these types of hormones.

Prohormone peptides are processed by peptidases which

3

commonly recognize and cleave at basic amino acids. These
enzymes are trypsin-like but have a much narrower range of
specificity. In the few cases which have been extensively
characterised, the proteins are a highly homologous group,
as judged by amino acid analysis and immunological cross-
reactivity (8,9), and form a distinct subgroup of serine
proteases, closely resembling trypsin in their primary
structure. They have been grouped by their ability to cleave
synthetic esters of arginine (e.g. benzoyl arginine ethyl
ester, BAEE) and subsequently termed arginyl-esteropeptidases. Their *in vivo* substrate(s) is not, in many cases,
known (9). Studies on the epidermal growth factor (EGF) and
nerve growth factor (NGF) processing enzymes suggest that
each prohormone may have its own specific peptidase (9).
Further, in both of these cases the peptidase remains bound
to the hormone and therefore is required in stoichiometric
amount with the hormone molecule. This apparent substrate
specificity suggests that a family of proteases are synthesized in a co-ordinate, tissue-specific manner with their
particular prohormone substrate. For example, synthesis of
EGF must always be accompanied by the synthesis of EGF-
binding protein, as also occurs for NGF and its γ-subunit.
Other prohormone molecules are not as clearly defined, in
that they appear not to remain bound to their protease after
processing. However, very similar proteases appear to be
involved (e.g. proinsulin is activated *in vitro* by kallikrein) (10). Both EGF-binding protein and the γ-subunit of
NGF are synthesized in the submaxillary glands where they
are found as members of a homologous group of arginyl
esteropeptidases (8) expressed in this tissue. Other
characterized members of this group are NGF-endopeptidase
which has kallikrein activity (i.e. it catalyzes the release
of kallidin a potent vasodilator, from kininogen) and γ
-renin, (11), a serine protease which cleaves angiotensin
II, a potent vasoconstrictor, from angiotensinogen ("normal"
renin activity, as found in the kidney, is an acid
protease). For the sake of simplicity we have called this
homologous group of enzymes the KAE family as the first
coding sequence was identified by homology to pancreatic
kallikrein and most characterised members have some kallikrein activity.

Detailed studies on the biosynthesis of EGF and NGF (9)
have shown quite clearly that members of the KAE family are
involved in their processing. *In vitro* studies have demonstrated the potential of various members of the KAE family
to specifically and correctly cleave angiotensinogen,

prorenin, proinsulin, pro-opiomelanocortin and kininogen.
These observations, and the demonstration that the normal
processing cleavage site of many prohormones is composed of
basic amino acids, are consistent with a general involvement
of the KAE family in prohormone processing.

A major site of expression for this gene family is the
male mouse submaxillary gland. This gland produces a wide
range of hormones and growth factors, which is consistent
with the high level of expression of the KAE gene family in
this tissue (8).

The very similar chemical and physical properties of the
mouse submaxillary gland arginyl-esteropeptidases have
hindered their characterization to the extent that it has
been difficult to determine just how many of these genes are
expressed in this tissue. Heterogeneity observed by gel
electrophoresis can be due to protein modification (e.g.
proteolysis and glycosylation) as well as separate gene
products (12). Different members of this group have been
found to cross-react immunologically and this complicates
the use of immunological techniques. Purification has been
afforded by utilization of the substrate specificity and
binding properties of these enzymes (e.g. 7S nerve growth
factor contains two α subunits, as yet uncharacterized,
two β subunits, with neurite growth promoting activity and
two γ subunits which are the arginyl-esteropeptidase
processing enzymes specific for the β subunit.

In an attempt to characterize several genes expressed in
the male mouse submaxillary gland, in particular the
kallikrein-arginyl-esteropeptidases, we have constructed
cloned cDNA banks using mRNA isolated from this tissue.
Analysis of the clone banks by hyridization and nucleic acid
sequence analysis led to the identification of recombinant
plasmid pMK-1, (13). This plasmid contains cDNA coding for a
peptide sequence with homology to porcine pancreatic
kallikrein and three murine submaxillary gland arginyl
esteropeptidases, NGF-γ subunit, EGF-BP and γ-renin
(11,12,14,15). Although the protein encoded by the pMK-1
insert sequence is clearly a member of this enzyme family it
is not completely homologous with any of the known peptide
sequences.

To address the question of how these sequences are
distributed in the genome we used the KAE cDNA probe on DNA
isolated from Chinese Hamster/Mouse hybrid cell lines

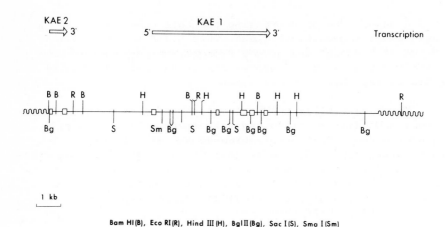

FIGURE 1.   *Molecular anatomy of arginyl esteropeptidase*
             genes

The organisation of the two genes contained in the
recombinant clone λ MSP-1 was initially determined by
restriction endonuclease digestion and Southern blotting and
confirmed by direct DNA sequencing. λMSP-1 was isolated
from a partial MboI library of mouse genomic DNA cloned in
bacteriophage λ charon 28.

constructed by Dr. David Cox (UCSF). These cell lines
contain all the Chinese Hamster chromosomes and various
assortments of the mouse chromosomes. Southern blot

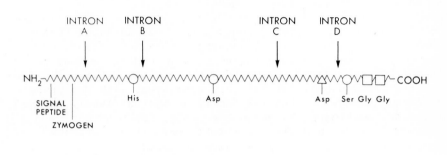

FIGURE 2. Position of introns in arginyl esteropeptidase genes in relation to the functional groups in the enzyme protein.

experiments indicate that all the cross-reacting sequences are located on chromosome 7 in the mouse, providing the first evidence that the members of this gene family are physically linked.

The second approach, that of isolation and character-ization of genomic DNA sequences, has shown that the genes are very closely linked. One clone of 13 kb of chromosomal DNA contained within it one complete gene and the 3' half of another (Fig. 1). We have sequenced most of these two genes and their flanking regions and have mapped the positions of introns or intervening sequences in the genes. As shown in Fig. 2, the introns interrupt the protein coding sequences at sites which separate the component amino acids of the active site of serine proteases. This suggests that the individual components of the active site may have been brought together on individual exons during the evolution of an ancestral serine protease gene.

Isolation of further genomic clones has revealed that gene linkage is common. DNA sequences obtained from some of the genes reveal homology in the protein coding regions of 75-85%. These figures, compared with homology between pMK-1 and the γ-NGF subunit of 77%, are consistent with the probe detecting members of the KAE family at the DNA level. We are therefore confident the genes for γ-NGF, EGF-binding protein, γ-renin and glandular kallikrein itself will be amongst the genomic clones we have isolated.

Hybridization histochemistry experiments on several different tissues have been performed by John Coghlan and Jenny Penschow at the Howard Florey Institute. Hybridization was localized only to those cells of the male mouse submaxillary gland and kidney which are known (by immuno-fluorescent methodology) to contain kallikreins, i.e. granular, tubular and striated duct cells in the salivary gland (data not shown) and the distal convoluted tubules in the kidney. We hope to use this technique on other tissues in order to determine the extent of expression of the KAE gene family.

Using recombinant DNA technology we hope to assign protease specificity to particular hormone or growth factor activation. This will involve the characterization of individual genes, their reintroduction into mammalian cells in culture, isolation of the expressed KAE protease and its identification by enzyme assays on various prohormone substrates. Further characterization using these techniques will allow a determination of which tissues express these genes and the nature of the correlation between the expression of a prohormone gene and that of a particular processing enzyme.

REFERENCES

1.   Seeburg, P.H., Shine, J., Martial, J.A., Baxter, J.D. and Goodman, H.M. (1977). *Nature* 270, 486.

2. Shine, J. Seeburg, P.H., Martial, J.A., Baxter, J.D. and Goodman, H.M. (1977). *Nature* 270, 494.
3. Cooke, N.E., Coite, D., Shine, J., Baxter, J.D. and Martial, J.A. (1981). *J. Biol. Chem.* 256, 4007.
4. Fiddes, J. and Goodman, H.M. (1979). *Nature* 281, 351.
5. Fiddes, J. and Goodman, H.M. (1980). *Nature* 286, 648.
6. Wallis, M. (1980). *Nature* 284, 512.
7. Chretien, M. and Seidah, N.G. (1981). *Molec. Cell. Biochem.* 34, 101.
8. Bothwell, M.A., Wilson, W.H. and Shooter, E.M. (1979). *J. Mol. Chem.* 254, 7287.
9. Server, A.C. and Shooter, E.M. (1976). *J. Biol. Chem.* 251, 165.
10. ole-MoiYoi, O., Pinkus G.S., Spragg, J. and Austen, K.F. (1979). *N. Engl. J. Med.* 300, 1289.
11. Bennett, C.D. and Poe, M. pers. comm.
12. Thomas, K.A., Baglan, N.C. and Bradshaw, R.A. (1981). *J. Biol. Chem.* 256, 9156.
13. Richards, R.I., Catanzaro, D.F., Mason, A.J., Morris, B.J., Baxter, J.D. and Shine, J. (1982). *J. Biol. Chem.* 257, 2758.
14. Tschesche, H., Mair, G., Godec, G., Fiedler, F., Ehret, W., Hirschauer, C., Lemon, M., Fritz, H., Schmidt-Kastner, G., and Kutzbach, C., in "Kinins II: Biochemistry, Patho-physiology and Clinical Aspects (Fujii, S., Moriya, H. and Suzuki, T. eds.), p. 245, Plenum Press, New York, (1979).
15. Silverman, R.E. (1977), PhD thesis, Washington University, St. Louis.

2
Immunoscreening of Expression Clones Using Antibodies to Renin, EGF and NGF

D. F. Catanzaro, A. R. Baker, G. R. Cam and B. J. Morris

Department of Physiology, The University of Sydney,
N.S.W., Australia

The mouse submandibular gland (MSG) expresses in abundance the genes coding for a number of physiologically important proteins including renin (2% of gland protein), epidermal growth factor (EGF)(0.1%), nerve growth factor (NGF)(0.3%) and several serine proteases, including kallikrein (5%). In order to determine the base sequence coding for these polypeptides we isolated MSG mRNA for use in construction of clone banks by recombinant DNA techniques. Base sequencing of cDNA from one colony chosen at random from the 500 colonies of a bank containing MSG cDNA at the $Hind$III site of pBR322 indicated a 500 base pair cDNA insert having 57% homology with the C-terminal 149 amino acids of pig kallikrein (Morris et $al.$, 1981; Richards et $al.$, 1982). Since then attempts have been made to identify bacterial colonies containing cDNA coding for renin, EGF and NGF. However, because amino acid sequence data for renin was not available and because approaches using synthetic oligonucleotide probes had proved unsuccessful in the identification of EGF and NGF cDNA, we decided to screen expression clones with antibodies directed at renin, EGF and NGF.

The in $situ$ immunoassay of Kemp and Cowman (1981) was used to screen a clone bank of MSG cDNA inserted at the PstI site of pBR322 (provided by P. Hudson of the Howard Florey Institute, Melbourne). Colonies were grown on nitrocellulose filters, lysed and polypeptide contents were bound to cyanogen-bromide activated filter papers. Affinity purified antirenin (1:75), antiEGF (1:150) or antiNGF (1:150) were applied and later localized with ^{125}I-labelled protein A from $Staphylococcus$ $aureus$ and autoradiography. Immunoreactivity was found in 2/500 colonies using antirenin and in 1/1000 colonies using antiEGF or antiNGF. Background was very low and the clear positive spots had a characteristic 'tail'.

MANIPULATION AND EXPRESSION
OF GENES IN EUKARYOTES
ISBN 0 12 513780 x

The intensity corresponded to 60, 1 and 1 fmol standard/10^6
cells for renin, EGF and NGF, respectively. Plasmid DNA was
isolated and agarose gel electrophoresis indicated cDNA
inserts of 700-1000 base pairs for the antirenin colony and
300-500 for the others. Using *Pst*I the cDNAs were cut out
and when used in hybrid-selected translation experiments
yielded polypeptides similar in size to the biosynthetic
precursors, in the case of renin and EGF. Renin cDNA
provided by J. Mullins of the University of Leicester
hybridized to renin-immunoreactive colonies. Base sequencing,
using 3',5'-dideoxy chain terminating inhibitors after subcloning in M13mp7.1 gave a sequence similar to a MSG renin
cDNA sequence published recently (Panthier *et al.*, 1982).
MSG renin cDNA probe was then used to screen a human genomic
clone bank and sequencing of positive colonies is in progress.
    The present study has therefore indicated the usefulness
of an immunoscreening procedure for the identification of
specific cDNAs in banks of expression clones.

REFERENCES

Kemp, D.M., and Cowman, A.F. (1981) *Proc. Natl Acad. Sci.
    USA 78*, 4520.
Morris, B.J., Catanzaro, D.F., Richards, R.I., Mason, A.,
    and Shine, J. (1981) *Clin. Sci. 61*, 351s.
Panthier, J.-J., Foote, S., Chambraud, B., Strosberg, A.D.,
    Corvol, P., and Rougeon, F. (1982) *Nature 298*,90.
Richards, R.I., Catanzaro, D.F., Mason, A.J., Morris, B.J.,
    Baxter, J.D., and Shine, J. (1982) *J. Biol. Chem. 257*,
    2758.

3

Structure of the Human Prothrombin Gene and a Related Gene[1]

Sandra J. Friezner Degen, Ross T. A. MacGillivray[2]
and Earl W. Davie

Department of Biochemistry
University of Washington
Seattle, Washington

Human prothrombin is a single-chain plasma glycoprotein (MW 70,000) composed of 579 amino acids. It is synthesized in the liver and then secreted into the plasma. During the coagulation process, prothrombin is converted to thrombin by factor Xa in the presence of factor Va, Ca^{++} and phospholipid. Upon conversion to thrombin, a large profragment is released from the amino-terminal region of the protein resulting in the formation of a two-chain molecule held together by a disulfide bond. Thrombin (MW 39,000) then converts fibrinogen to fibrin in the final stages of blood coagulation.

In previous work from our laboratory, we have described the isolation and sequence of a cDNA clone coding for bovine prothrombin mRNA (1). More recently, two human prothrombin cDNA clones have been isolated from a human liver cDNA library of 5000 clones. The cDNAs are 610 bp and 1288 bp in length. They have been mapped by restriction enzymes and sequenced. The 1288 bp cDNA codes for amino acids 200 to 579 and contains a non-coding region of 97 bp followed by a poly(A) stretch of 14 bp. Fourteen bp upstream from the poly(A) tail, there is a typical sequence of AATAAA which

[1]This work was supported in part by Grant HL 16919 from the National Institutes of Health.

[2]Present address: Department of Biochemistry, Univ. of British Columbia, Vancouver, B.C., Canada.

MANIPULATION AND EXPRESSION
OF GENES IN EUKARYOTES
ISBN 0 12 513780 x

may play a role in the processing of mRNA (2). When the
predicted amino acid sequence from the DNA sequence is compared with that by conventional protein chemistry (3-5), a
Glu-Glu dipeptide at positions 266 and 267 has been deleted.
A number of other differences have also been observed, particularly changes involving amide-acid assignments.

We have also screened a human genomic DNA library to
study the structure of the gene coding for human prothrombin.
The library was constructed from *Alu* I and *Hae* III digests
of human fetal liver DNA and was kindly provided by Dr. Tom
Maniatis (6, 7).

Twelve positive phage were isolated that hybridized with
the probe to varying degrees. Each has been partially characterized by restriction mapping and three of the phage have
been shown by DNA sequencing to contain coding sequences for
human prothrombin. One of the other strongly hybridizing
phage was found to code for a gene related to prothrombin in
the "kringle region" of the protein.

To date, about 5500 bp of DNA sequence has been obtained
for the human prothrombin gene. All sequencing has been
performed by the procedures of Maxam and Gilbert (8). Sixty-
four percent of the sequence has been obtained on both of
the strands and 91% has been sequenced two or more times.
This represents 63% of the coding region of the gene including amino acid residues 144 through 509. The coding
sequences for this part of the gene include six exons that
are interrupted by seven intervening sequences (Figure 1).
The intervening sequences occur between amino acid residues
143-144, 248-249, 291-292, 334-335, 390-391, 447-448, and
509-510 employing the numbering system for human prothrombin.
The first and second intervening sequences are on each side
of an exon that codes for the second kringle structure of
prothrombin. The placement of these intervening sequences
is identical with respect to each kringle and gives further
evidence that the structures are the result of a gene duplication event (9).

The DNA sequence of the coding region of the gene is in
excellent agreement with that obtained from the cDNA and
shows greater than 85% identity to the bovine cDNA for prothrombin (1). At the amino acid level, human prothrombin
is also 85% identical to the bovine molecule.

The intervening sequences vary in size from 84 bp to
over 1955 bp in length. Each has the traditional GT-AG
recognition sequence (10) for splicing on the 5' and 3' end,
respectively. The two largest intervening sequences have
been found to contain two copies each of *Alu* repetitive DNA
sequence. The first two *Alu* sequences occur in an exact

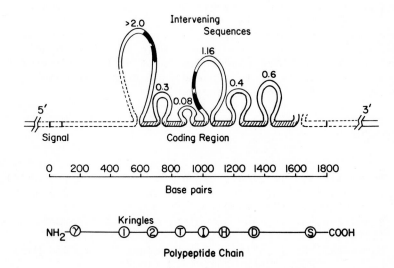

FIGURE 1. Schematic representation of the human prothrombin gene from the DNA sequence known at the present
time. The four Alu repetitive DNA sequences present in the
two largest intervening sequences are represented by solid
arrows. The six exons are hatched and the uncharacterized
portions of the gene are dashed. The coding region for
human prothrombin is approximately 1800 bp (by amino acid
sequence). The mature polypeptide chain is represented on
the bottom line indicating the active site amino acids (H =
His, D = Asp, and S = Ser), amino-terminal residues at the
activation sites (T = Thr and I = Ile), the two kringle
regions, and the ten Glu residues (γ) in the amino-terminal
region.

tandem repeat orientation with no extraneous DNA in between.
They are flanked on the extreme 5' and 3' ends by direct
repeats of 15 bp. The other two *Alu* repetitive DNA sequences are in an inverted repeat orientation with 141 bp
of DNA in between. These *Alu* sequences are also flanked by
direct repeats of 17 bp and 16 bp. All four sets of direct
repeats are not similar to any other reported direct repeat
sequences. When the *Alu* repetitive DNA sequences from the
human prothrombin gene are compared to a reported consensus
sequence for *Alu* repetitive DNA (11), there is 85% identity
and when compared to each other there is 78% identity.

The prothrombin related gene has been partially characterized and has been extensively mapped by restriction enzyme analysis and partially sequenced. It contains a sequence that codes for at least one kringle region, but it
is different from any of the prothrombin and plasminogen
kringles. Whether this gene codes for a protein containing
at least one kringle or is a pseudogene to prothrombin or
plasminogen is not known at present.

REFERENCES

1. MacGillivray, R. T. A., Degen, S. J. Friezner, Chandra,
    T., Woo, S. L. C., and Davie, E. W., *Proc. Natl. Acad.
    Sci. USA 77*, 5153 (1980).
2. Proudfoot, N. J., and Brownlee, G. G., *Nature 263*, 211
    (1976).
3. Butkowski, R. J., Elion, J., Downing, M. R., and Mann,
    K. G., *J. Biol. Chem. 252*, 4942 (1977).
4. Walz, D. A., Hewett-Emmett, D., and Seegers, W. H.,
    Proc. Natl. Acad. Sci. USA 74, 1969 (1977).
5. Thompson, A. R., Enfield, D. L., Ericsson, L. H.,
    Legaz, M. E., and Fenton, J. W., II, *Arch. Biochem.
    Biophys. 178*, 356 (1977).
6. Maniatis, T., Hardison, R. C., Lacy, E., Lauer, J.,
    O'Connell, C., Quon, D., Simm, G. K., and Efstratiadis,
    A., *Cell 15*, 687 (1978).
7. Lawn, R. M., Fritsch, E. F., Parker, R. C., Blake, G.,
    and Maniatis, T., *Cell 15*, 1157 (1978).
8. Maxam, A., and Gilbert, W., *Methods Enzymol. 65*, 499
    (1980).
9. Hewett-Emmett, D., McCoy, L. E., Hassouna, H. I.,
    Reuterby, J., Walz, D. A., and Seegers, W. H., *Thromb.
    Res. 5*, 421 (1974).
10. Breathnach, R., Benoist, C., O'Hare, K., Gannon, F.,
    and Chambon, P., *Proc. Natl. Acad. Sci. USA 75*, 4853
    (1978).
11. Deininger, P. L., Jolly D. J., Rubin, C. M., Friedmann,
    T., and Schmid, C. W., *J. Mol. Biol. 151*, 17 (1981).

4

Direct Selection of Human Leukocyte Interferon Genes from a Genome Library Using Synthetic Oligonucleotides

Graeme C. Woodrow, Mark Murphy, Gabrielle L. McMullen
and Anthony W. Linnane

Department of Biochemistry, Monash University, Victoria,
Australia

Synthetic oligodeoxyribonucleotides are finding wide application in molecular biology. Radioactively end-labelled oligonucleotides as short as 11 bases have been used to directly screen cDNA libraries cloned in *E. coli* for the presence of specific genes. When used as specific primers with mRNA as template, oligonucleotides can direct the synthesis of long cDNA probes. Oligonucleotides have also been used to screen genome libraries of relatively simple organisms. However, the use of oligonucleotides to screen human genome libraries has not always been successful. As part of our work on the human interferons, we have employed a hexadecanucleotide (16-mer) to screen a lambda human genome library for leukocyte interferon genes. Since these genes lack introns, they can be used directly for insertion into expression vectors as the first step in the production of large amounts of a given interferon.

RESULTS

Synthesis of Oligonucleotides

We have used the synthesis procedure of Duckworth *et al.* (1981) to produce three oligonucleotides which are complementary to the 3'-terminal sequence of the known leukocyte interferon genes (Goeddel *et al.*, 1981; Streuli *et al.*, 1980). In three interferon genes there is a common sequence as follows:

5' ATTAAGGAGGAAGGAATGA 3'

In a fourth gene the sequence is the same except that the penultimate G at the 3'-end is replaced by an A. The final

17

MANIPULATION AND EXPRESSION
OF GENES IN EUKARYOTES
ISBN 0 12 513780 x

three nucleotides constitute the termination codon.   We synthesised the following oligonucleotides complementary to
these sequences, as follows:

16-mer	3' TAATTCCTCCTTCCTT 5'
19-mer-(a)	3' TAATTCCTCCTTCCTTACT 5'
19-mer-(b)	3' TAATTCCTCCTTCCTTATT 5'

The 16-mer is absolutely complementary to the 3'-terminal
sequence of the above four interferon genes, the 19-mer-(a)
is complementary to three of the genes while the 19-mer-(b)
is absolutely complementary to the remaining gene.   In
addition, each of these oligonucleotides is complementary to
at least six other interferon genes with mismatches ranging
between one and three.

Screening of Human Genome Library with 16-mer

The human genome has been cloned as a series of randomly
generated 20 Kbp DNA fragments in the coliphage lambda,
λCh4A (R.M. Lawn *et al.*,1978).   A sample of this library was
used to infect *Escherichia coli* on 15 cm plates to give
approximately 6,000 plaques per plate.   The phage were
transferred to nitrocellulose filters and the phage DNA
hybridised with the 16-mer, radioactively labelled at the 5'-
end, by the procedure of Benton and Davis (1977).   Of 70,000
plaques screened, four (designated F,J,K and X) showed weakly
positive hybridisation.   On a second round screen, this time
using the plaque transfer procedure of Woo (1979) each of
these clones produced a strongly positive signal when hybridised with the 16-mer.

Lysates of Clones Contain Interferon Activity

In order to determine whether the four clones contained
a complete interferon gene, lysates of *E. coli* infected with
each of the clones were tested for interferon activity.   A
10 ml lysate of *E. coli* was centrifuged to remove cell debris
and then passed through a small (0.2 ml) interferon antibody
affinity column.   This column consists of an interferon
monoclonal antibody (Secher and Burke, 1980) coupled to
Sepharose and is highly specific for leukocyte interferons.
Adsorbed and subsequently eluted material (concentrated
approx. 20-fold) was assayed in a standard interferon bio-
assay for its·ability to protect cells against viral infection.   Anti-viral activity was present in lysates of all four
clones (Table I).   In addition, the anti-viral activity
conferred by the material eluted from the antibody affinity

TABLE I. *Interferon Activity in Lysates of Clones*

Lambda clone	Interferon level in eluate of antibody column[a] (International Units/ml)
F	16
J	3
K	5
X	2

[a] *Interferon activity was measured as an inhibition of cytopathic effect of Semliki Forest virus infection of monolayer calf testis fibroblasts.*

column for clone K was completely neutralised by either a sheep leukocyte interferon antiserum (from Dr. K. Cantell, Finland) or a mouse monoclonal leukocyte interferon antibody produced in our laboratory, confirming that the anti-viral activity was indeed due to the presence of interferon.

Small Differences in Oligonucleotide Probes Can Distinguish between Different Clones

Phage DNA from each of the four clones was hybridised *in situ* at 28°C to each of the three oligonucleotides described above. As shown in Table II the clones could be differentiated into three classes on the basis of their hybridisation with the three oligonucleotides.

Purified DNA from each of the lambda clones was prepared, digested with a variety of restriction endonucleases, run on an agarose gel and transferred to either nitrocellulose or diazotised aminothiophenol paper for hybridisation with the

TABLE II. *Hybridisation pattern of clones with oligonucleotides[a]*

Oligonucleotide	Hybridisation to clone			
	F	J	K	X
16-mer	+	+	+	+
19-mer-(a)	+	+	+	-
19-mer-(b)	+	+	-	-

[a]*Hybridisations were carried out at 28°C using the procedure of Woo (1979).*

appropriate end-labelled oligonucleotides. Autoradiograms from these "Southern" experiments showed distinct and different patterns of hybridisation for each clone. This indicates that the four clones are indeed different from each other, confirming the hybridisation results obtained above. These experiments demonstrate that oligonucleotide probes with minor base differencescan be used to differentiate between related genes.

SUMMARY

We have demonstrated that a hexadecanucleotide is sufficiently long to select particular genes from a human genome library. In order to maximise the signal to noise ratio in screening a lambda library, the plaque screening procedure of Woo (1979) is recommended. Four different leukocyte interferon genes have so far been isolated in this work and nucleotide sequencing experiments are currently in progress to determine the structure of these genes.

ACKNOWLEDGEMENTS

We thank Dr. Kari Cantell for the gift of sheep anti-human leukocyte interferon serum and Dr. T. Maniatis for sending us a sample of his lambda human genome library. Throughout this work we were ably assisted by Tricia Shevenan, David Pye, Karen Pluck and Ian Nisbet.

REFERENCES

Benton, W.D., and Davis, R.W. (1977) *Science 196*, 180-182.
Duckworth, M.L., Gait, M.J., Goelet, P., Hong, G.F., Singh, M., and Titmas, R.C. (1981) *Nucleic Acids Res., 9*, 1691-1706.
Goeddel, D.V., Leung, D.W., Dull, T.J., Gross, M., Lawn, R.M., McCandliss, R., Seeburg, P.H., Ullrich, A., Yelverton, E., and Gray, P.W. (1981) *Nature 290*, 20-26.
Lawn, R.M., Fritsch, E.F., Parker, R.C., Blake, G., and Maniatis, T. (1978) *Cell 15*, 1157-1174.
Secher, D.S., and Burke, D.C. (1980) *Nature 285*, 446-450.
Streuli, M., Nagata, S., Weissmann, C. (1980) *Science 209*, 1343-1347.
Woo, S.L.C. (1979) *Methods in Enzymology, Vol. 68*, 389-395.

5

The Application of Hybridization Histochemistry for the Analysis of Gene Expression in Endocrine Tissues

Peter Hudson, Jenny Penschow, Fred Stauder, Hugh Niall
and John Coghlan

Howard Florey Institute of Experimental
Physiology and Medicine,
University of Melbourne, Parkville, Victoria

THE TECHNIQUE

A procedure, termed hybridization histochemistry, has
been developed to locate in specially prepared whole sections
of tissue, areas which contain specific mRNA populations [1].
Cell populations or tissue regions binding ^{32}P-cDNA probes
are identified by autoradiography. Hybridization histochemistry is thus similar in principle to the widely used
immunohistochemical procedures for protein localization.
An endorphin cDNA probe consistently labelled the
rat pituitary pars intermedia which is known to be particularly rich in the corresponding mRNA. Likewise the growth
hormone probe specifically labelled the anterior pituitary.
Control tissues were not labelled by either probe and
provides further evidence of the specificity of the procedure. These results, which are highly reproducible, indicate
that the mRNA species for endorphin and growth hormone are
present in whole sections of pituitary in a physical state
which leaves them accessible to cDNA probes.
One obvious application of this technique is in the
semi-quantitative analysis of gene expression, i.e., for
determining relative amounts of mRNA either within one
tissue or in different tissues. We have used a ^{32}P-labelled
cDNA probe containing structural gene sequences for rat
relaxin to evaluate the level of gene expression during pre-
nancy in the rat. Relaxin is synthesized in the ovaries

21

MANIPULATION AND EXPRESSION
OF GENES IN EUKARYOTES
ISBN 0 12 513780 x

and the mRNA levels are found to increase sharply following
the 7th-9th day of pregnancy (term is 21 days).

In an improvement in the methodology, we have used ^{3}H
and ^{125}I-labelled calcitonin cDNA probes to localize the
calcitonin secreting parafollicular (C-cells) in the rat
thyroid at higher resolution than was possible with ^{32}P-cDNA
probes. (Figure 1). These C-cells are detected only in the
central region of the thyroid lobe and occur either as isolated cells or in small clusters. This enables a comparison to be made between resolution of this <u>in situ</u> hybridization technique and the current immunohistochemical procedures to localize proteins.

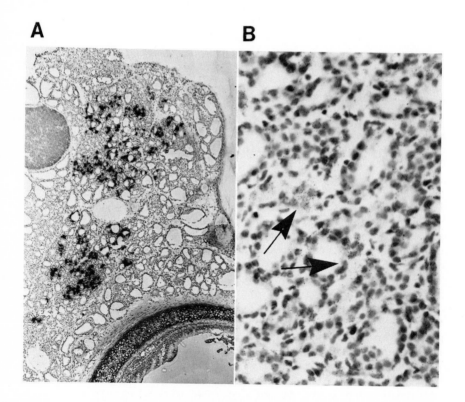

Figure 1.   *Stripping film autoradiographs of the rat thryroid.*
A) ^{32}P-calcitonin probe (magnification x 40)
B) ^{3}H-calcitonin probe (magnification x 240)

OUTLINE OF PROCEDURE

. Fresh tissues placed in mould of O.C.T. gel. (Lab-Tek
  products, Naperville, Ill, USA)
. Rapidly frozen in hexane/dry ice.
. Embedded in block of O.C.T.
. 5μ sections cut at -20°C (cryostat).
. Thawed onto unchilled gelatinized glass slides.
. Cooled to -40°C for 5 min.
. Fixed in Carnoy's or 4% glutaraldehyde solution.
. Pre-hybridized (50% Formamide; 5 x SSC, Denhardt's)
. Incubated at 40°C for 2 hours.
. Ethanol rinsed: air dried. (slides can be stored).
. ^{32}P-cDNA probe added in droplet of hybridization buffer.
. Sealed with cover slip and wax.
. Incubated in bags at 40°C for 1-4 days.
. Washed in limiting concentrations of SSC at 40°C.
. Rinsed in absolute ethanol: air dried.
. Autoradiographed (X-ray film or emulsions).
. Tissue stained (e.g. haematoxylin and eosin).

High Resolution Autoradiography

Coat slides with AR-10 stripping film (Kodak) by the
standard procedure (Kodak leaflet Nos. Pl. 1157 and SC-10)
with one exception - the water bath for floating of film is
filled with 1 x SSC.
Liquid emulsions are an alternative method to stripping
film. Pre-coating sections with 2% gelatine at 40°C assists
in obtaining an even emulsion layer, thereby minimising
background irregularities with high energy (^{32}p) isotopes.

cDNA probes and Hybridization Conditions

Radioactively labelled cDNA was prepared by random
primers under similar conditions to those described by
Taylor et al (Biochim Biophys ACTA 442, 324-330 (1976)). The
alternative procedure of nick-translation would be suitable
since both methods generate short cDNA copies of the double
stranded DNA excised from the plasmid.
The critical elements of hybridization (4 x SSC, 50%
formamide, 40°C, 48 hrs) appear to be optimal for our cDNA
probes which have been prepared from cloned DNA fragments of
150-3000 bp. Lower temperatures or decreasing formamide
concentration result in higher background labelling in
tissue.

Higher temperatures will cause the sections to fall off
during the washing procedures.

Specific hybridization appears to be essentially
complete after four days. Further incubation results in a
slightly enhanced signal, but at the expense of higher background labelling caused by evaporation under the cover slip.

Using probes prepared from DNA still attached to the
plasmid vector (pBR322) gave high backgrounds due to entrap-
ment of non-homologous bacterial DNA sequences in the tissue.
Probes made from M13 vectors are currently under evaluation.

The presence of tails (e.g. 20-30 bp poly dG.dC from
the cloning procedure; or longer poly-A tails from the mRNA)
definitely increases background labelling in some tissues.

CONCLUSIONS

The present work shows clearly that hybridization histochemistry, with further refinement, should have wide applicability in the localization and semi-quantitative analysis of
intracellular mRNA in whole frozen sections of tissue. The
technique is complementary to immunohistochemical procedures
in that the site of biosynthesis can be distinguished from
storage sites. Furthermore, there may be circumstances where
hybridization histochemistry will prove useful in clinical
diagnosis. Specific cDNA probes could be used to detect
either viral RNA/DNA or virus specific mRNA in infected
tissues.

ACKNOWLEDGMENTS

We are particularly grateful to Dr. John Shine (ANU)
for providing rat growth hormone and mouse endorphin cDNA
clones and Dr. Jack Jacobs (San Antonio, Texas) for the rat
calcitonin cDNA clone. Mr. John Walsh kindly provided the
rat ovaries and Dr. Larry Eddie the rat pituitaries.

REFERENCES

Hudson, P., Penschow, J., Shine, J., Ryan, G., Niall, H. and
    Coghlan, J. (1981) *Endocrinology, 108*, 353.

6

Characterisation of Cloned cDNAs for Bovine α_{s1}-, β- and κ-Caseins and β-Lactoglobulin

I. M. Willis, A. F. Stewart, A. Caputo, A. R. Thompson
and A. G. Mackinlay

School of Biochemistry, University of N.S.W., Australia

Bovine milk has six major protein components; four caseins α_{s1} (30% of total milk protein), α_{s2} (9%), β (30%) and κ (10%) and two whey proteins; β-lactoglobulin (9%) and α-lactalbumin (2%) (1). Several genetic variants of each of the caseins are known and segregation studies on the progeny of bulls heterozygous for these have indicated close linkage between the four genetic loci (2). β-lactoglobulin and α-lactalbumin are not implicated in this proposed gene family, however all are coordinately induced during lactogenesis as the result of multiple interactions of peptide and steroid hormones.

To examine the organisation of the casein genes and the mechanism of co-ordinate induction during lactogenesis, we have cloned cDNAs, made from polyA$^+$ RNA isolated from lactating bovine mammary gland, by self-priming of the second strand and G-C tailing. Clones corresponding to α_{s1}, β and κ caseins and β-lactoglobulin have been identified by comparison of DNA sequence with the known protein sequences.

α_{s1}- and β-caseins are of similar size (199 and 209 amino acids respectively) and the leader peptide sequences from ovine α_{s1}- and β-caseins are virtually identical (1). Both proteins contain a strongly conserved sequence of 8 amino acids which encompass a cluster of 4 serine residues. Apart from this only minor amino acid homologies between the two can be found. Our present data shows that the α_{s1}-casein mRNA is 1,140 nucleotides long with 40 nucleotides of 5' non-coding sequence and 460 3' non-coding sequence. For β-casein these figures are 1,100, 40 and 400. The similar size and organisation between these two mRNAs suggests an evolutionary or functional relationship. Blackburn et al., (3) have

MANIPULATION AND EXPRESSION
OF GENES IN EUKARYOTES
ISBN 0 12 513780 x

published the complete nucleotide sequence for rat β-casein cDNA. This sequence contains 1154 nucleotides which includes 696 nucleotides of coding sequence, 52 nucleotides of 5' non-coding and 406 nucleotides of 3' non-coding sequence. Comparisons between the rat β-casein cDNA sequence and our partial bovine β-casein cDNA sequence reveal approximately 60% homology both in the coding and 3' non-coding sequences.

The caseins are phosphorylated at specific serine residues in order to chelate Ca^{2+} for transport to the growing infant. The post-translational phosphorylation of the caseins is well documented and a mechanism has been proposed (4) for the sequential incorporation of all the phosphate moieties by an enzyme which recognizes either glutamic acid or serine phosphate at position n+2 relative to an available serine residue. However, Sharp and Stewart (5) found a minor bovine mammary gland seryl-tRNA that could be phosphorylated by crude bovine mammary gland protein extracts. This raises the possibility that some of the phosphorylated serines in the caseins may be incorporated during translation via a phosphoseryl tRNA. If this were the case it would be necessary for a specific codon to be reserved for recognition by the phosphoseryl-tRNA. We have sequenced 32 serine codons from our casein clones, 9 of which were for phosphoserine residues. Only one codon, UCG, codes for phosphoserine but not for non-phosphorylated serine residues. This codon occurs at only one position, amino acid 66 of α_{s1}-casein. This may be a reserved codon but is more likely to reflect the low usage of the CG dinucleotide found in eukaryotes.

The four bovine cDNA clones described above have been used to select bovine genomic fragments from a bovine Charon 30 library.

REFERENCES

1. Mercier, J-C., and Gaye, P., *Ann. N.Y. Acad. Sci. 343*, 232 (1980).

2. Grosclaude, F., Joudrier, P., and Mahe , M-F., *J. Dairy Res. 46*, 211 (1979).

3. Blackburn, D.E., Hobbs, A.A. and Rosen, J.M., *Nuc. Acid Res. 10*, 2295 (1982).

4. Mercier, J-C., Grosclaude, F. and Ribadeau-Dumas, B., *Eur. J. Biochem. 23*, 41 (1971)

5. Sharp, S.J. and Stewart, T.S., *Nuc. Acid Res. 4*, 2123, (1977).

7

Analysis of the Molecular Heterogeneity of Phenylalanine Hydroxylase by Cell-Free Translation of Rat Liver RNA

J. F. B. Mercer, S. M. Hunt, A. Grimes and R. G. H. Cotton

Birth Defects Research Institute, Royal Children's Hospital, Parkville, Victoria 3052, Australia

The enzyme phenylalanine hydroxylase (PH) catalyses the conversion of phenylalanine to tyrosine and is principally found in the liver of mammals. Analysis of the purified enzyme from human and monkey livers has shown that they contain multiple polypeptides that differ in both charge and molecular weight (Choo et al., 1979). The two molecular weight forms (higher and lower molecular weight, H- and L-forms respectively), have also been found with rat liver PH (Kaufman and Fisher, 1970). Evidence from previous studies favoured the hypothesis that the L-form derives from the H-form by proteolysis (Choo et al., 1979).

We have been investigating the molecular heterogeneity of phenylalanine hydroxylase produced by translation of rat liver RNA in a rabbit reticulocyte lysate. The PH produced by cell-free translation was specifically immunoprecipitated and analysed by one dimensional SDS/polyacrylamide gel electrophoresis. As expected from the proteolysis hypothesis for generation of the L-form, the RNA from adult male rat livers translated to yield a polypeptide with the same molecular weight as the H-form of PH. Surprisingly, when we examined the immunoprecipitated translation products of RNA from neonatal rats (days 10-15 of age), the PH comigrated with the L-form of the enzyme isolated from adult male rat livers. This suggests that two distinct messenger RNAs code for the two forms of PH.

To establish whether the neonatal rat livers contained a distinct form of PH corresponding to the smaller product detected by cell-free translation, we isolated the enzyme from livers of 12-15 day-old rats using the procedure of Shiman et al. (1979) which was developed for adult rat liver

27

MANIPULATION AND EXPRESSION
OF GENES IN EUKARYOTES
ISBN 0 12 513780 x

PH. The PH from the neonatal rat liver contained mainly
one molecular weight form when examined by SDS/
polyacrylamide gel electrophoresis and this had the same
mobility as the L-form of the protein isolated from the
adult. The neonatal and adult forms of PH are very similar
when examined by peptide mapping and are antigenically
similar suggesting that the proteins may have the same amino
acid sequence, but differ by a C- or N-terminal extension.
This situation has been reported in a number of gene systems,
with tissue specific levels of expression (Young et al.,
1981) or isozyme type (Noguchi et al., 1982) or with
secreted and non-secreted forms (Perlman et al., 1982;
Rogers et al., 1981).

Two-dimensional polyacrylamide gel analysis of the PH
produced by cell-free translation of rat liver RNA revealed
that it consisted of several differently-charged species.
The pattern of spots was found to be less complex than the
rat liver enzyme, but very similar to the polypeptides
isolated from rat hepatoma cells using a PH-specific
monoclonal antibody. The possible reasons for this charge
heterogeneity are at present being examined. PH is known
to be phosphorylated so this is one possibility, but
glycosylation can be disregarded since there is no evidence
for carbohydrate attachment to rat liver PH. Further work
will be needed to establish whether the charge heterogeneity
is entirely due to post-translational modification of the
protein formed in the reticulocyte lysate or whether
different genes code for some of the polypeptides.

REFERENCES

Choo, K.H., Cotton, R.G.H., Danks, D.M. and Jennings, I.G.
    (1979). *Biochem. J. 181*, 285-294.
Kaufman, S. and Fisher, D.B. (1970). *J. Biol. Chem. 245*,
    4745-4750.
Noguchi, T. and Tanaka, T. (1982). *J. Biol. Chem. 257*,
    1110-1113.
Perlman, D., Halvorson, H.O. and Cannon, L.E. (1982). *Proc.
    Natl. Acad. Sci. US 79*, 781-785.
Rogers, J., Early, O., Carte, G., Calame, K., Bond, M.,
    Wall, R. and Hood, L. (1980). *Cell 20*, 313-319.
Shiman, R., Gray, D.W. and Pater, A. (1979). *J. Biol. Chem.
    254*, 11300-11306.
Young, R.A., Hagenbüchle, O. and Schibler, U. (1981). *Cell
    23*, 451-458.

8

Heavy Chain Immunoglobulin Gene Rearrangement and Expression in Early B-Lymphocytes

Steven D. Gerondakis, A. Boyd, Ora Bernard, Suzanne Cory,
Alan W. Harris, Elizabeth A. Webb and Jerry M. Adams

The Walter and Eliza Hall Institute of Medical Research,
Melbourne, Australia

Immunoglobulin genes comprise multiple coding elements in the germline which rearrange during B-lymphocyte development to form a functional transcription unit. An assembled heavy (H) chain variable region sequence represents a fusion of separate variable (V_H), diversity (D_H) and joining (J_H) elements to form a contiguous V_H-D_H-J_H coding sequence. V_H-D_H-J_H sequences are first expressed as μ H-chain polypeptide in early B-cells, but in more mature cells, the V_H-D_H-J_H coding region can be switched to associate with other H-chain types by recombination events between the flanking sequences of different C_H genes (Adams & Cory, in press).

What is the relationship between rearrangement and the expression of V_H-D_H-J_H sequences? Does rearrangement automatically activate H-chain synthesis? To address the question of H-chain rearrangement and expression, virally transformed immature lymphocytes or "pre-B" cells appear to provide the best model. Although these cells show J_H rearrangement at both loci, a large proportion lack μ chain expression (Alt et al., 1981). The rearrangements could be incomplete D_H-J_H recombinations or complete V_H-D_H-J_H joins. If they are V_H-D_H-J_H rearrangements, is the absence of μ-chain expression due to a block in transcription or translation?

We have examined two virally induced pre-B lymphomas, ABLS 8 (A8) and Avrij 1.3 (A1), which are rearranged at both J_H alleles (Fig. 1), but do not synthesize μ-chains (Walker & Harris, 1980). Both rearranged loci from the two cell lines were molecularly cloned. Sequence analysis

MANIPULATION AND EXPRESSION
OF GENES IN EUKARYOTES
ISBN 0 12 513780 x

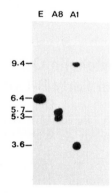

Fig. 1.   Rearrangement of J_H loci in the pre-B lymphomas A1 and A8.   DNA from embryos and the pre-B lymphomas A1 and A8 was cleaved with EcoRI and analyzed by the Southern procedure using the J_H probe.

in the vicinity of the J_H coding regions revealed that each rearrangement represented an assembled V_H-D_H-J_H.   In each line, V_H, D_H and J_H were in phase on one allele, but out of phase on the other, thereby excluding synthesis of a functional polypeptide.   Thus aberrant rearrangement on the second allele accounts for allelic exclusion, whereby only one allele is involved in H-chain synthesis.   As expected, different V_H genes have recombined at the two alleles in A1. Surprisingly, in A8 however the V_H genes on the two alleles are identical; each has recombined with different D_H and J_H sequences.   This observation argues strongly for a mechanism of random assortment of V_H, D_H and J_H sequences within a cell during the assembly process.

To determine if these V_H-D_H-J_H rearrangements are transcribed we performed RNA blot hybridization (Alwine et al., 1977) using as probes the cloned V_H-D_H-J_H sequences from the correctly joined alleles of both tumors.   Surprisingly, only trace amounts of 2.4 and 2.7 kb H-chain transcripts corresponding in size to the μ membrane and secreted mRNAs were detected (Fig.2).   These transcripts are in the order of 0.1 →0.2 molecules per cell and probably represent transcription within a very minor sub-population of cells.   These levels compare with approximately 20-50 molecules of μ mRNA per cell for a typical pre-B cell (Perry & Kelley, 1979). Although the aberrantly joined alleles would not be able to synthesize a functional polypeptide, they should nevertheless be capable of directing transcription at levels com-

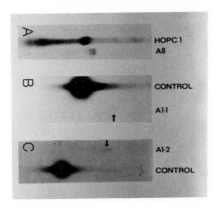

Fig. 2. Detection of mRNA species in A1 and A8 hybridizing to the correctly and aberrantly rearranged V_H-D_H-J_H
sequences from these cell lines. In A, the A8 track shows
hybridization to the membrane and secreted mRNA species,
with HOPC 1 mRNA being a positive control. In B and C, the
arrows indicate very faint hybridization to mRNA species in
A1 with the aberrantly (A1.1) and correctly joined (A1.2)
V_H-D_H-J_H alleles. Positive DNA hybridization controls for
both A1.1 and A1.2 are also shown in B and C respectively.

parable to the correctly rearranged allele, as found for
excluded kappa and lambda light chain genes (Alt et al.,
1980). However, RNA hybridization with the aberrantly
joined V_H-D_H-J_H sequences also revealed only trace amounts
of H-chain transcript, of the order seen for the correctly
joined sequences. Thus the absence of μ-chain expression in
A1 and A8 appears to be due to a block in transcription.
This transcription block could be due either to faulty V_H
promoters, or to genetic control of transcription. A
promoter lesion cannot be the explanation for the blocked
A1.2 allele because it has the same V_H sequence as that
expressed abundantly in plasmacytoma MOPC 141 (Sakano
et al., 1980).
    Somatic cell hybridization between B-cells and a plasma-
cytoma line greatly augments the expression of the B-cell
immunoglobulin genes. Based on this finding, we attempted
to amplify immunoglobulin gene expression in A1 and A8 by
fusion to the myeloma line NS1. No detectable induction of
immunoglobulin synthesis upon cell fusion was found, ruling
out the need for a trans-acting factor as an activation requirement.

The block in A1 and A8 appears to be different to that described for another pre-B line, 18-81, in which long term tissue culture abolishes μ-chain synthesis. While lipopolysaccharide (LPS) restores μ mRNA and protein synthesis in 18-81 (Alt et al., 1982), it had no effect on either A1 or A8.

We propose that the pre-B cells A1 and A8 represent a stage in early B-cell development at which V_H-D_H-J_H assembly has occurred, but subsequent event(s) are required to activate the transcription unit. Since cell fusion has ruled out the need for a trans-acting factor, the transcription block must be a cis-acting mechanism such as a change in methylation or chromatin packaging.

ACKNOWLEDGMENTS

We thank D. Kemp for helpful discussion. This work was supported by the N.H.&M.R.C. (Australia), the Drakensberg Trust and the NCI (USA) (CA29634 and CA12421).

REFERENCES

Adams, J., and Cory, S., in "Eukaryotic Genes: Their Structure, Activity and Regulation" (N. Maclean, S. Gregory, and R. Flavell, eds.). Butterworth, Lond.

Alt, F.W., Enea, V., Bothwell, A.L.M., and Baltimore, D. (1980). Cell 21, 1.

Alt, F.W., Rosenberg, N., Lewis, S., Thomas, E., and Baltimore, D. (1981). Cell 27, 391.

Alt, F.W., Rosenberg, N., Casonova, R.J., Thomas, E., and Baltimore, D. (1982). Nature 296, 325.

Alwine, J.C., Kemp, D.J., and Stark, G.R. (1977). Proc.Natl.Acad.Sci.USA 72, 3962.

Perry, R.P., and Kelley, D.E. (1979). Cell 18, 1333.

Sakano, H., Maki, R., Kurosawa, Y., Roeder, W., and Tonegawa, S. (1980). Nature 288, 676.

Walker, I.D., and Harris, A.W. (1980). Nature, 288, 2901.

9

Processing of Immunoglobulin Heavy Chain Gene Transcripts

David J. Kemp, Alan F. Cowman, Steven D. Gerondakis,
Brett M. Tyler, G. Morahan, Elizabeth A. Webb, Ora Bernard,
Alan W. Harris, Suzanne Cory and Jerry M. Adams

The Walter and Eliza Hall Institute of Medical Research
Melbourne, Australia

Functional expression of immunoglobulin heavy chain genes in a B lymphoid cell requires DNA rearrangements which assemble the V_H, D_H and J_H segments (reviewed in ref. 1). Later in development the B cell can "switch" the assembled V_H-D_H-J_H segment from the $C\mu$ gene to another of the C_H genes, arranged in the order $(V_H$-D_H-$J_H)$-$C\mu$-$C\delta$-$C\gamma3$-$C\gamma1$-$C\gamma2b$-$C\gamma2a$-$C\varepsilon$-$C\alpha$ (1). As well as these rearrangements, transcriptional regulation and RNA processing play vital roles in heavy chain gene expression, enabling the simultaneous production of multiple polypeptides bearing the same V region and modulating the levels of expression. We discuss transcription and processing here with emphasis on recent work from our own laboratory.

Alternative Heavy Chain mRNAs and Transcription Units

Transcripts of assembled μ genes are processed into two distinct μ mRNA species which differ at their 3' termini, encoding secreted (μS) and membrane-bound (μM) polypeptides (2-4). These and other transcripts discussed below are shown in Fig. 1. Recently we have demonstrated that γ heavy chain transcripts from switched chromosomes bear analogous S-and M-termini (5,6). We identified γM gene segments 3' to the $\gamma2a$ and $\gamma3$ genes (6) and similar segments have now been identified on $\gamma2b$ (7), δ (8) and α genes (9). Hence S and M processing pathways are probably a feature of all immunoglobulin heavy chain genes.

MANIPULATION AND EXPRESSION
OF GENES IN EUKARYOTES
ISBN 0 12 513780 x

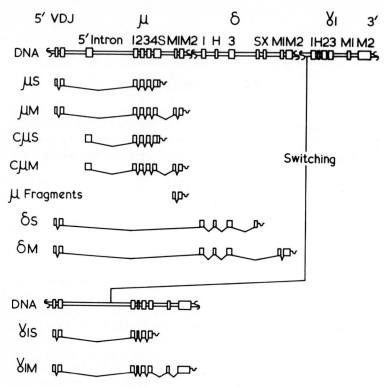

Fig. 1   Some polyA RNAs from the heavy chain locus. DNA molecules bearing a rearranged V_H-D_H-J_H sequence are shown before (top) and after switch rearrangement, RNA exons (boxes) are aligned below the DNA and splicing is indicated. Additional Cµ, Cγ and δ species also exist, as well as γ3, γ2a, γ2b, ε and α species.

As well as alternative processing pathways for a single gene, complex transcription units encode mRNAs for more than one heavy chain gene. While early B cells express only µ polypeptides, mature B cells simultaneously express µ and δ membrane receptors. The Cµ and Cδ genes comprise segments of a longer transcription unit (8,10) arranged (V_H-D_H-J_H) -Cµ-Cδ, which can simultaneously produce µS, µM, δS and δM mRNAs bearing the same V_H region (Fig. 1). The data suggest that in early B cells only the µ region is transcribed (11, and see below), while in later cells transcription procedes through both the µ and δ regions. Cells bearing both µ and

γ or μ and ε immunoglobulins have also been described (12), and so the entire C_H locus may constitute a giant transcription unit 180kb long, subject to intricate developmental control mechanisms.

Early Events in Transcriptional Activation of the Heavy Chain Locus.

Rearrangement is not the first event in activation of heavy chain gene transcription. We detected a transcriptionally competent state of the $C\mu$ gene which does not require J_H rearrangement in immature B lymphoid as well as in T lymphoid and myeloid cells (13,14). The chromatin of the J_H-$C\mu$ region is decondensed and the cells produce $C\mu$ RNA transcripts (13,15). The $C\mu$ RNAs (Fig. 1) are equivalent to μ mRNAs throughout the $C\mu$ region and each is terminated by μS or μM (16). However, the J_H loci are either unrearranged or bear incomplete D_H-J_H rearrangements; hence the $C\mu$ RNAs do not contain V_H regions and have no known function (18,19). Instead the $C\mu$1 exon is spliced to a segment of the J_H-$C\mu$ intron (17,18). Transcription of unrearranged $C\kappa$ genes is also common in plasmacytomas (20) but not in immature cells. There is no evidence for transcription of unrearranged V_H genes in any cell line (21).

Later in B cell development, V_H-D_H-J_H rearrangements permit expression of functional μ mRNAs. We conclude that transposition of a V_H gene into the competent J_H-$C\mu$ region is necessary to permit transcription from the V_H promoter region. Similar conclusions for activation of κ genes have been reached (22).

The Cδ Gene is not Activated Early in Development

Because the μ and δ genes comprise a complex transcription unit we investigated whether transcriptional competence of the J_H-$C\mu$ region extends to the adjacent $C\delta$ gene. We screened $C\mu$ RNA-producing lymphoid lines for δ RNA production (Fig. 2). No δ RNA was seen in any of these lines although 1 molecule per cell would have been detected. Hence the $C\delta$ gene is not activated at this early stage of lymphoid development. Presumably transcription from the $C\mu$ gene terminates 5' to the $C\delta$ gene in these cells, as in a B lymphoma (11).

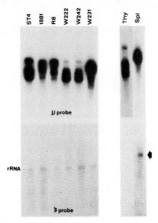

Fig. 2   A search for δ RNAs in lymphoid cells. RNA samples from T (ST4, W222, W242), pre-B (1881, RAW8), and B (W231) lymphoma, thymus (Thy) and spleen (Spl) cells were analysed by the Northern procedure (13) with μ and δ probes. The arrow indicates δ mRNA in spleen RNA.

Demethylation of DNA Accompanies Early Stages of J_H-Cμ Activation

The widespread correlation of DNA demethylation with transcription holds for immunoglobulin genes, at least in relatively mature cells. The δ and γ genes are methylated in μ producing cells while the μ gene is methylated in non-lymphoid tissue (11; but see 23). To compare DNA methylation in cells producing Cμ RNA with mature IgM-producing cells we examined their J_H-Cμ regions. Fig. 3 shows clearly that methyl groups in embryo DNA are not present in IgM-producing plasmacytoma HPC76. Similarly the DNA is demethylated on at least one allele of T lymphoma ST4 (Fig. 3) a Cμ RNA-producing line with incomplete rearrangements on both alleles (18). We conclude that demethylation accompanies early stages of transcriptional activation in the heavy chain locus.

Transcriptional Amplification and the Control of S:M Ratios in Mature Cells

On maturation of B cells to plasma cells, the abundance of μ mRNA and the relative abundance of μS:μM termini both increase by up to two orders of magnitude ("transcriptional

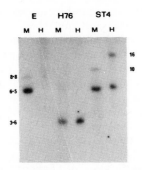

Fig. 3 Demethylation of the J_H-Cμ region. DNA from embryos, plasmacytoma HPC76 and T lymphoma ST4 cells was cleaved with MspI or HpaII as indicated and analysed by the Southern procedure, using the 5' intron probe (18).

amplification") (3,4,24). If the cell has switched to expression of, for example, the γ1 gene, a similar amplification of γ_1 transcription occurs upon maturation (6,7). In B lymphoma-plasmacytoma somatic hybrid cell lines, transcripts of functional μ (or γ) genes from the B lymphoma are present at the amplified level and are predominantly terminated by μS (or γS) (3,4,7,24). Hence plasmacytomas most likely provide trans-acting molecules which control transcription and determine the 3' processing path.

We examined the mechanisms of transcriptional amplification and control of S:M ratios by fusing T lymphoma cells to cells of plasmacytoma NS1, which contains no Cμ genes. In the fused lines an increase in the ratio of μS:μM terminated Cμ RNAs was observed, without a concomitant increase in the abundance of CμRNAs (25). Because the T lymphoma Cμ genes transcribed here lack V_H genes, we infer that the plasmacytoma factors which control transcriptional amplification act upon sequences within or 5' to V_H genes and hence cannot amplify Cμ RNAs. This contrasts with transcriptional competence, which is a function of the J_H-Cμ region and does not require V_H sequences.

Two models exist for production of μS and μM (3). A common precursor transcript may give rise to both μS and μM, specific cleavage of this precursor being required to generate μS. Alternatively, μS termini may result from termination of transcription 5' to the μM exons. In the T lymphoma-plasmacytoma hybrids the increased ratio of μS:μM terminated Cμ RNAs was accompanied by 0.47 kb μM fragments (Fig. 1) bearing only sequences from exons μM1 and μM2 (25). The μM

fragments were present in about equimolar ratios with the
μS-terminated Cμ RNAs, suggesting that they are a byproduct
of μS production, and terminated near the M1 splice boundary
(25). This is consistent with cleavage of a common precursor
to produce μS but not with the termination model.

Hence we conclude that control of μS:μM ratios (and presumably S:M ratios of all heavy chain RNAs) is mediated by a
trans-acting factor which specifically cleaves a common precursor to the S and M species (25). In contrast, the control
of δ production in B cells is mediated by termination of
transcription 5' to the δ exons (11). It is therefore
probable that at least two different general mechanisms
operate to control production of various mRNAs from this
complex transcription unit.

ACKNOWLEDGMENTS

We thank K. Easton, J. Mitchell, P. Bruglieri and
S. Carson for technical assistance. This work was supported
by the N.H.&M.R.C. and the Drakensberg Trust (Australia),
the NCI (USA) (CA29634 and CA12421).

REFERENCES

1. Adams, J., and Cory, S., in "Eukaryotic Genes: Their
    Structure, Activity and Regulation" (N. Maclean,
    S. Gregory, and R. Flavell, eds.). Butterworth, Lond.
2. Rogers, J., Early, P., Carter, C., Calame, K.,
    Bond, M., Hood, L., and Wall, R. Cell 20, 303-312
    (1980).
3. Early, P., Rogers, J., Davis, M., Calame, K.,
    Bond, M., Wall, R., and Hood, L. Cell 20, 313-319
    (1980).
4. Alt, F., Bothwell, A., Knapp, M., Siden, E.,
    Mather, E., Koshland, M., and Baltimore, D. Cell 20,
    293-301 (1980).
5. Tyler, B., Cowman, A., Adams, J., and Harris, A.
    Nature 293, 406-408 (1981).
6. Tyler, B., Cowman, A., Gerondakis, S., Adams, J., and
    Bernard, O. Proc. Natl. Acad. Sci. USA 79, 2008-2012
    (1982).

7. Rogers, J., Choi, E., Souza, L., Carter, C., Word, C., Kuehl, M., Eisenberg, D., and Wall, R. Cell 26, 19-27 (1981).
8. Cheng, H., Blattner, F., Fitzmaurice, L., Mushinski, J., and Tucker, P. Nature 296, 410-415 (1982).
9. Kikutani, H., Sitia, R., Good, R. and Stavnezer, J. Proc. Natl. Acad. Sci. USA 78, 6436-6440 (1981).
10. Maki, R., Roeder, W., Traunecker, A., Sidman, C., Wabl, M., Raschke, W., and Tonegawa, S. Cell 24, 353-365 (1981).
11. Rogers, J., and Wall, R. Proc. Natl. Acad. Sci. USA 78, 7497-7501 (1981).
12. Yaoita, Y., Kumagai, Y., Okumura, K., and Honjo, T. Nature 297, 697-699 (1982).
13. Kemp, D., Harris, A., Cory, S., and Adams, J. Proc. Natl. Acad. Sci. USA 77, 2876-2880 (1980).
14. Perry, R., and Kelley, D. Cell 18, 1333-1339 (1979).
15. Storb, U., Wilson, R., Selsing, E., and Walfield, A. Biochem. 20, 990-996 (1981).
16. Kemp, D., Wilson, A., Harris, A., and Shortman, K. Nature 286, 168-170 (1980).
17. Kemp, D., Harris, A., and Adams, J. Proc. Natl. Acad. Sci. USA 77, 7400-7404 (1980).
18. Cory, S., Adams, J., and Kemp, D. Proc. Natl. Acad. Sci. USA 77, 4943-4947 (1980).
19. Alt, F., Rosenberg, N., Enea, V., Siden, E., and Baltimore, D. Mol. Cell. Biol. 2, 386-400 (1982).
20. Perry, R., Kelley, D., Coleclough, C., Seidman, J., Leder, P., Tonegawa, S., Matthyssens, G., and Weigert, M. Proc. Natl. Acad. Sci. USA 77, 1937-1941 (1980).
21. Kemp, D., Adams, J., Mottram, P., Thomas, W., Walker, I., and Miller, J. J. Exp. Med. (submitted).
22. Van Ness, B., Weigert, M., Coleclough, E., Mather, D., Kelley, D., and Perry, R. Cell 27, 593-602 (1981).
23. Yagi, M., and Koshland, M.E. Proc. Natl. Acad. Sci. USA 78, 4907-4911 (1981).
24. Perry, R., Kelley, D., Coleclough, C., and Kearney, J. Proc. Natl. Acad. Sci. USA 78, 247-251 (1981).
25. Kemp, D., Morahan, G., Cowman, A., and Harris, A. Nature (submitted).

10

Non-Immunoglobulin DNA Region That Rearranges Frequently in Lymphoid Tumours[*]

Jerry M. Adams, Steven D. Gerondakis, Elizabeth A. Webb, Ora Bernard and Suzanne Cory

The Walter and Eliza Hall Institute of Medical Research, Melbourne, Australia

DNA rearrangements manifest in mouse plasmacytomas have assembled immunoglobulin variable (V) and constant (C) region genes (reviewed by Adams and Cory, 1982). For the heavy (H) chain locus, on chromosome 12, these include fusion of V_H, D_H and J_H elements into a complete V_H-D_H-J_H gene, and its subsequent "switch" from C_μ to another C_H gene, via recombination between switch regions 5' to the two C_H genes. Less well understood are the translocations in plasmacytomas that join the end of chromosome 15 to the region of 12 where the H locus lies (Ohno et al., 1979). We are studying a sequence that represents a candidate for such chromosomal events. In several plasmacytomas the same unknown DNA region has recombined 5' to the C_α gene (Kim and Hood, personal comm; Kirsch et al., 1981). We have cloned such a rearranged C_α gene, fused to a region designated here as LyR (Lymphoid Rearranging) DNA. We sequenced the recombination region and have searched for LyR rearrangement and transcription in diverse tumor lines and in normal cells. We find that most tumors of mature B cells contain rearranged LyR and that rearrangement alters expression of LyR encoded genes, as will be more fully documented elsewhere (Adams et al., 1982).

[*] This work was supported by the N.H. & M.R.C. (Canberra, the U.S. NCI (CA12421), the American Heart Association and the Drakensberg Trust.

41

MANIPULATION AND EXPRESSION
OF GENES IN EUKARYOTES
ISBN 0 12 513780 x

LyR has Fused to the J558 Sα Region within an Sα Repeat Unit

We cloned from plasmacytoma J558 a 14.3 kb EcoRI fragment bearing an aberrantly rearranged Cα gene. Its restriction map showed that the incoming (LyR) DNA had recombined 1.85 kb 5' to the Cα gene, within the Sα switch region. We sequenced a 417 bp fragment spanning the LyR-Sα recombination point. The 5' 344 residues did not derive from Sα or Sµ. From the Sα side, the LyR-Sα fusion looks like C_H switch recombination. LyR has entered Sα within a run of three of the GAGCT sequences found in most S_H repeat units and the recombination point lies within the 16th of 17 ~32 bp repeat units implicated in Sα switch recombination (Davis et al., 1980). Moreover, a TAGGCTG 9 bp 5' to the recombination point in germline Sα approximates the consensus YAGGTTG found within 20 bp 5' of nearly all switch sites (Marcu et al., 1982). In contrast, the LyR region sequenced lacked switch region features: no repeat unit was evident, GAGCT and GGGGT were not frequent, and no YAGGTTG occured within 140 bp 5' to the recombination point.

LyR Rearrangement is Lymphoid Tumor Associated

We analysed Southern blots of DNA from 63 tumor lines. Twenty one of 23 plasmacytomas had, as well as the germline EcoRI fragment of ~21.5 kb, a new fragment, typically 14 to 17 kb long; these included lines secreting µ, δ, γ3, γ1, γ2a, γ2b and α. Of 5 B lymphomas, which express surface immunoglobulin, 3 had rearranged LyR. All 11 Abelson murine leukemia virus-induced pre-B lymphomas, however, had only germline LyR. This striking difference from other tumors of B lineage might relate to retroviral induction, or to dormant switch machinery in the pre-B lines.

To assess normal T and B cells, we examined peripheral T cells and spleens of nude mice. No rearrangement was detected; control experiments suggested that 10-20% rearrangement would have been detectable within such polyclonal populations. To test monoclonal B cells, we analyzed four hybridomas. The only rearranged fragment in three lines was that of the NS1 parent. The fourth had a new fragment, but it may represent a secondary rearrangement of the tumor parent. No LyR rearrangement due to the normal B cell was found in 8 other hybridomas (Harris et al., 1982). Thus LyR rearrangement appears to be confined largely, if not exclusively, to lymphoid tumors, primarily of mature cells of B lineage.

Most rearranged LyR fragments were labeled by a 5' Cα cDNA probe, including 17 of the 21 plasmacytomas, and 2 of the 3 B lymphomas but not the three T lymphomas. In contrast to these BALB/c tumors, none of 11 NZB lines with rearranged LyR had recombined near Cα (Harris et al. 1982). Thus the targets for LyR rearrangement are genetically determined.

Altered LyR Transcription

Gel blot analysis showed that most lymphoid and non-lymphoid tumor lines (46/50) had polyadenylated LyR transcripts. Lines with only germline LyR DNA displayed partially resolved 2.3 and 2.4 kb species. These "germline (G)" RNAs were found in 6 non-lymphoid lines, in 11 T lymphomas and in all 7 pre-B lines examined. In striking contrast, B lymphoid lines with a rearranged LyR allele had prominent new RNA species of 2.0 and 1.85 kb. The "R transcripts" were seen in 16 of 17 plasmacytomas and B lymphomas with rearranged LyR, including lines expressing diverse H chains and with LyR rearranged elsewhere than Sα. This excellent correlation argues that the R transcripts derive from the rearranged LyR allele.

LyR Sequences Appear to Arise Outside the H Chain Locus, Probably from a Separate Chromosome

The LyR restriction map excluded its arising from the J_H-C_H locus or the known D locus. Much of the V_H locus and the entire V_H-J_H region can be excluded because pre-B lymphomas with V_H-D_H-J_H joins on both alleles have deleted those regions but retain germline LyR. The V_H locus also appears unlikely because all 6 V_H probes that we have tested reveal restriction site polymorphism but none was detected with LyR probes in mouse strains. Evidence that LyR is not closely linked to the H locus is that plasmacytomas M104E and S117 appear haploid for the H locus yet exhibit both germline and rearranged LyR. Conversely, an ST1 subline with both germline and rearranged J_H alleles exhibits one rearranged but no germline LyR fragment. To test linkage to the H chain locus further, we exploited the tendency of cell hybrids to discard chromosomes. We examined a clone from a TIKAUT x A/J lymphocyte fusion because EcoRI digests distinguish the Cμ genes of TIKAUT (AKR strain) and A/J, and TIKAUT LyR from A/J LyR. Significantly, C126 retained some A/J Cμ but no

detectable A/J LyR. The simplest interpretation is that the
C126 population had lost all copies of the A/J chromosome
bearing LyR but not all copies of A/J chromosome 12.

Conclusions

LyR is a transcriptionally active region of mouse germline DNA (>21 kb long) that has rearranged in most tumors of
mature B lymphoid cells. The polyadenylated RNAs in B lines
with rearranged LyR DNA differ from the germline transcript.
If, as seems likely, these potential mRNAs arise from the
rearranged LyR allele, rearrangement must have activated a
new promoter, or mode of splicing. Discordance between LyR
and C_H copy number in four lines suggests that LyR does not
derive from chromosome 12. Thus LyR may be the first region
undergoing interchromosomal translocation to be characterized molecularly. LyR-Sα fusion may correspond to the
chromosome 15:12 shift in plasmacytomas or to a previously
undetected translocation.

REFERENCES

Adams, J.M., Gerondakis, S., Webb, E., Mitchell, J.,
    Bernard, O. and Cory, S. (1982) Proc.Natl.Acad.Sci.USA.
    (submitted).
Adams, J.M. and Cory, S. (1982). In "Eukaryotic Genes: Their
    Structure, Activity and Regulation", ed. McLean, N.,
    Gregory, S., and Flavell, R. Butterworth Press (London)
    in press.
Davis, M.M., Kim, S.K. and Hood, L. (1980). Science, 209,
    1360.
Harris, L.J., Lang, R.B. and Marcu, K.B. (1982). Proc.Natl
    Acad.Sci.USA. 79, 4175-4179.
Kirsch, I.R., Ravetch, J.V., Kwan, S-P., Max, E.E., Ney, R.L.
    and Leder, P. (1981). Nature (London) 293, 585.
Marcu, K.B., Lang, R.B., Stanton, L.W., and Harris, L. (1982).
    Nature (in press).
Ohno, S., Babonits, M., Wiener, F., Spira, J., Klein, G.,
    and Potter, M. (1979). Cell 18, 1001.

11
Expression of DNA Repair Genes in Mammalian Cells[1]

Ann K. Ganesan, Graciela Spivak and Philip C. Hanawalt

Department of Biological Sciences
Stanford University
Stanford, California

Studies designed to obtain information about the chromosomal location, organization and expression of genes controlling excision repair (ER) in cultured mammalian cells, particularly human cells, have been performed in several laboratories, including our own. Most of the current ideas about
DNA repair in mammalian cells are based upon results obtained
from studies of the bacterium, *Escherichia coli*, from which a
wide variety of mutants defective in repair has been isolated.
Biochemical and genetic analyses of some of these mutants
have provided evidence that ER requires at least four steps:
incision of the damaged DNA strand, removal of the damage together with a stretch of adjacent nucleotides, repair synthesis to replace the excised nucleotides, and finally, ligation
to restore the continuity of the DNA. By comparison with *E.
coli* the variety of available mutants of cultured mammalian
cells is limited, and none of the mutants has been well characterized either genetically or ʰiochemically. However, accumulated data indicate that ER in mammalian cells resembles
that in *E. coli*, at least with regard to its principal features (reviewed in Hanawalt, *et al.*, 1979; Ganesan, *et al.*,
1982).

The most extensively studied repair deficient mammalian

[1] *Supported by grant NP161 from the American Cancer
Society and NATO Research Grant No. 943 to A.G.*

MANIPULATION AND EXPRESSION
OF GENES IN EUKARYOTES
ISBN 0 12 513780 x

cells are those derived from humans suffering from xeroderma
pigmentosum (XP), a rare disease inherited as an autosomal
recessive allele and characterized by hypersensitivity of the
skin to sunlight, usually expressed as neoplasms in exposed
areas of the skin.   Some affected individuals also have neuro-
logical abnormalities (reviewed in Bootsma, 1978).   Analyses
of unscheduled DNA synthesis (UDS) in heterokaryons formed by
fusing cells from different XP patients have defined 7 complementation groups for the "classical" form of XP in which
the mutation appears to affect an early step in ER.   (In the
"variant" form of XP no ER deficiency is evident.)   Compared
to cells cultured from normal individuals, XP cells show diminished capacity to incise cellular DNA, to excise pyrimidine dimers (the major form of damage resulting from UV irradiation), to perform repair synthesis, and to form colonies
after UV irradiation.   Furthermore, the introduction of purified preparations of T4 endonuclease V, a small enzyme coded
by the *denV* gene of bacteriophage T4, restores the capacity
for repair synthesis and enhances the survival of XP cells
after UV irradiation (Tanaka, *et al.*, 1977; Smith and
Hanawalt, 1978: Hayakawa, *et al.*, 1981).   This enzyme, which
specifically incises DNA containing pyrimidine dimers and
catalyses the first step in ER of UV irradiated T4 DNA, may
possess an activity analogous to one present in normal human
cells but deficient in XP cells, or it may provide a different activity that can substitute for the defective one.
In either case, XP cells appear to retain the enzymes needed
for the steps in ER subsequent to incision.
     Although the precise biochemical defect in XP cells has
not been identified, the correlation between diminished capacity for ER and hypersensitivity to UV irradiation provided
evidence for an important role for ER in human cells, and en-
couraged us and other investigators to attempt to identify
genetic determinants of ER.   For this purpose we have utilized two experimental approaches, analysis of interspecific
somatic cell hybrids, and intraspecific DNA-mediated gene
transfer.

INTERSPECIFIC SOMATIC CELL HYBRIDS

     In one type of study, hybrid cell lines formed by fusing
human cells with mouse cells have been tested for ER.   Typically, cell lines derived from mice or hamsters show less ER
than normal, repair proficient human cell lines.   Thus, it
seemed likely that examination of hybrids containing differ-

ent partial sets of human chromosomes would reveal a correlation between the capacity for normal human levels of ER and
the presence of a specific human chromosome or set of chromosomes. Our studies have included 3 kinds of mouse-human hybrid. (1) A hybrid derived from a chloramphenicol-resistant
mutant of ER-proficient HeLa cells: this hybrid retains a
full complement of chromosomes from the HeLa parent, but only
a partial set of mouse chromosomes (Wallace and Eisenstadt,
1979). (2) Hybrids derived from normal, ER-proficient human
cells: these hybrids contain different partial sets of human
chromosomes together with the entire murine complement (Craig,
et al., 1976; Solomon, *et al.*, 1976). (3) Hybrids derived
from ER-deficient XP cells belonging to complementation group
A (XP-A). To test for ER we monitored the excision of pyrimidine dimers with T4 endonuclease V (van Zeeland, 1978;
Ganesan, *et al.*, 1981). In some cases we also examined UDS
by autoradiography, and repair synthesis by density labeling
with bromodeoxyuridine followed by equilibrium sedimentation
in CsCl (Smith, *et al.*, 1981).

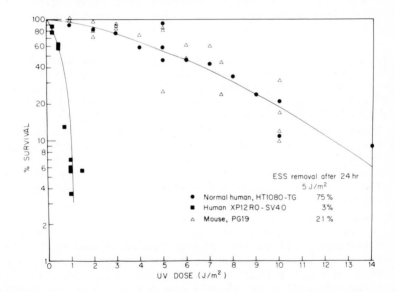

*FIGURE 1. Colony-forming ability of UV irradiated human
cells and mouse cells. Cells seeded in 10 cm culture dishes
at a density of 500-1000 cells per dish were allowed to attach. The medium was then removed, the cells were irradiated
and fresh medium was added. The dishes were then incubated
undisturbed for 2 weeks, after which colonies were stained
and counted.*

The behavior of the parental cell lines was consistent with results reported by other investigators. Within 24 hours after irradiation our ER-proficient human cells typically removed 50-80% of the endonuclease sensitive sites (ESS) produced by a dose of 5 J/m^2 at 254 nm. The colony-forming efficiency of cells irradiated with this dose was 50-90% of that of unirradiated cells (Fig. 1). Under the same conditions the XP-A cells removed 5% or less of the ESS and their colony-forming efficiency was reduced to an undetectable level (less than 1% of that of the unirradiated controls). They also showed little or no UDS or repair synthesis. Irradiated mouse cells showed significant UDS, repair synthesis and removal of ESS compared to XP-A cells, but less than ER-proficient human cells. For example, 24 hours after irradiation they usually had removed 10-30% of the ESS produced by a dose of 5 J/m^2 at 254 nm. Although ER appeared to be less efficient, or at least slower, in mouse cells than in normal human cells, this property was not correlated with lower survival. Instead, the colony-forming ability of mouse cells was similar to that of ER-proficient human cells (Fig. 1), indicating that mouse cells can survive a higher level of unrepaired damage than can human cells.

As expected, the colony-forming ability of the hybrid cell lines after UV irradiation was the same as that of mouse cells or ER-proficient human cells. However, their behavior with regard to ER was unanticipated. We had hoped that hybrids derived from ER-proficient human cells would fall into 2 categories: one showing low levels of ER like mouse cells, and one showing high levels of ER like normal human cells. However, all those tested, including the one containing an entire HeLa chromosome complement, showed low levels of ER indistinguishable from that observed in mouse cells (Table 1).

This result is consistent with findings recently reported by Collins and Waldren (1982) who used incision as a criterion for ER in various hamster-human hybrids. Although they did not examine a hybrid containing the full human complement of chromosomes, all of the hybrids tested showed ER levels equivalent to hamster cells and lower than human cells, leading them to suggest that certain hamster determinants can inhibit the expression of human genes for ER. Lin and Ruddle (1981) reported that a hybrid between mouse embryo fibroblasts and XP-A cells, which contained a full complement of human chromosomes together with a full complement of mouse chromosomes, showed a "normal" level of ER as measured by UDS. However, no quantitative data were presented, so it is not certain whether the level was the same as that in embryonic mouse cells or in normal human cells, and thus whether their

TABLE I. *Removal of ESS in irradiated cells of a mouse-human hybrid cell line. Cells irradiated with 5 J/m^2 at 254 nm were analyzed for ESS as described by Ganesan, et al., (1981)*

| Cell line | Expt. | ESS per $10^8 d$ | | % |
		0 hr	24 hr	removed
ROH8A[a]	1	5.0	3.3	34
	2a	4.0	3.1	23
	2b	7.8	4.1	47
296-1[b]	2a	5.6	1.4	75
	2b	7.8	1.4	82
	3a	9.1	1.5	84
	3b	5.7	1.3	77
RAG[c]	1	4.8	5.5	–
	3a	4.8	4.0	17
	3b	5.3	4.2	21

[a]*Hybrid containing full HeLa chromosome complement (Wallace and Eisenstadt, 1979).*

[b]*HeLa parent of ROH8A.*

[c]*Murine parent of ROH8A.*

observations differ from ours.

The decreased expression of normal human ER genes in human-rodent hybrids is probably different from the effect described by Bootsma, *et al.*(1982) who reported that UDS in the nuclei of normal human fibroblasts was depressed shortly after fusion of the fibroblasts with chicken erythrocytes. This effect was transient, however. By 8 days after fusion, UDS in the human nuclei occurred at nearly normal levels. In contrast, in the hybrid cells we examined, which had doubled many times since fusion, reduced expression of human ER genes is apparently a stable property.

The hybrids derived from XP-A cells that we tested behaved like those derived from ER-proficient human cells. Neither the rate of ESS removal nor colony-forming ability after UV irradiation was significantly different from that of mouse cells. This result appears to differ slightly from that of Goldstein and Lin (1972) who found that UV resistant hybrids produced by fusing XP cells (probably XP-A) with hamster

cells exhibited UDS levels slightly lower than that of the
hamster cells, although their survival levels were similar.
    Our results can be summarized as follows: (1) Fusing
XP-A with mouse cells did not *per se* produce hybrids with ER
levels significantly different than those typical of mouse
cells. (2) Fusing ER-proficient human cells with mouse cells
did not result in ER levels significantly higher than those
characteristic of mouse cells even when an entire chromosome
complement from the ER-proficient human cells was present.
This latter result, in particular, indicates that mouse-human
hybrid cell lines may not be suitable for determining the
chromosomal location of human genes required for ER; however,
further studies of mouse-human hybrids which retain complete
sets of human chromosomes while segregating mouse chromosomes
may provide insight into the effect of mouse determinants on
the regulation and expression of human genes required for ER,
as well as information about the location of murine genes required for DNA repair (Lin and Ruddle, 1981).

INTRASPECIFIC GENE TRANSFER

    An alternative approach to the characterization of human
ER genes was to utilize intraspecific gene transfer. For example, Ishizaki, *et al.* (1981) produced UV resistant derivatives of an XP-A cell line, XP20S(SV40), by fusion with
microcells obtained from normal human cells. However, in
that study the chromosome(s) that carried genes conferring UV
resistance could not be identified because of the complex
karyotype of the recipient cell line, which had been transformed with SV40. Efforts to obtain UV resistant fusion products from non-transformed XP-A cells with a normal karyotype
were not successful.
    We have undertaken experiments to introduce DNA purified
from ER-proficient human cells into XP-A cells by calcium
phosphate co-precipitation (Graham and van der Eb, 1973;
Wigler, *et al.*, 1979). As recipients we chose continuous
cell lines resulting from SV40 transformation. Although it
has been reported that some SV40-transformed cell lines respond differently to certain chemical agents than do their
non-transformed counterparts (Fujiwara and Tatsumi, 1976;
Sklar and Strauss, 1981), continuous lines are preferable to
finite lines for characterizing the products of gene transfer.
Since cell lines appear to differ with respect to their ability to act as recipients for DNA (Graf, *et al.*,1979;
Peterson and McBride, 1980; Lester, *et al.*, 1980; Corsaro and

Pearson,1981),we tested two XP-A cell lines for this property using DNA from the chimeric plasmid, pSV2-*gpt* (Mulligan and Berg;1980). This plasmid carries the *gpt* gene from *Escherichia coli* in a form that can be expressed and selected for in mammalian cells. Using the modification of the calcium phosphate coprecipitation procedure recommended by Mulligan and Berg(1981) we found that both XP-A lines tested, XP12RO(SV40) and XP20S(SV40),were able to act as recipients for the plasmid DNA. The frequency of Gpt$^+$ transferents (Table 2) was comparable for the two cell lines, and only slightly lower than values published for other types of cells, indicating that either cell line should be suitable for our experiments. We then performed gene transfer experiments with XP12RO(SV40) as the recipient for DNA purified from an ER-proficient diploid line, HT1080 (Croce, 1976) using the same calcium phosphate co-precipitation procedure for the cellular DNA that we had used for the pSV2 DNA. To select for UV resistant transferents we used several low UV doses (2-4 J/m^2) spaced several

TABLE II. XP-A cells as recipients for plasmid DNA. Recipient cells were treated with purified DNA (Parker and Stark, 1979) from the plasmid, pSV2-gpt (Mulligan and Berg, 1981). In some cases carrier DNA, purified from the recipient cells, was also inincluded. The indicated amount of DNA was added to each culture dish, which contained approximately 10^6 cells. The cells were reseeded after 3 days and Gpt$^+$ transferents were selected in medium containing hypoxanthine, xanthine, thymidine, methotrexate and mycophenolic acid.

Recipient	DNA		Gpt$^+$ transferents
	(μg per 10^6 cells)		(per 10^6 reseeded cells)
	pSV2	carrier	
XP12RO(SV40)	3.8	20	41
	3.8	0	17
	0	0	0
XP20S(SV40)	2.5	20	15
	2.5	0	5
	0	0	0

days apart. Reconstruction experiments with mixed populations of cells (XP12RO(SV40) being the major component and a
6-thioguanine resistant derivative of HT1080 the minor component) showed that this schedule resulted in more efficient
selection for the ER-proficient cells than did a single dose
of UV. The gene transfer experiments indicated that UV resistant derivatives could be isolated from the XP-A recipient.
Although some of the colonies that survived the UV selection
procedure when subcultured and tested were as UV sensitive as
the recipient cells, others were as resistant as the donor
cells (Fig. 2) and a few were of intermediate sensitivity.
However, the frequency of UV resistant derivatives was not

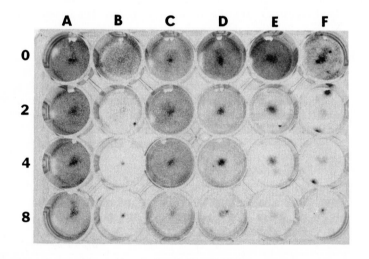

FIGURE II. *Relative UV sensitivity of several derivatives of XP12RO(SV40). Cells seeded at a density of approximately 10^4 per well in 24-well plates were allowed to attach.
The medium was then removed, the cells were irradiated, and
fresh medium was added. The plates were reincubated for 3
days, the medium was removed and the remaining cells stained.
Columns A and F contain ER-proficient cells (HT1080) and
XP12RO(SV40), respectively. Columns B through E contain 4
different derivatives of XP12RO(SV40). Rows labeled 0, 2, 4,
and 8 received 0, 2, 4 and 8 J/m^2, respectively. (Dark
centers in wells are due to non-uniform seeding of cells.)*

reproducibly greater among cells treated with DNA from HT1080
than among control cells treated with DNA from the recipient
cells or with no DNA at all. Thus, the UV resistant derivatives may be revertants, rather than transferents. We are
currently testing these derivatives for their ability to remove ESS after UV irradiation, and for other aspects of ER,
to determine whether their resistance to UV results from an
increased capacity for ER or from enhanced tolerance of un-
removed damage, similar to that seen in mouse cells.

Although we have not yet identified conditions suitable
for the transfer of UV resistance to XP12RO(SV40) by purified
DNA, Takano, *et al.*(1982) have recently reported success with
XP20S(SV40). Since they employed a modification of the calcium phosphate co-precipitation procedure that differs in
several respects from the one we used, we plan to compare the
two modifications using the same recipient cells, XP20S(SV40),
in an attempt to determine what factors affect the efficiency
of transfer or expression of human ER genes in XP cells.

REFERENCES

Bootsma, D. (1978) *In* "DNA Repair Mechanisms" (Hanawalt,
P. C., Friedberg, E. C. and Fox, C. F. eds.), pp. 589-
601, Academic Press, New York.
Bootsma, D., Kleijer, W., van der Veer, E., Rainaldi, G. and
de Weerd-Kastelein, E. A. (1982) *Exptl. Cell. Res. 137*,
181-189.
Collins, A. R. S. and Waldren, C. A. (1982) *J. Cell Sci.* in
press.
Corsaro, C. M. and Pearson, M. L. (1981) *Somat. Cell. Genet.
7*, 603-616.
Craig, I., Tolley, E. and Bobrow, M. (1976) *Cytogenet. and
Cell Genetics 16*, 114-117.
Croce, C. M. (1976) *Proc. Natl. Acad. Sci. U.S.A. 73*, 3248-
3252.
Fujiwara, Y. and Tatsumi, M. (1976) *Mut. Res. 37*, 91-110.
Ganesan, A. K., Cooper, P. K., Hanawalt, P. C. and Smith,
C. A. (1982) *Prog. in Mutation Res. 4*, 313-323.
Ganesan, A. K., Smith, C. A. and van Zeeland, A. A. (1981)
In "DNA Repair: a laboratory manual of research
procedures" (Friedberg, E. C. and Hanawalt, P. C. eds.),
pp. 89-97, Marcel Dekker, Inc., New York.
Goldstein, S. and Lin, C. C. (1972) *Nature New Biology 239*,
142-145.
Graf, L. H., Urlaub, G. and Chasin, L. A. (1979) *Somat. Cell.
Genet. 5*, 1031-1044.

Graham, F. L. and van der Eb, A. J. (1973) *Virology 52,* 456-467.

Hanawalt, P. C., Cooper, P. K., Ganesan, A. K. and Smith, C. A. (1979) *Ann. Rev. Biochem. 48,* 783-836.

Hayakawa, H., Ishizaki, K., Inoue, M., Yagi, T. Sekiguchi, M. and Takebe, H. (1981) *Mutat. Res. 80,* 381-388.

Ishizaki, K., Shima, A., Sekiguchi, T. and Takebe, H. (1981) *Jpn. J. Genet. 56,* 105-115.

Lester, S. C., LeVan, S. K., Steglich, C. and De Mars, R. (1980) *Somat. Cell. Genet. 6,* 241-259.

Lin, P. F. and Ruddle, F. H. (1981) *Nature 289,* 191-194.

Mulligan, R. C. and Berg, P. (1980) *Science 209,* 1422-1427.

Mulligan, R. C. and Berg, P. (1981) *Proc. Natl. Acad. Sci. U.S.A. 78,* 2072-2076.

Parker, B. A. and Stark, G. R. (1979) *J. Virol. 31,* 360-369.

Peterson, J. L. and Mc Bride, O. W. (1980) *Proc. Natl. Acad. Sci. U.S.A. 77,* 1583-1587.

Sklar, R., Strauss, B. (1981) *Nature 289,* 417-420.

Smith, C. A. and Hanawalt, P. C. (1978) *Proc. Natl. Acad. Sci. U.S.A. 75,* 2598-2602.

Smith, C. A., Cooper, P. K. and Hanawalt, P. C. (1981) *In* "DNA Repair: a laboratory manual of research procedures" (Friedberg, E. A. and Hanawalt, P. C. eds.) pp. 289-305, Marcel Dekker, Inc., New York.

Solomon, E., Bobrow, M., Goodfellow, P. N., Bodmer, W. F., Swallow, D. M., Povey, S. and Noël, B. (1976) *Somat. Cell Genet. 2,* 125-140.

Takano, T., Noda, M. and Tamura, T. (1982) *Nature 296,* 269-270.

Tanaka, K., Hayakawa, H., Sekiguchi, M. and Okada, Y. (1977) *Proc. Natl. Acad. Sci. U.S.A. 74,* 2958-2962.

Van Zeeland, A. A. (1978) *In* "DNA Repair Mechanisms" (Hanawalt, P. C., Friedberg, E. C. and Fox, C. F. eds.), pp. 307-310, Academic Press, New York.

Wallace, D. C. and Eisenstadt, J. M. (1979) *Somat. Cell. Genet. 5,* 373-396.

Wigler, M., Sweet, R., Sim, G. K., Wold, B., Pellicer, A., Lacy, E., Maniatis, T., Silverstein, S. and Axel, R. (1979) *Cell 16,* 777-785.

12

A Novel Pleiotropic Mutation in Baby Hamster Kidney Cells Causing Increased Resistance to Arabinosyladenine, Sensitivity to Adenosine and Adenosine Kinase Deficiency[1]

V. L. Chan, F. Meffe, S. Guttman, P. Juranka and S. M. Archer

Department of Microbiology
University of Toronto
Toronto, Ontario, Canada

INTRODUCTION

Recently we (Chan and Juranka, 1981) reported the isolation of 3 distinct phenotypic groups of arabinosyladenine (araA)-resistant mutants of baby hamster kidney (BHK 21/C13) cells all of which are cross-resistant to deoxyadenosine (dAdo). The major group (Class I) of araA^r mutants are adenosine Kinase (AK) deficient. The second class of araA^r/dAdo^r mutants, probably has an altered ribonucleoside diphosphate reductase (RDR) because RDR activity in these cells showed an increased resistance to inhibition by araATP and dATP (Chan et al., 1981). The Class III mutants are also araA^r/dAdo^r but unlike those of Class I and II exhibit extreme adenosine sensitivity (Chan and Juranka, 1981). All the Ado^s/araA^r mutants have an altered level of AK activity relative to the wild type BHK cells. Some are totally deficient while one of these Ado^s/araA^r mutants, ara-S10d, which was derived spontaneously, possesses 35% AK activity with an altered pH optimum. In this communication we wish to report the results that implicate the involvement of a single polar

[1]This work was supported by the National Cancer Institute (NCI) and the Medical Research Council (MRC) of Canada. P.J. has a studentship from NCI. V.L.C. was an NCI Research Scholar. S.M.A. had an MRC Studentship.

MANIPULATION AND EXPRESSION
OF GENES IN EUKARYOTES
ISBN 0 12 513780 x

or pleiotropic mutation in the production of the multiple
phenotypic properties, AK-alteration, mutator activity, Ado
sensitivity, araA and dAdo resistance.

RESULTS AND DISCUSSION

We reasoned that if the multiple apparent lesions (araA^r/
dAdo^r/Ado^s/ AK-deficiency) of the Class III araA^r mutants
were due to a single mutation, then one should be able to
isolate Ado^r revertants with alterations in araA and/or dAdo
resistance and perhaps AK level. For these studies we utilized the Ado^s/araA^r mutant, ara-S10d, which was originally
isolated without mutagenic treatment. A number of subclones
of ara-S10d were used in the isolation of single-step Ado^r
revertants to ensure that we obtain independently derived
revertants. Two of these Ado^r revertants, C9f1R and C7a
showed a similar increase in sensitivity to araA as compared
to the ara-S10d parent (See Fig. 1, only data for C9f1R are
presented). The isolation of two independent single-step
Ado^r revertants with a concurrent change in sensitivity to
araA suggests that a single mutation in ara-S10d may be
responsible for the Ado^s/araA^r properties. If a single mutation was responsible for the partial AK-deficiency of ara-
S10d cells, then C9f1R, C7a and possibly other Ado^r

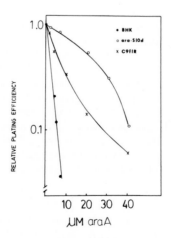

FIGURE 1. *Relative plating efficiency of BHK, ara-S10d
and C9f1R in increasing concentration of araA.*

revertants would also show an alteration in AK level. As shown in Table 1, all seven revertants tested contained non-detectable levels of AK activity unlike the parental ara-S10d cells which had 35% level of wild type AK activity at pH 5.6. Deficiency in AK could not be attributed to the presence of an inhibitor as no inhibition of AK activity was observed in wild type and revertant mixed extracts (data not included).

In order to examine if the 35% level of the AK activity in ara-S10d is due to a mutation which alters the structure of the enzyme we determined the pH optimum of the AK activity in crude extracts of the wild type BHK cells and that of ara-S10d cells. As can be seen in Fig. 2, the pH optimum of the AK activity of BHK cells is 5.7 as observed by other workers. In contrast the AK activity of the Ado[s] mutant, ara-S10d, has a pH optimum of 7.0, thus suggesting a structural change of the AK in ara-S10d cells.

TABLE I.   Relative Adenosine Kinase Activity

Cell Strain	*AK activity (nmole AMP formed/mg protein/min)*	*% of Wild Type*
BHK (Wild type)	*2.91 ± 0.55*	*100*
ara-S10d	*1.01 ± 0.22*	*35*
Ado[r] revertants (R202A, R203A, C7b, C8a, C7a, C7c, C9f1R)	*<0.01*	*<0.5*

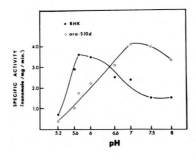

FIGURE 2.   Effect of pH on AK activity in extracts of BHK and ara-S10d cells.

The observation that the reversion from Ado^S to Ado^r results in a concurrent dramatic reduction in AK activity, and the isolation of Ado^r revertants C9f1R and C7a which not not only showed alteration in AK activity but also partially reverted to araA sensitivity supports the thesis that the ara-S10d phenotype, $araA^r/dAdo^r/Ado^S$ and the partial AK-deficiency were all attributed to a single mutation event. The single mutation model is also supported by the relatively high frequency of occurrence of $Ado^S/araA^r$ mutants with altered AK. The overall araA resistant mutation frequency for non-mutagenized and mutagenized cells were 6×10^{-5} and 3.5×10^{-4} respectively (Chan and Juranka, 1981) and the Class III Ado^S mutants represent 6 out of a total of 168 mutants isolated. The $Ado^S/araA^r$ mutants are at least as frequent as the Class II $araA^r$ RDR mutant (2 out of 168).

All the Ado^S mutants are associated with an AK-deficiency (Table 1, Chan and Juranka, 1981). However, the AK-deficiency in itself will not produce an Ado^S phenotype since the Class I $araA^r$ AK-deficient mutants do not exhibit the Ado^S phenotype (Chan and Juranka, 1981). Therefore the Ado^S and AK-deficiency are determined by two separate genes. The Ado sensitivity has been shown to be associated with the mutator activity (Chan *et al.*, 1981). Thus, the phenotype of Ado sensitivity, $araA^r$, $dAdo^r$, mutator activity and an altered AK activity is due to a pleiotropic or polar mutation, the nature of which is presently unknown. It is conceivable that the AK gene and the gene that determines Ado sensitivity are adjacent genes. A possible model would involve integration at the AK gene of a transposable element that possesses a promotor or a sequence that enhances the transcription of the adjacent promoter. Integration of such an element would result in an AK-deficiency or altered AK enzyme and at the same time alter the transcription rate and/or the control of the adjacent gene which determines Ado sensitivity. According to this model all the Ado^r revertants isolated were presumably due to excision or rearrangement of the transposable element resulting in a deletion mutation in the AK gene.

REFERENCES

Chan, V.L., Guttman, S. and Juranka, P. (1981). *Mol. Cell. Biol.*, 1, 568.

Chan, V.L., and Juranka , P. (1981). *Somat. Cell Genet. 7,* 147.

13

Alterations in Gene Expression during Phytohemagglutinin Induced Transformation of Human Peripheral Blood Lymphocytes

Eric McCairns, George E. O. Muscat, Dale Fahey
and Peter B. Rowe

Children's Medical Research Foundation
University of Sydney
Camperdown, New South Wales, Australia

Phytohemagglutinin (PHA) stimulation of the normally quiescent human lymphocyte results in increased DNA, RNA and protein synthesis. Torelli et al. (1981) found that in PHA transformed lymphocytes the complexity of total cellular poly (A)$^+$RNA is decreased and that many sequences increase in their abundance with respect to resting lymphocytes. Willard and Anderson (1980) showed by two-dimensional gel electrophoresis that concanavalin A transformed lymphocytes do not produce any unique proteins during the first 18 hours of transformation.

We have found that the translational activity of cytoplasmic poly(A)$^+$RNA from resting cells was approximately 20% of that from transformed cells in a rabbit reticulocyte lysate assay. Co-translation of each poly(A)+RNA fraction with rabbit globin mRNA showed that both preparations were essentially free of inhibitors of peptide synthesis. There was no difference in translational efficiency between the polysomal and non-polysomal poly(A)+RNA fractions of resting cells.

The size range of cytoplasmic poly(A)$^+$RNA from resting and transformed lymphocytes was similar, and was in the range 5S to 28S. The number average size of mRNA was 800 nucleotides in resting cells and 950 in transformed cells, while the corresponding average poly(A) tail lengths, were 45 and 50 residues respectively. Hybridisation with [^{3}H]poly(U) showed that both poly(A)$^+$RNA fractions were essentially pure.

MANIPULATION AND EXPRESSION
OF GENES IN EUKARYOTES
ISBN 0 12 513780 x

Both cytoplasmic poly(A)+RNA fractions were capable of directing the synthesis of peptides in the molecular weight range 15,000 to 90,000 in the cell free assay. Two dimensional gel patterns of total proteins from resting and transformed lymphocytes were qualitatively similar and indicated predominantly quantitative changes, implying changes in the relative abundance of cytoplasmic mRNA species on transformation. As individual mRNA's may have different translational efficiencies, changes in the relative amounts of individual mRNA's could affect the overall translational efficiency of the total mRNA population.

The yield of cytoplasmic RNA increased 4-fold on transformation, while the proportion of the RNA that was polyadenylated increased by a factor of 1.6 giving a net 6-fold increase in mRNA per cell on transformation. Hybridisation of cDNA synthesised from cytoplasmic poly(A)+RNA from transformed cells with its homologous and heterologous mRNA templates indicated that approximately 5-6% of the mRNA from transformed cells was transformation specific. Amplication of this by isolating the nonhybridised cDNA from a cross-hybridisation on hydroxylapatite and using this in homologous and heterologous hybridisations indicated that 4% of the mRNA of transformed lymphocytes is unique to the transformed state. The analogous experiments using the cDNA synthesised from cytoplasmic poly-(A)+RNA from resting cells indicated that there were no unique resting cell mRNA sequences.

The data in this report imply that changes in only a small proportion of the number of expressed genes are necessary for the transition from the resting to the growing state in spite of a large increase in the total amount of mRNA per cell. Whether the changes in gene expression found are due to transcriptional or post-transcriptional events will be further investigated by probing the nuclei of these cells with the transformation specific cDNA's and by examining the expression of specific mRNA sequences.

REFERENCES

Torelli, G., Ferrari, S., Donelli, A., Cadossi, R.,Ferrari,S., Bosi, P., and Torelli, U. (1981). Nuc. Acids Res. 9, 7013.
Willard, K.G., and Anderson, N.L. (1980). In "Electrophoresis '79" (B.J. Radola, ed.), p. 415. Walter de Gruyter and Co., Berlin.

14

Identification of the Murine Ly-6.2 Antigen and Precursor Proteins

Claudia A. Mickelson, Vivien R. Sutton, Glen H. Tobias
and Ian F. C. McKenzie

Research Centre for Cancer and Transplantation
Department of Pathology
University of Melbourne
Parkville, Victoria, Australia

1. INTRODUCTION

The murine *Ly-6* determinant was first detected using an alloantisera (McKenzie et al, 1977). The antigen was characterized as a single lymphocyte differentiation marker, present in high levels on peripheral and activated T cells and absent or in low amounts in thymocytes and B cells (Potter et al, 1980). This report describes the use of an anti-Ly-6.2 monoclonal antibody (α-Ly-6.2 Mab) to characterize the Ly-6.2 cell surface antigens and precursor polypeptides.

2. METHODS

The monoclonal antibody used throughout, 5041-24.2 was produced from the fusion of spleen cells from CBA/H mice, immunized with C57BL/c-H-2^{bml2}spleen and lymph node cells to P3-NSI-1-AG-4 myeloma cells. The cells used for ^{35}S-methionine labelling and RNA isolation were EL4 and BW5147, both B6 T-lymphoma cell lines. Poly A^{+}RNA was isolated using oligo-d T cellulose and translated in an mRNA dependent rabbit reticulocyte lysate. The procedures used for cell labelling, immune precipitations and .1% SDS-10% acrylamide gel electrophoresis (SDS-PAGE) have been previously described (Hogarth et al, 1982).

MANIPULATION AND EXPRESSION
OF GENES IN EUKARYOTES
ISBN 0 12 513780 x

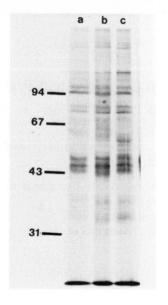

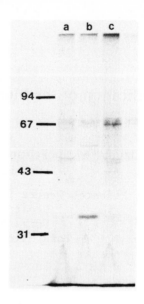

*Figure 1. SDS gel electro-
phoresis of immune
precipitates of ^{35}S-methionine
labelled proteins from EL4
lysates*
(a) Normal Mouse Serum
(b) α-H-2 Mab
(c) α-Ly-6.2 Mab.

*Figure 2 SDS gel electro-
phoresis of immune precipitates
of proteins synthesized from
EL4 poly A$^+$RNA*
(a) α-H-2 Mab
(b) α-Ly-6.2 Mab
(c) α-thymocyte Mab.

3.   RESULTS AND DISCUSSION

The 5041-24.2 monoclonal antibody was used to precipitate
the Ly-6.2 antigen from ^{35}S-methionine labelled EL4 cells.
SDS-PAGE analysis of the immune precipitate showed that the
antibody specifically precipitated two bands at 57,000 and
34,000 (Fig. 1).   To determine whether one protein was a
cleavage product of the other or merely co-precipitated as
part of a cell surface complex, the poly A$^+$RNA from EL4 cells
was translated using an mRNA dependent reticulocyte lysate.
The translation products were immune precipitated and
analysed by SDS gel electrophoresis.   Two protein bands were
observed at 54,000 and 36,000 (Fig. 2).   It is unlikely that
the two bands were the result of post-translational cleavage
occurring in the *in vitro* system as the amount of 54,000
protein relative to 36,000 protein does not alter with time

during the translation assay, (data not shown). The simplest interpretation of the *in vitro* translation data is that the two proteins each carry the Ly-6.2 determinant and are encoded by distinct mRNA molecules. We conclude that the 54,000 band observed in the *in vitro* translation is the unglycosylated precursor of the 57,000 surface glycoprotein, likewise the 36,000 translation product is the precursor for the 34,000 surface protein.

4. REFERENCES

Hogarth, P.M., Crewther, P.E., and McKenzie, I.F.C. (1982). Eur. J. Immunol. *12,* 374–379.
McKenzie, I.F.C., Cherry, M., and Snell, G.D. (1977). Immunogenetics. *5,* 25–32.
Potter, T.G., McKenzie, I.F.C., Morgan, G.M., and Cherry, M. (1980). J. Immunol. *125,* 541–545.

15

Avian Keratin Genes: Organisation and Evolutionary Inter-relationships

Keith Gregg, Stephen D. Wilton, George E. Rogers

Department of Biochemistry
University of Adelaide
Adelaide, South Australia

Peter L. Molloy

C.S.I.R.O. Molecular and Cellular Biology Unit
North Ryde, New South Wales

The keratins synthesized by birds occur in a wide range of structural tissues namely the surface of the epidermis, the beak, claws, scales and feathers. Protein studies upon keratinized tissues have led to the identification of a large number of similar but non-identical proteins and hybridization studies upon DNA and mRNA have indicated that such proteins arise from large families of homologous genes (Kemp, 1975).

The most interesting recent development in avian keratin gene studies involves the evolutionary relationship of the different genes and the way in which genome modification may contribute to the development of new genes and stabilization of existing ones. Findings of this nature are the major subject of this report.

Information on the DNA sequence of chicken feather and scale genes has been obtained from both mRNA derived clones (Saint *et al.*, manuscript in preparation) and from genomic DNA clones selected from a Charon 4A library prepared by J. Dodgson and D. Engel (Cal. Tech.). The feather genes in the clone λCFK1 are five evenly spaced genes, A, B, C, D and E, which are all transcribed in a single direction (see

MANIPULATION AND EXPRESSION
OF GENES IN EUKARYOTES
ISBN 0 12 513780 x

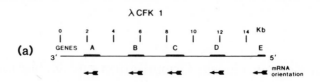

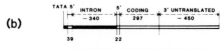

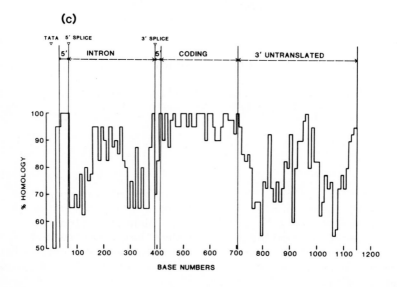

FIGURE 1. (a) Organization of feather keratin genes within the genomic clone λCFK1. (b) General structure of a feather keratin gene. Numbers denote length of each region in nucleotide bases. (c) DNA sequence homology among the five feather keratin genes.

Figure 1 (a) for details) and which all appear to have the same general structure (Figure 1(b), Molloy et al., 1982). The deduced amino acid sequences agree closely with the peptide sequences obtained previously from a number of bird species (O'Donnell & Inglis, 1974; Walker and Rogers, 1976; Walker and Bridgen, 1976) and define these genes as coding for embryonic feather keratins. There is considerable homology among the known sequences of feather keratin genes, as shown in Figure 1(c).

Clearly, the protein coding region is excellently conserved, reflecting the strong homology of amino acid sequences. However, other regions are also remarkably well conserved, most notably the 5′ untranslated region and splice sites and the sequence around the polyadenylation site. There are other small regions within the untranslated parts of the gene, which have highly conserved sequence but do not have a known function.

Sequencing of scale keratin gene clones has provided the information shown in Figure 2(a) on the arrangement which bears an overall similarity to the organisation of λCFK1. The scale gene structure known so far is shown in Figure 2(b) and also bears a resemblance to the feather gene structure. The known scale gene homologies are shown in Figure 2(c).

Using the sequence divergence between eukaryotic genes of several kinds, it has been possible to estimate their evolutionary relationship to one another. The correlation between DNA sequence divergence and evolutionary separation has been calculated previously for globin genes and for preproinsulin genes (Perler *et al.*, 1980). In those studies it was concluded that the most reliable "evolutionary clock" was provided by the DNA sequence coding for protein structure, considering specifically those positions in which base changes lead to amino acid replacements. Non-coding regions were found to diverge more rapidly and less consistently, presumably because there have been fewer constraints upon them. We have made calculations of the percentage divergence between coding and non-coding regions of the keratin genes and all calculations have been corrected for multiple events as outlined by Perler *et al.* (1980). As is the case with other genes, our calculations show that the interrelationships of protein coding regions do not agree with the calculation based upon non-coding regions (see Table 1). One possible explanation for this anomaly is that the more ancient separation is probably closer to the true evolutionary relationship while, more recently, the apparently closer homologies have resulted from intergenic exchange by events such as gene conversion. A strongly supportive observation lies in the relationship between the 3′ untranslated regions of genes B, C and E. While both genes B and C have a similar degree of divergence from gene E throughout the 3′ untranslated region (37.5% and 31.2%, respectively), the sequences can be

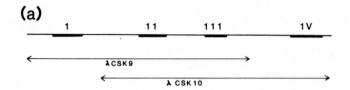

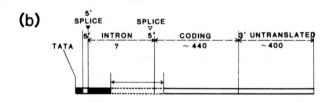

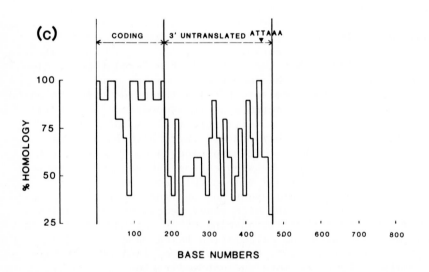

FIGURE 2. (a) Organization of scale keratin genes within the genomic clones λCSK9 and λCSK10. (b) General structure of a scale keratin gene. Note: The presence of an intron in the scale keratin gene is inferred from DNA sequence of the clone λCSK2. This clone was isolated using scale cDNA probes and has features in common with other scale genes, but has not been unequivocally identified. (c) DNA sequence homology between two scale keratin genes.

TABLE I. *Percentage divergence of protein coding (upper section) and untranslated regions (lower section) of feather keratin genes.*

	A	B	C	D	E
		Protein Coding Regions			
A		5.5	7.8	7.0	16.0
B	*58.0		1.3	1.3	16.0
C	*44.0	#34.0		0.9	16.0
D	*69.0	#54.0	#53.0		16.0
E	--	‡37.0	‡31.0	‡37.0	

Untranslated Regions

*Intron only; #Intron + 3' untranslated; ‡3' untranslated only.

divided into distinct regions of greater and lesser homology (figure 3); i.e., from base 1 to base 240 the divergence of gene B is less than that of gene C (28% compared with 45%) and from base 250 to base 440 divergence of gene B is greater (60.7% compared with 22.6%). Interestingly, the point of homology crossover is at a 19 base sequence which is

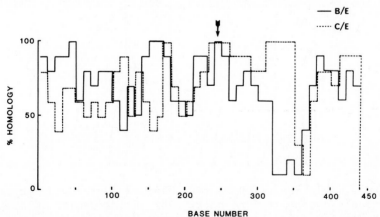

Figure 3. An example of sequence homology crossover when comparing the 3' untranslated region of genes B and C with that of gene E. The arrow denotes a 19 base sequence which is perfectly conserved in 5 out of 6 genes.

perfectly conserved in 5 out of the 6 genes sequenced over this region. The sixth has one base substitution. This might be regarded as a hot-spot for recombination events and the homology crossovers, which occur in several places, as indications of conversion-like events.

The scale genes show less evidence of intergenic exchange events (data not shown), although this may be because less sequence information is available for scale genes. It is also possible that the smaller number of embryonic scale genes might lead to their being less readily involved in exchange reactions. In the embryo there are believed to be 12 scale genes expressed (S. Wilton, manuscript in preparation) compared with about 30 expressed feather keratin genes (Kemp, 1975).

Taking into account the protein coding regions only, an evolutionary tree can be constructed as in figure 4(a). However, this appears to be an over-simplification and a more likely pattern, taking into account the probable conversion events, is shown in figure 4(b). When the large number of genes is considered the true picture must be still more complex. Both figures indicate a common ancestor for avian scales and feathers, a relationship which has been postulated for many years (Spearman, 1964). In figure 5 we show that this putative ancestry for feather genes is supported by the homology which can be demonstrated between genes for scale and feather keratins. Clearly there has

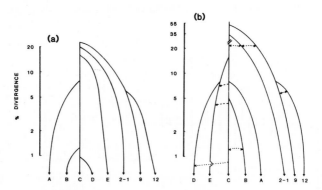

Figure 4. (a) Evolutionary tree derived from the divergence of protein coding regions only. (b) Relationship of the avian keratin genes, taking into consideration the divergence of untranslated region and coding regions. Broken lines indicate untranslated region exchange. Dotted lines indicate coding region exchange. Feather (A-E), scale (2-1, 9 & 12).

been much modification since the original divergence,
particularly in the deletion of a direct repeat sequence of
39 bases which is present as five copies in the scale gene
depicted in figure 5. However, the very well conserved
region clearly demonstrates the relationship between these
two gene types.

Divergence analysis of the coding regions which exhibit
in-phase homology, showed 23% corrected divergence. Since
the development of birds from reptile ancestors is believed
to have occurred approximately 200 million years ago, it is
calculated that a 1% change in sequence occurs within 8.7
million years (MY) compared with 10 MY for globin genes and
4.8 MY for preproinsulin genes.

Continuing work on the avian keratin genes, centred
largely around the better known feather genes, is aimed at
establishing the functional significance of the family
characteristics such as the intron. One possibility is that
there might be an alternative promoter region for each gene
which can be spliced to give rise to an alternative 5′
leader sequence on the mRNA in another tissue or at another
stage of development, analogous to the mouse α-amylase
system (Young *et al.*, 1981). Investigations are also
proceeding on the arrangement of nucleosome complexes and

*Figure 5. Sequence comparison between an avian scale
keratin gene and a feather keratin gene. Base homology is
represented by the dark line. S, scale gene; F, feather
gene.*

the patterns of DNA methylation which occur through developmental stages of keratin synthesis. It is hoped that we might ultimately be able to define the signals which lead to the characteristically sudden and overwhelming induction of the keratin gene products in avian skin appendages.

ACKNOWLEDGMENTS

This project was supported, in part, by a research grant from the Australian Research Grants Scheme.

REFERENCES

Kemp, D.J. (1975). *Nature 254*, 573-577.
Molloy, P.L., Powell, B.C., Gregg, K., Barone, E.D. and
    Rogers, G.E. (1982). *Nucleic Acids Res.* (submitted)
O'Donnell, I.J. and Inglis, A.S. (1974). *Aust. J. Biol. Sci.*
    27, 369-382.
Perler, F., Efstratiadis, A., Lomedico, P., Gilbert, W.,
    Kolodner, R. and Dodgson, J. (1980). *Cell 20*, 555-565.
Saint, R.B., Morris, C.P., Crowe, J.S., Kuczek, E.S.A. and
    Rogers, G.E. (manuscript in preparation).
Spearman, R.I.C. (1964). *In* Symp. Zool. Soc. Lond. vol.12,
    p.67. Acad. Press, London.
Walker, I.D. and Bridgen, J. (1976). *Eur. J. Biochem. 67*,
    283-293.
Walker, I.D. and Rogers, G.E. (1976). *Eur. J. Biochem. 69*,
    341-350.
Young, R.A., Hagenbrickle, O. and Schibler, V. (1981). *Cell
    23*, 451-458.

16
Chicken Histone Genes: Analysis of Core, H1, H5 and Variant Sequences

Julian R. E. Wells, Leeanne S. Coles, Richard D'Andrea,
Richard P. Harvey, Paul A. Krieg, Allan Robins
and Jennifer Whiting

Department of Biochemistry
University of Adelaide
Adelaide, South Australia

Developmental control of eukaryotic gene expression must eventually account for mechanisms which initiate and maintain sets of genes in an active or inactive state for long periods. We are investigating the control of histone genes whose products are intimately associated with packaging DNA in eukaryotic cells and which may be involved in gene activity. There is little doubt that histones influence the packaging of DNA. Primary sequence variants of histones exist (except H4) and these, together with post-translational modifications may influence DNA/histone interactions (Isenberg, 1979).

We are especially interested in tissue-specific histone genes and their biological function. Avian red blood cells contain a prominent tissue-specific histone, H5, which is H1-like and which accumulates during red cell maturation. This highly basic protein is present in early dividing erythroblasts (Appels *et al*., 1972) and so its presence *per se* does not prevent DNA replication and cell division. However, increasing levels of this histone can be correlated with increasing condensation of nuclei and eventual template inactivity (Weintraub, 1978). If H5 plays a role in this process, it may be an example of a more general phenomenon,

73

MANIPULATION AND EXPRESSION
OF GENES IN EUKARYOTES
ISBN 0 12 513780 x

namely that tissue-specific histone variants are involved in chromatin re-modelling.

I.   CHICKEN HISTONE GENES ARE CLUSTERED, BUT DISORDERED

All studies so far are consistent with the proposition that chicken histone genes are clustered but not ordered (Harvey *et al.*, 1981; Engel & Dodgson, 1981). In this sense there is an obvious difference between the organisation of these vertebrate genes and the tandemly repeated clusters typified by embryonic sea urchin histone genes and *Drosophila* genes (Hentschel & Birnstiel, 1981).

Examination of overlapping chicken histone genomal clones has not revealed any long-order pattern (if such exists). Figure 1a shows the positions of histone genes found in clones overlapping λCH-01. While there is conservation of gene order around the single H3 gene of λCH-01, the spacing of genes and restriction sites is not maintained. There are long stretches devoid of histone genes (up to 9 kb between the H4 and H1 genes in λCH-01) which contrasts with the close association of genes in other regions of the cluster. Particularly close are the divergent H2A/H2B pair in λCH-01, in which respective TATA boxes are only separated by 77 base-pairs of DNA. This presumably sets some limits for 5′ transcriptional control regions assuming both genes are expressed. It may also be a site for an origin of DNA replication if the proposal made by Smithies turns out to be correct (Smithies, 1982).

The embryonic sea urchin histone genes are highly reiterated and organised into conserved, tandemly repeated units (Hentschel & Birnstiel, 1981). Conversely, the chicken histone genes are reiterated about ten-fold (Crawford *et al.*, 1979) and, as discussed above, are disorganised. In the former case it is likely that gene correction mechanisms will maintain homogeneity of the gene cluster, whereas with the chicken system, the gene copy number has dropped sufficiently that the frequency of correctional events will not maintain such homogeneity. In *Xenopus* an intermediate situation is evident. Some histone genes exist in tandem arrays (Zernick *et al.*, 1980), but the order of genes is not maintained in all repeats. It is significant that in the newt, where reiteration frequency is again high, histone genes appear once again in tandem arrays (Stephenson *et al.*, 1981).

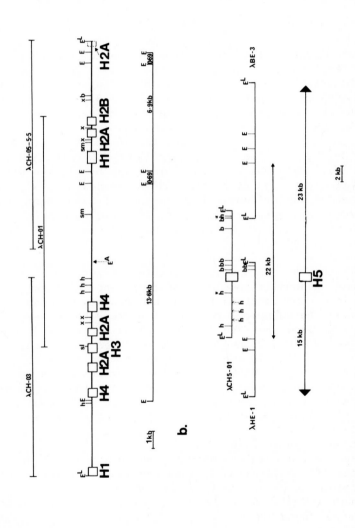

FIGURE 1.   *Location of chicken histone genes in overlapping genomal clones.*   *(a) Core and H1 genes.   (b) H5 gene.   (Note scale change.)   E*L, *EcoRI linkers; E, EcoRI; E*A, *an allelic EcoRI site; b, BamHI; h, HindIII; sl, SalI; sm, SmaI; x, XhoI.*

II. AN UNLINKED H5 GENE

H5 is found in red cells of birds, reptiles amphibians and fish. This tissue-specific histone gradually (but not completely) replaces H1 during maturation of red blood cells. A functional counterpart of H5 in mammalian tissues is the H1^{o} protein which increases in cells with a decreased rate of cell division (Perhson & Cole, 1980).

We are investigating regulation of the chicken H5 gene and its relationship to other histone genes. An extended synthetic primer was used to detect H5-coding sequences from chicken cDNA recombinants (Krieg *et al.*, 1982) and a cloned probe was then used to isolate a genomal H5 gene from a Charon 4A library by standard procedures.

The genomal clone λCH5-01 contains 16 kb of insert DNA and the H5 gene is almost centrally located within this (Figure 1b). When DNA from this clone was hybridised with ^{32}P-labelled core and H1 histone DNA sequences, no hybridisation was evident. In a reciprocal experiment, H5-coding sequences were hybridised to DNA from some 50 recombinants of chicken histone genes in λ Charon 4A. No hybridisation was detected. These results are consistent with absence of close linkage of H5 and other histone genes, but other interpretations are possible. First, the 50 independent genomal clones may not represent the total histone gene complement in the chicken genome. Secondly, we have already noted spacer regions of up to 9 Kb between histone genes and thus the 7.5 Kb regions on each side of the H5 gene in λCH5-01 needed to be extended for further analysis. The data in Figure 1b show the positions of clones overlapping λCH5-01. Together these three inserts correspond to some 39 Kb of contiguous DNA. None of these inserts gave positive hybridisation signals with homologous core or H1 histone gene probes and it therefore appears that the isolated H5 gene is not closely linked to histone genes of the type shown in Figure 1a.

In summary, a complete H5 gene has been isolated and sequenced. It contains no introns and is not clustered with other histone genes of the type seen in the recombinant λCH-01. Southern blot analysis indicate that the sequence found in λCH5-01 is the major chicken H5 gene in the genome.

III.   RECOGNITION SEQUENCES IN CHICKEN HISTONE GENES

A. 5' Sequences

DNA sequence comparisons of a number of genes transcribed by polymerase II have identified two promoter elements. The 'TATA box' at -30 corresponds to a selector sequence since, in its absence, initiation occurs from several sites; the 'CAT box' at -80 acts as a modulator of gene expression (Grosschedl & Birnstiel, 1980). Chicken histone genes, in common with many other polymerase II transcribed genes, contain these sequences.

The H2A/H2B gene pair of λCH–01 (Figure 1a) was referred to previously because of the close proximity of the 5' ends of these divergently transcribed genes. Similar organisation has been described in histone genes of other species (Hentschel & Birnsteil, 1981), for silk moth chorion genes (Jones & Kafatos, 1980), and *Drosophila* heat shock genes (Keene *et al.*, 1981). It has been suggested that coordinate expression of these gene pairs may occur via shared promoter sequences. The inter-gene region of the chicken H2A/H2B pair has been sequenced and each contains a conserved 'TATA box'. Between these there are 77 base pairs of DNA containing only one obvious 'CAT box' oriented as a modulator for the H2B gene, but located almost centrally in the inter-gene region. The lack of any obvious H2A gene 'CAT box' and the striking symmetry of the overall arrangement suggest that the H2B 'CAT' sequence may in fact be shared by both genes. It may be particularly instructive to manipulate this intergene region for *in vivo* or *in vitro* transcription studies.

Thus far only well established 5' promoter elements have been considered. However, our analysis of isocoding genes in the chicken genome shows that coding regions are highly conserved but flanking regions are immediately and obviously divergent. This situation of independently evolving genes within the same species coding for the same protein presents an ideal opportunity to look for conserved gene specific regions within the flanking sequences.

The 5' leader regions of the H2B genes found in λCH–01 and λCH–02 (Harvey *et al.*, 1981) have been examined in detail. In addition to the CAT box and TATA box, a third element of 13 base pairs was found in both genes located between the two boxes. The sequence, 5' CTCATTTGCATAC 3' was then found in all other H2B genes (but not other histone genes) for which sequence data is available. These include

H2B genes of sea urchin species, *P. miliaris* (Busslinger
et al., 1980) and *S. purpuratus* (Sures *et al.*, 1978) and for
Xenopus (Moorman, personal communication).   It seems likely
that this conserved sequence will be found as an H2B gene
specific element in histone genes of other species.   Its
location and conservation strongly suggests that this 13
base-pair element   may be involved in the regulation of H2B
gene expression.

B.   *3' Sequences*

     A hyphenated dyad symmetry element in the 3' non-coding
region of sea urchin histone genes (Busslinger *et al.*, 1979)
was also found in a chicken H2A sequence (D'Andrea *et al.*,
1981) and subsequently in a variety of histone genes from
many sources (Hentschel & Birnstiel, 1981).   All chicken
core histones so far examined contain this element with the
exception of an unusual H3 gene (Engel *et al.*, 1982).   It is
also present in H1 genes (our unpublished data), but not in
the tissue-specific H5 gene.   The absence of the histone
gene-specifc element in H5 is one obvious difference between
it and H1 genes.   The hyphenated dyad element is involved in
termination of transcription but insufficient on its own for
this function (Birchmeier *et al.*, 1982).
     In summary, examination of 5' chicken histone gene
sequences has revealed a candidate for an H2B gene-specific
signal.   In 3' sequences, H5 is unusual in that it does not
contain the highly conserved hyphenated-dyad symmetry
element associated with termination of transcription.

IV.   AN EXTREME VARIANT CORE HISTONE GENE

     We have constructed cDNA recombinants from 5 day chicken
embryo RNA which had been enriched for histone mRNA
(Crawford *et al.*, 1979).   Colonies were screened with gene-
specific core histone probes.   The intensity of
hybridisation to different clonies with either H2A or H2B
gene probes was strikingly different and not explained by
concentration or insert length effects.
     An insert from the most weakly hybridising H2A gene
plasmid has been studied in more detail.   Translation of the
DNA sequence into a polypeptide shows that this expressed
sequence is an extreme H2A variant since the conserved
sequence of amino acids found in all H2A proteins so far

studied (^{21}ala-gly-leu-gln-phe-pro-val-gly-arg^{29}) is
present. If the variant sequence is aligned with the known
amino acid sequence of the major chicken H2A.1 protein,
other blocks of amino acid homology are evident throughout
the sequence. However, the striking result is that there is
some 40% mis-match in amino acid sequence, ranging from
single residues up to blocks of seven amino acids. Because
the amino acid sequence data for variant histones is sparse,
it is difficult to assign this sequence to a particular
protein. Peptide mapping data from West and Bonner (1980
and personal communication) indicate that it is similar to
H2A.z. Whereas the abundant H2A.1 protein contains an
N-terminal serine residue which is subject to
phosphorylation *in vivo*, both H2A.z and the variant
described here lack this residue.

Preliminary 'Northern' blot analysis suggest that the
RNA is polyadenylated and is restricted in tissue
distribution. It will be of interest to see if such a
variant core histone modifies nucleosome structure in a
biochemically significant fashion.

In addition to the H2A, we also have preliminary
evidence for an H2B variant. Other unusual histone genes
include the H5 which is expressed only in red cells and a
chicken H3 histone gene with introns (Engel *et al*., 1982).
These examples suggest that there is a great deal more to be
learnt about vertebrate histone genes and the way in which
their products package DNA.

V. SUMMARY

1. Chicken histone genes are clustered but no repeat
    pattern is observed. Spacer regions between the genes
    vary from 9 Kb down to 0.13 Kb. Transcription occurs
    from both strands.
2. The red cell-specific histone H5 gene is not closely
    linked with other histone genes (39 Kb examined). It
    lacks the histone gene specific 3' terminator sequence.
3. A 13 base-pair H2B gene specific 5' sequence found in
    chicken genes is present in all other H2B genes
    examined (*Xenopus* and sea urchin).
4. An extreme H2A variant has been deduced from DNA
    sequence analysis of a cDNA clone.

REFERENCES

Appels, R., Wells, J.R.E. and Williams A.F. (1972). *J. Cell Sci. 10*, 47–59.

Birchmeier, C., Grosschedl, R. and Birnstiel, M.L. (1982). *Cell, 28*, 739–745.

Busslinger, M., Portmann, R., Irminger, J.C. and Birnstiel, M.L. (1980). *Nucl. Acids. Res. 8*, 957–977.

Busslinger, M., Portmann, R. and Birnstiel, M.L. (1979). *Nucl. Acids Res. 6*, 2997–3008.

Crawford, R.J., Krieg, P., Harvey, R.P., Hewish, D.A. and Wells, J.R.E. (1979). *Nature, 279*, 132–136.

D'Andrea, R., Harvey, R.P. and Wells, J.R.E. (1981). *Nucl. Acids Res. 9*, 3119–3128.

Engel, J.D. and Dodgson, J.B. (1981). *Proc. Natl. Acad. Sci. U.S.A. 78*, 2856–2860.

Engel, J.D., Sugarman, B.J. and Dodgson, J.B. (1982). *Nature, 297*, 434–436.

Grosschedl, R. and Birnstiel, M.L. (1980). *Proc. Natl. Acad. Sci. U.S.A. 77*, 1432–1436.

Harvey, R.P., Krieg, P.A., Robins, A.J., Coles, L.S. and Wells, J.R.E. (1981). *Nature, 294*, 49–53.

Hentschel, C.C. and Birnstiel, M.L. (1981). *Cell, 25*, 301–313.

Isenberg, I. (1979). *Ann. Rev. Biochem., 48*, 159–191.

Jones, C.W. and Kafatos, F.C. (1980). *Cell, 22*, 855–867.

Keene, M.A., Corces, V., Lowenhaupt, K. and Elgin, S.C.R. (1981). *Proc. Natl. Acad. Sci. U.S.A., 78*, 143–146.

Krieg, P.A., Robins, A.J., Gait, M.J., Titmas, R.C. and Wells, J.R.E. (1982). *Nucl. Acids Res., 10*, 1495–1502.

Pehrson, J. and Cole, R.D. (1980). *Nature, 285*, 43–44.

Smithies, O. (1982). *J. Cell Physiol. Supp., 1*, 137–143.

Stephenson, E.C., Erba, H.P. and Gall, J.G. (1981). *Cell, 24*, 639–647.

Sures, I., Lowry, J. and Kedes, L.H. (1978). *Cell, 15*, 1033–1044.

Weintraub, H. (1978). *Nucl. Acids Res., 5*, 1179–1188.

West, H.P. and Bonner, W.M. (1980). *Biochemistry, 19*, 3238–3245.

Zernick, M., Heintz, N., Boime, I. and Roeder, R. (1980). *Cell, 22*, 807–815.

Part II
Amphibians, Fish and Insects

Introduction

Research in development in animals has made use of species from different phyla. Some of the classical experimental systems are still used to contribute to knowledge at the molecular level of regulation of gene expression in eukaryotes. For example, the large size of amphibian oocytes has enabled them to be used in detailed analyses of expression of gene segments derived from many different animal sources.

One of the best analysed systems of gene expression in higher animals is the 5S ribosomal RNA genes of *Xenopus*. The unique feature of these genes is the transcription control properties of a sequence in the middle of the coding region; in addition, some regulatory proteins are known that interact with this region of the gene. The differential expression of somatic and oocyte 5S RNA genes is another gene regulation event which is well established, but its control mechanisms have not yet been exposed by recombinant DNA techniques. Chapter 17 combines classical and molecular approaches to the problem of regulation of transcription of these genes.

Drosophila has also been important in developmental studies, but the recent development of the gene transfer system based on mobile gene elements has vastly increased the options for research into the development of this organism. This application of the P element as a gene vector in *Drosophila* will lead to many advances in our understanding of control of gene action. Chapter 20 demonstrates the use of specific gene transposons in analysing chorion gene amplification and expression.

Another special feature of *Drosophila* and other dipteran insects which is useful in genetic and developmental analyses is the polyteny of their chromosomes in some tissues. Chapter 21 reports advances in understanding salivary polytene chromosome structure and Chapter 22 discusses gene expression in polytene chromosomes using *Rhynchosciara* which provides remarkable examples of gene amplification. The control and mechanism of these two cases of localized gene amplification will ultimately be analysed to a level equivalent to that of the *Drosophila* chorion genes, and it may be desirable to develop a gene transfer system for the sciarids.

We can expect that in the next few years many of the complex regulatory mechanisms behind ordered development will come to be understood at a molecular level.

17

Transcription of *Xenopus* 5S Ribosomal RNA Genes in Somatic Nuclei Injected into *Xenopus* Oocytes

Laurence Jay Korn, Jennifer Price

Department of Genetics
Stanford University School of Medicine
Stanford, California 94305 U.S.A.

John B. Gurdon

MRC Laboratory of Molecular Biology
University Medical School
Hills Road, Cambridge, CB2 2QH
England

This article reviews the work in our laboratories on transcription of the *Xenopus* 5S ribosomal RNA genes, focusing on the developmental regulation of transcription of these genes. There are two types of 5S RNA (Brown and Sugimoto, 1973), the oocyte-type (Wegnez *et al.*, 1972; Ford and Southern, 1973) and the somatic-type (Ford and Southern, 1973). In *Xenopus laevis* there are 20,000 copies of the oocyte-type 5S RNA genes and 400 copies of the somatic-type 5S RNA genes per haploid genome (Peterson *et al.*, 1980). Transcripts from these two types of genes are the same length but differ in sequence by six nucleotides, allowing them to be distinguished (Ford and Brown, 1976; Korn and Gurdon, 1981). In oocytes (immature, unfertilized eggs), both the oocyte- and somatic-type 5S

This work was supported by NSF, Grant No. PCM 8104522 and March of Dimes Birth Defects Foundation, Grant No. 5-323 to L.J.K.

MANIPULATION AND EXPRESSION
OF GENES IN EUKARYOTES
ISBN 0 12 513780 x

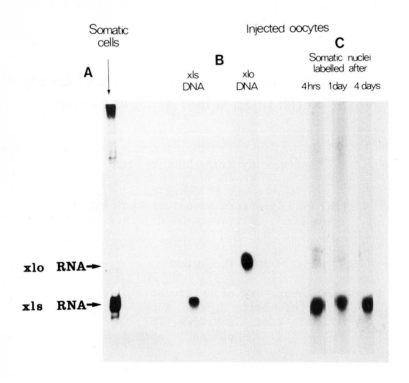

FIGURE 1.   *5S RNA transcription from 5S DNA and somatic nuclei injected into* Xenopus *oocytes.   (A) Cultured somatic cells were labelled and RNA extracted by the method of Brown and Littna (1966).   RNA was electrophoresed on a non-denaturing 17% polyacrylamide gel (Korn and Gurdon, 1981).   (B) 5 ng DNA (pXlo8-pBR322 containing 4 repeat units of* X. laevis *oocyte-type 5S DNA-Brown and Gurdon, 1978, or pXls11-pBR322 containing 1 repeat unit of* X. laevis *somatic-type 5S DNA-Peterson* et al., *1980) and 1 μCi α-³²P-GTP were injected into the germinal vesicle of oocytes.   The oocytes were incubated at 20°C for 24 hrs and the RNA extracted and electrophoresed as above.   X.* laevis *oocyte-type 5S RNA (Xlo);* X. laevis *somatic-type 5S RNA (Xls).   (C) Approximately 500 somatic nuclei (Gurdon, 1976) were injected into individual oocytes, aiming for the germinal vesicle, followed by a second injection of 1 μCi α-³²P-GTP at 4 hrs, 1 day or 4 days after the first injection.   Half the RNA from individual oocytes was electrophoresed in each lane of the gel.*

RNA genes are expressed (Wegnez *et al.*, 1972; Ford and
Southern, 1973). In contrast, in somatic cells, the somatic-type 5S RNA genes are expressed, but the oocyte-type
genes are transcriptionally inactive (Fig. 1A). It is this
developmental control of transcription, the switching off
of oocyte-type 5S RNA synthesis, that we wish to understand on a molecular level.

The transcription of the 5S RNA genes has been studied
extensively and has been recently reviewed (Brown, 1981;
Korn, 1982; Korn and Bogenhagen, 1982). When injected into
the nucleus (germinal vesicle) of *Xenopus* oocytes or incubated in a cell-free extract, 5S DNA is accurately and
efficiently transcribed (Brown and Gurdon, 1977; 1978;
Gurdon and Brown, 1978; Birkenmeier *et al.*, 1978; Korn *et*
al., 1979; Brown *et al.*, 1979; Ng *et al.*, 1979; Weil *et*
al., 1979). These transcription systems have helped
identify the DNA sequences and other macromolecular components required for faithful synthesis of 5S RNA. Unfor-
tunately, however, it has not yet been possible to demonstrate any developmental regulation of transcription,
using purified DNA. When DNA is isolated from somatic
cells (in which the oocyte-type genes are inactive) and
injected into *Xenopus* oocytes, the oocyte- and somatic-
type 5S RNA genes are both transcribed efficiently (Brown
and Gurdon, 1977). Similarly, both cloned oocyte- and
somatic-type 5S DNA are transcribed when injected into
oocytes (Fig. 1B; Brown and Gurdon, 1978; Gurdon and
Brown, 1978). We are therefore studying the transcription
of the 5S RNA genes as part of intact chromosomes within
somatic cell nuclei when such nuclei are injected into the
germinal vesicle of *Xenopus* oocytes (Korn and Gurdon, 1981).
(Due to its very large size, it is possible to inject 500-
1000 somatic nuclei into one germinal vesicle.) Analysis
of RNA transcripts from nuclei injected into oocytes of
certain frogs, shows that, as part of chromosomes, the 5S
RNA genes do maintain their regulated state. The oocyte-
type genes in the somatic cell chromosomes remain quiescent
while the somatic-type genes are expressed as efficiently
as they are in somatic cells (Fig. 1C; Korn and Gurdon,
1981). This finding leads to two important conclusions.
First, 5S RNA genes as part of intact chromosomes behave
differently from purified 5S RNA genes with respect to
regulation of transcription. Second, the developmentally
regulated transcription pattern of 5S RNA genes can be
maintained when genes are transferred from one cell to
another as part of chromosomes.

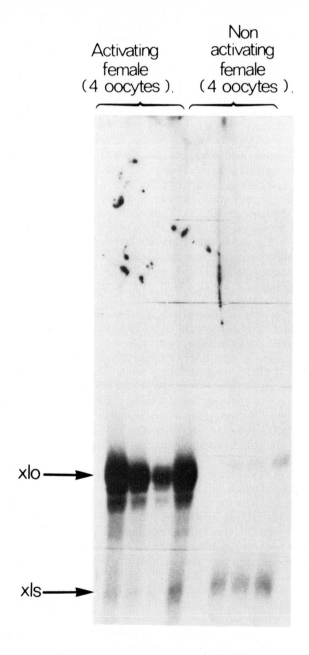

FIGURE 2. *Activating and non-activating oocytes.*
About 500 somatic cell nuclei (Gurdon, 1976) were injected

with 1 µCi α-32P-GTP into individual oocytes from a X.
laevis *activating female (lanes 1-4) or non-activating*
female (lanes 5-8), aiming for the germinal vesicle.
Labelling was for 24 hrs. The RNA was extracted and elec-
trophoresed on a polyacrylamide gel as described in the
legend to Fig. 1. The positions of X. laevis *oocyte- and*
somatic-type 5S RNA, prepared as described in Figure
Legend 1 are indicated by arrows. It should be recalled
that there are 50 times more oocyte- than somatic-type
5S RNA genes (Peterson et al., 1980).

This is not the case with oocytes from all frogs, however. In several experiments, we have found that the
oocyte-type 5S genes are in fact reactivated (Korn and
Gurdon, 1981). Figure 2 shows the result of injecting
samples of the same suspension of somatic nuclei into
oocytes from two female frogs. In the oocytes of one female, oocyte-type 5S genes were activated (activating
oocytes); in oocytes of the other, they remained quiescent
(non-activating oocytes). Oocytes taken from the same
female at different times over a one year period show the
same "activating" or "non-activating" pattern as initially
determined (Korn and Gurdon, 1981). Furthermore, the
synthesis of somatic-type 5S RNA in injected oocytes is
approximately the same, whether or not the oocyte-type
5S RNA genes are activated (Fig. 2). Somatic-type 5S RNA
therefore provides an essential control, eliminating the
possibility that the inactivity of oocyte-type 5S RNA genes
could be due to the inviability of the injected nuclei in
these oocytes. We conclude that the developmental inactivation of oocyte-type 5S genes takes place by a process
that is readily reversible by components of the oocytes of
some, but not all, females.

We have further analyzed the properties of these different kinds of oocytes. Extracts prepared from activating
or non-activating oocytes have been used to search for
factors which affect the transcription of the 5S RNA genes.
Somatic nuclei were injected, either alone or after mixing
with extracts prepared from either activating or non-
activating oocytes, into the germinal vesicles of both
activating and non-activating oocytes. Labelled transcripts were extracted (Brown and Littna, 1966) and electrophoresed on polyacrylamide gels to differentiate between
somatic- and oocyte-type 5S RNA (Korn and Gurdon, 1981).
When somatic nuclei were mixed with extracts prepared from

activating oocytes and injected into non-activating oocytes,
the previously inactive oocyte-type 5S RNA genes in the
somatic nuclei were now transcribed as efficiently as when
such nuclei were injected into activating oocytes (Korn,
Price and Gurdon, unpublished results). On the other hand,
extracts from non-activating oocytes, injected with somatic
nuclei into activating oocytes did not diminish the expression of the oocyte-type 5S RNA genes. Mixing the
somatic nuclei with extracts prepared from the same kind
of oocyte into which they were then injected gave the same
result as when somatic nuclei were injected alone into
either activating or non-activating oocytes. Since activating extracts exert their influence in non-activating
oocytes and not vice versa, it appears that at least one
factor controlling the reactivation of oocyte 5S genes
regulates 5S transcription in a positive fashion. In
addition, we have shown the activating factor to be heat
labile and protease sensitive, indicating that a protein
component is involved in activation (Korn, Price and
Gurdon, unpublished results).

A key question in modern biology is how gene expression
is regulated during development. The molecular mechanisms
of activation or inactivation are not yet known for any
gene in a higher eukaryotic system. We have shown for a
particular gene system, the oocyte-type 5S RNA genes of
Xenopus, that specific, although so far unidentified,
protein factors can activate transcription. We are currently attempting to purify and analyze these factors.

ACKNOWLEDGMENTS

We would like to thank P. Botchan, P. Farnham, C.L.
Queen and J. Warwick for their helpful suggestions in the
preparation of this manuscript and J. Warwick for typing
this manuscript.

REFERENCES

Birkenmeier, E.H., Brown, D.D. and Jordan, E. (1978). *Cell*
  15, 1077.
Brown, D.D. (1981). *Science* *211*, 667.
Brown, D.D. and Gurdon, J.B. (1977). *Proc. natn. Acad. Sci.*
  U.S.A. *74*, 2064.

Brown, D.D. and Gurdon, J.B. (1978). *Proc. natn. Acad. Sci. U.S.A. 75*, 2849.

Brown, D.D., Korn, L.J., Birkenmeier, E., Peterson, R. and Sakonju, S. (1979). *ICN-UCLA Symp. 14*, 511.

Brown, D.D. and Littna, E. (1966). *J. molec. Biol. 20*, 95.

Brown, D.D. and Sugimoto, K. (1973). *Cold Spring Harb. Symp. quant. Biol. 38*, 501.

Ford, P.J. and Brown, R.D. (1976). *Cell 8*, 485.

Ford, P.J. and Southern, E.M. (1973). *Nature new Biol. 241*, 7.

Gurdon, J.B. (1976). *J. Embryol. exp. Morph. 36*, 523.

Gurdon, J.B. and Brown, D.D. (1978). *Devl. Biol. 67*, 346.

Korn, L.J. (1982). *Nature 295*, 101.

Korn, L.J., Birkenmeier, E.H. and Brown, D.D. (1979). *Nucleic Acids Res. 7*, 947.

Korn, L.J. and Bogenhagen, D.F. (1982). *In "The Cell Nucleus"* (H. Busch and L. Rothblum, eds.), Vol. XII. Academic, New York.

Korn, L.J. and Gurdon, J.B. (1981). *Nature 289*, 461.

Ng, S.Y., Parker, C.S. and Roeder, R.G. (1979). *Proc. natn. Acad. Sci. U.S.A. 76*, 136.

Peterson, R.C., Doering, J.L. and Brown, D.D. (1980). *Cell 20*, 131.

Wegnez, M., Monier, R. and Denis, H. (1972). *FEBS Lett. 25*, 13.

Weil, P.A., Segall, J., Harris, B., Ng, S.Y. and Roeder, R.G. (1979). *J. biol. Chem. 254*, 6163.

18

Characterization of a Cloned Repeat Element Derived from Polyadenylated Nuclear RNA from Gastrula Stage of *Xenopus laevis*

Wolfgang Meyerhof, Maliyakal E. John and Walter Knöchel

Institut für Molekularbiologie und Biochemie
Freie Universität Berlin
Berlin, Germany

To investigate a possible role of transcribed repeat elements in gene expression we have prepared a cDNA clone bank from polyadenylated nuclear RNA of gastrula stage (1) and screened with cDNA prepared from template RNA as well as with nick-translated nuclear DNA. 108 clones hybridized strongly to the cDNA probe and 9 clones to nick-translated DNA. None of the clones had strongly hybridized to both labelled probes. This means we have not detected sequences that are highly reiterated within the genome and also abundant in polyadenylated gastrula nuclear RNA. Plasmid DNAs from 5 clones containing repeat elements were isolated and digested with various restriction enzymes (Table I). Comparison of restriction enzyme recognition sites enables to distinguish between four out of the 5 repeat elements but fails to differentiate between clone 1 and 9. Since DNA from clone 9 does not hybridize to DNA from the other four clones (not shown) all five repeat elements are assumed to belong to different repeat families.

To investigate the genomic organization of the repeat element of clone 9 we have digested nuclear DNA from adult liver with different restriction enzymes (Fig. 1A). After gel electrophoresis and Southern transfer (2), digests were hybridized to DNA of clone 9. Hybridization signals cover a broad range from high to low molecular weights. Whereas sequence cluster would be digested to only a limited number of restriction fragments, the ubiquitous hybridization signals indicate

91

MANIPULATION AND EXPRESSION
OF GENES IN EUKARYOTES
ISBN 0 12 513780 x

Table I. Characteristics of cloned repeats

clone	enzyme cleavage sites[a]	insert lengths (bp)	reiteration frequency[b]
1	Bam HI, Sau 96I	180	24 000
2	none	210	6 000
3	Hae III, Alu I	230	3 000
6	Eco RI, Pst I, Hae III	700	3 500
9	Bam HI, Sau 96I (2x)	178	15 000

[a] enzymes tested: Bam HI, Eco RI, Hind III, Alu I, Hae III, Pst I, Hha I, Sau 96I
[b] calculated on the basis of % nuclear DNA hybridized and Xenopus genome size of 3.1×10^9 bp (3).

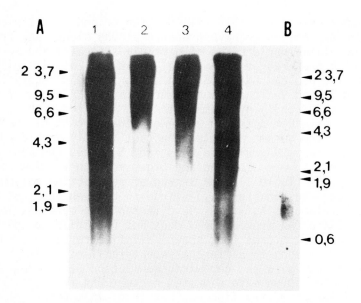

Fig. 1. (A) Hybridization of restricted nuclear DNA to nick-translated DNA of clone 9; Hind III (1), Eco RI (2), Hpa II (3), Pst I (4). (B) Hybridization of labelled plasmid DNA from clone 9 to size fractionated renatured and S_1 trimmed DNA. Hind III fragments of λ DNA as size markers.

that clone 9 repeat element is scattered throughout the genome. To investigate length distribution of genomic repeat elements complementary to clone 9 repeat, we performed a S_1-trimming experiment (4). As can be seen from fig. 1B most intense autoradiographic signals cover a size range of about 1.5 KBP to 0.8 KBP. Thus we conclude that the sequence of clone 9 is part of a repetitive genomic DNA element, which is considerably longer than the cloned sequence.

The representation of repeat element of clone 9 among RNA transcripts is investigated by Northern blot analysis (5). It can be seen from fig. 2 that clone 9 sequence hybridizes with transcripts of different sizes. This result implies clone 9 repeat to be part of different transcription units. Fig. 2 also demonstrates very clearly that transcription of the clone 9 repeat is regulated in the course of embryogenesis, as we can detect presence of this repeat element only at neurula and tadpole stages. However, we would suggest this sequence to be present at low copy numbers also at gastrula stage, since our cDNA clone bank is derived from RNA of that stage. Furthermore.

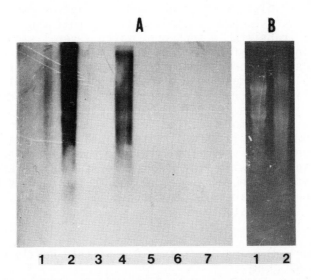

Fig. 2. Northern blot hybridization of labelled plasmid DNA from clone 9. Left: RNA from tadpole polysomes (1), tadpole nuclei (2), neurula polysomes (3), neurula nuclei (4), gastrula polysomes (5), gastrula nuclei (6), unfertilized eggs (7) were size fractionated and after transfer hybridized to nick-translated DNA from clone 9. Right: acridine orange stain·ing of gastrula nuclear RNA (1) and gastrula polysomal RNA (2)

the clone 9 repeat element hybridizes only with nuclear RNAs but not with polysomal RNAs. We cannot yet decide whether the various transcripts containing the repeat element are entirely confined to the nucleus or whether the repeat element is removed in the event of cytoplasmic export of the associated sequences.

The nucleotide sequence of the cloned Xenopus repeat element is shown in fig. 3. As we could not orientate the sequen-

```
5' ...GGG GGTATTGGCT TTTAATGTTT ACATGATTTT CTGGTAGACA AGGTATGA
3' ...CCC CCATAACCGA AAATTACAAA TGTACTAAAA GACCATCTGT TCCATACT
AG ATCCTAATTG CAGAAGGATC CGTTGTCTGG AAAGCCCCAG GTCCCGAGCA TTCT
TC TAGGATTAAC GTCTTCCTAG GCAACAGACC TTTCGGGGTC CAGGGCTCGT AAGA
GGATAA CAGGTCCCCT ACCTGTACTA TTTTTCTGAC CCCC... 3'
CCTATT GTCCAGGGGA TGGACATGAT AAAAAGACTG GGGG... 5'
```

Fig. 3. Nucleotide sequence of clone 9 repeat

ce with respect to template RNA, both strands are shown. The sequence exhibits start codons and exon/intron donor site consensus elements. However it is unlikely that the repeat element is a translatable sequence, since several stop codons are present in 5 out of 6 reading frames (3 on each strand). Presence of stop codons in repeats has also been reported by others (6). Provided there are protein coding single-copy sequences linked to repeat elements containing stop codons, then in these instances removal of repeat elements seems to be a necessary step before translation.

In summary it is shown that a cloned repeat element is transcribed stage specifically during early embryogenesis and it is associated with many RNA molecules of different size. This repetitive sequence is confined to the nucleus and due to the presence of various stop codons, it is unlikely that this element is translatable.

1. Knöchel, W., and Bladauski, D., Wilhelm Roux's Archives 190, 97 (1981).
2. Southern, E. M., J. Mol. Biol. 94, 51 (1975).
3. Dawid, I. B., J. Mol. Biol. 12, 581 (1965).
4. Pearson, W. R., and Morrow, J. F., Proc. Natl. Acad. Sci. USA 78, 4016 (1981).
5. Thomas, P. S., Proc. Natl. Acad. Sci. USA 77, 5201 (1980).
6. Posakony, J. W., Scheller, R.H., Anderson, D.M., Britten, R.J., and Davidson, E.H., J. Mol. Biol. 149, 41 (1981).

19

A Model for the Regulation of Transcription of the Preprosomatostatin I Gene from Anglerfish

Robert Crawford,[+][*] Peter Hobart[*] and W. J. Rutter[*]

[*]Department of Biochemistry
University of California
San Francisco

[+]Howard Florey Institute
University of Melbourne
Victoria

Somatostatin is a 14 amino acid peptide that plays a central role in homeostasis by regulating the secretion of several hormones in a variety of tissues. For example it inhibits growth hormone secretion in the hypothalamus (Vale et al., 1972) and insulin and glucagon secretion in the endocrine pancreas (Koerker, D. et al., 1974). Somatostatin has also been implicated in neurotransmission (Cohn, M.L. & Cohn, M. 1975).

Recently we cloned and sequenced two related prepro-somatostatin cDNAs from the Anglerfish endocrine pancreas (Hobart et al., 1980). Preprosomatostatin 1 cDNA contained sequence coding for a 14 amino acid somatostatin identical to the peptide detected in mammals. Preprosomatostatin II cDNA codes for a peptide that differs by 2 residues within the somatostatin region of the prepro-hormone. The discovery of a second preprosomatostatin was unexpected and suggests the activity previously attributed to somatostatin may be due to a family of related somatostatins.

We have subsequently cloned both Anglerfish preprosoma-tostatin 1 and preprosomatostatin II genomic sequences in order to determine the structural and evolutionary relationship between the genes, and to compare their regulatory sequences. This paper reports the sequence of the 5' flanking region of the preprosomatostatin 1 gene

95

MANIPULATION AND EXPRESSION
OF GENES IN EUKARYOTES
ISBN 0 12 513780 x

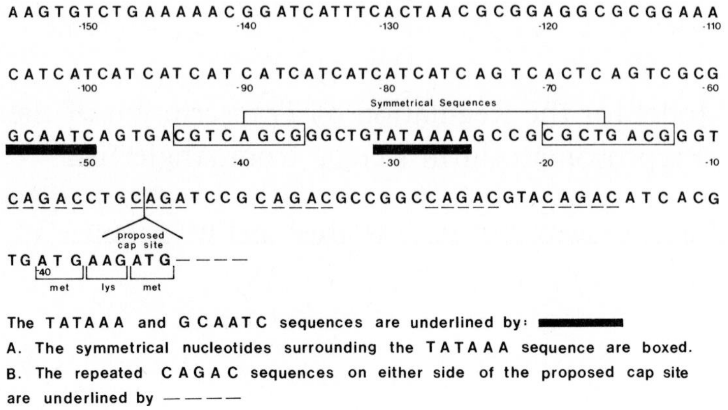

A A G T G T C T G A A A A A C G G A T C A T T T C A C T A A C G C G G A G G C G C G G A A A
    -150   -140   -130   -120   -110

C A T C A T C A T C A T C A T C A T C A T C A T C A T C A G T C A C T C A G T C G C G
    -100   -90   -80   -70   -60

Symmetrical Sequences

G C A A T C A G T G A CGT C A G C G GGC T G T A T A A A A G C C G C G C T G A C G G G T
    -50   -40   -30   -20   -10

C A G A C C T G C A G A T C C G C A G A C G C C G G C C A G A C G T A C A G A C A T C A C G

proposed
cap site

T G A T G A A G A T G — — — —
   40
   met   lys   met

The **T A T A A A** and **G C A A T C** sequences are underlined by: ■■■■■■■
A. The symmetrical nucleotides surrounding the **T A T A A A** sequence are boxed.
B. The repeated **C A G A C** sequences on either side of the proposed cap site
are underlined by — — — —

Fig. 1. Nucleotide sequence of the 5' regulatory region of the anglerfish preprosomatostatin I gene.

and discusses a possible mechanism for regulating the transcription of this gene.

An anglerfish genomic library was prepared from DNA partially digested with Sau 3A, using the Bam Hl arms of λ Charon 28 as vector. Phage containing either preprosomatostatin 1 (λAFGS1) or preprosomatostatin II (λAFGS2) sequences were detected and plaque purified. A 5.4 Bam Hl fragment within λAFGS1, containing preprosomatostatin 1 cDNA sequences, was isolated for sequence analysis using the dideoxy nucleotide chain termination method (Sanger et al, 1977).

The nucleotide sequence of the 5' flanking region is shown in Fig 1. It shares several features with the regulatory sequences of other eukaryote genes transcribed by RNA polymerase II. The TATAAA sequence (generally necessary for correct initiation and efficient transcription) is located in a similar position at -25 to -30 to that in other genes (Corder, J. et al 1980). However the GCAAT sequence, which apparently promotes transcription within in vivo systems is positioned differently at -50 to -56. This sequence is generally located at -70 to -80 in other genes (Efstradiatis, et al 1980).

The significance of the repeated CAT sequences from -72

to -102, and the CAGAC sequence repeated on both sides of
the proposed capping point is not clear.

Perhaps the most striking sequences within the 5'
flanking region are the two sets of 8 base-pair symmetrical
sequences centered around the TATAAA box. The symmetrical
nature of this sequence is two fold. Firstly the sequence
itself is symmetrical. In addition it is symmetrical in 3
dimensional structure - the central nucleotide of each 8
base-pair sequence is positioned on the same face of the
DNA (based on 10 nucleotides/turn in B form DNA).

Within prokaryotes, symmetry appears to be central to
interactions between several regulatory proteins and their
operator sequences. For example the cro protein of λ
bacteriophage probably interacts as a dimer, with the symmetrical nucleotide sequence of the cro operator (Anderson,
et al 1981) - this interaction blocks transcription of the
cro operon.

A similar interaction is suggested here, to regulate
the transcription of the Anglerfish preprosomatostatin I
gene. The TATAAA sequence appears to be necessary for
correct transcription of most polymerase II transcribed
genes. It seems reasonable to propose that correct transcription is dependent on an interaction between the TATAAA
sequence and a positive regulatory factor. It follows that
if this interaction was blocked then transcription would not
proceed. We propose that the symmetrical sequence
surrounding the TATAAA box of the Anglerfish preprosomatostatin gene is involved in blocking the promotion of
transcription by the TATAAA sequence. The proposed mechanism
is illustrated in Fig 2. Essentially, if the symmetrical
sequence interacted with a dimeric regulatory protein, then
the positive regulatory factor might be physically excluded
from interacting with the TATAAA box. This model proposes
that binding of the dimeric protein to the symmetrical
sequence is controlled by a separate interaction between
the protein dimer and additional regulating components.

One might alternatively propose that the proposed
dimeric protein is the positive regulatory factor that
generally interacts with the TATAAA box to promote transcription. In our view this proposal is less likely. It seems
improbable that the general role that the TATAAA sequence
plays is dependent of a symmetrical sequence that is so far
unique to the Anglerfish preprosomatostatin 1 gene.

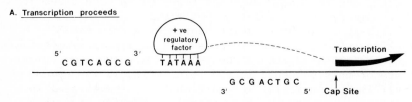

A. Transcription proceeds

Transcription is dependant on an interaction between the TATAAA box and a +ve regulatory factor.

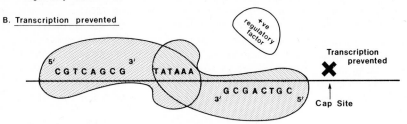

B. Transcription prevented

A dimeric protein interacts with the symmetrical sequences and physically prevents the +ve regulatory factor complexing with the TATAAA sequence.

Fig. 2.   Model for regulating the transcription of angler-fish preprosomatostatin I gene.

REFERENCES

Anderson, W.F., Ohlendorf, D.H., Takeda, Y. & Matthews, B.W. *Nature,* 290, 754-758 (1981).

Cohn, M.L. & Cohn, M. *Brain Res.* 96, 138-141, (1975).

Corden, J., Wasylyk, B., Buchwalder, A., Sasson-Corsi, P. Kedinger, C. & Chambon, P. *Science,* 209, 1406-1414 (1980).

Efstradiatis, A. et al. *Cell,* 21, 653 (1980).

Hobart, P., Crawford, R., Shen, Lu Ping, Pictet, R. & Rutter, W.J.   *Nature,* 288,137-141 (1980).

Koerker, D. et al. *Science,* 184, 482-484 (1974).

Sanger, F., Nicklen, S. & Coulson, A.R. *Proc. Natl. Acad. Sci.. U.S.A.* 74, 5643, (1977).

Vale, W. et al. *C.R. Legd. Seanc. Acad. Sci. Paris,* 275, 2913, (1972).

20

The Analysis of *Drosophila* Chorion Gene Amplification and Expression by P Element-mediated Transformation

Diane de Cicco, Laura Kalfayan, Joseph Levine, Suki Parks,
Barbara Wakimoto and Allan C. Spradling

Department of Embryology
Carnegie Institution of Washington
Baltimore, Maryland

I. THE PROBLEM OF DEVELOPMENTAL PROGRAMMING

During development, the genome of a higher eukaryote functions in much the same way as a computer during the execution of its program. Complex regulatory mechanisms, at present poorly understood, result in the establishment of precise patterns of genomic replication and transcription that are appropriate to the various parts of the developing organism. The timing of steps in a computer program can ultimately be related to the cycles of its clock, while program branches are known to depend on simple manipulations of the particular instruction in the instruction register. During development analogous timing and branching activities occur as particular genes are replicated and expressed at specific times in certain cell lineages. However, so little is known about the mechanisms underlying developmental timing and tissue specific branching that it is now diffi-cult to even formulate plausible models for its control based on known genomic elements. Consequently, our strategy for approaching these basic questions of developmental pro-gramming has been to concentrate on a relatively simple and isolated program segment in an organism highly favorable for

99

MANIPULATION AND EXPRESSION
OF GENES IN EUKARYOTES
ISBN 0 12 513780 x

genetic and molecular studies. Over the last several years, we have found that the amplification and expression of the genes coding for eggshell proteins in Drosophila melanogaster provides such a system.

II. DROSOPHILA CHORION GENES

Drosophila chorion proteins are abundantly synthesized by the ovarian follicle cells during characteristic, brief intervals near the end of oogenesis (Mahowald and Kambysellis, 1980). Each protein and its corresponding mRNA is produced over a specific, unique interval of several hours duration, and are found only in ovarian follicle cells. The genes for many of these proteins are located in a large cluster on the X and on the third chromosome (Spradling, 1981). Despite their close proximity, however, the transcription of each gene appears to be independent (Spradling et al., 1981; Griffin-Shea et al., 1982). The replication of many chorion genes is also subject to developmental regulation. Just prior to the time of gene expression, the genes in both clusters as well as 80-100 kb of flanking DNA sequences undergo 16- or 60-fold amplification in the follicle cells (Spradling and Mahowald, 1980). Available molecular and genetic evidence supports the view that amplification results from multiple rounds of initiation at specific origin sequences located in the chorion gene clusters, followed by bidirectional replication and random termination (Spradling, 1981; Spradling and Mahowald, 1981). Thus both the replication and transcription of chorion genes dramatically exhibit temporal and tissue-specific developmental regulation.

III. P ELEMENT-MEDIATED TRANSFORMATION

It is currently possible to generate mutations in cloned genomic segments in vitro with much greater frequency and specificity than corresponding lesions could be produced by classic genetic procedures (Shortle et al., 1981).

For this approach to be used in studying chorion gene regulation, however, methods are required for introducing mutated DNAs back into Drosophila germline cells, so that their ability to function can be tested in subsequent generations. An efficient transformation method has been developed recently based on the properties of a family of transposable elements ("P elements") which are responsible for the "hybrid dysgenesis" syndrome in Drosophila (Engels, 1981). When introduced into appropriate cleavage stage embryos, cloned P elements frequently transpose from the injected plasmid DNA into the chromosomes of germline cells (Spradling and Rubin, 1981). If element-encoded functions are supplied by a complementing complete P element, even a transposition-defective P element containing arbitrary DNA sequences inserted between its termini is capable of germline integration (Rubin and Spradling, 1982).

IV. CONSTRUCTION OF CHORION GENE TRANSPOSONS

     To begin analyzing sequences involved in controlling developmental timing and tissue specific expression it was first necessary to construct P element transposons containing a specific chorion gene. Figure 1 shows the construction of such a transposon starting from a 4.7 kb genomic Eco RI fragment containing the gene for protein s38-1 (Spradling, 1981).

     Two problems had to be overcome in designing these experiments. First, since mutant strains which lacked a normal s38-1 gene were not available as hosts for transformation, the s38 transposon would have to be introduced into cells containing a normal gene. To allow the activity of the exogenously added gene to be assayed, s38 "minigenes" were constructed. A small internal region of the gene was deleted with nuclease Bal-31 prior to cloning in the P element. Such a minigene is expected to make an mRNA of reduced length which could be separated from the normal mRNA by electrophoresis.

     The second problem involved the detection of those strains following the transformation procedure which have incorporated the chorion transposon. For this prupose a Sal I fragment containing the rosy (ry) gene was inserted into the P element along with the s38 minigene. Host strains containing mutant rosy genes have an easily detected abnormal eye color. Strains in which the eye color is wild type following transformation have incorporated the P element transposon and thus contain the chorion gene of interest (Rubin and Spradling, 1982).

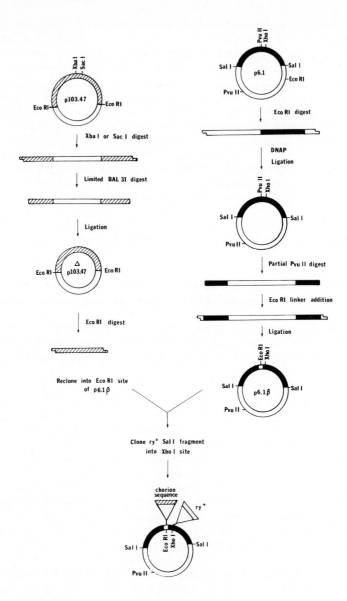

FIGURE 1.   Construction of composite P element transposons containing modified chorion genes. Starting material consisted of cloned genomic segments containing the chorion gene s38-1 (Spradling, 1981) and the defective P element-bearing clone p6.1 (Rubin et al., 1982). P element sequences are shown in black, s38-1 sequences are shaded, and <u>rosy</u> sequences are stippled.

V. LOCALIZATION OF AMPLIFICATION ORIGINS

Our initial studies have focused on defining sequences
required for the specific amplification of chorion genes. If
specific origin sequences exist, then the introduction of
such a sequence at a chromosomal site should result in its
specific amplification in the follicle cells of the transformed strain. Since the DNA sequences flanking the inserted
DNA will differ from those at its normal chromosomal position,
the amplification behavior of both normal and inserted DNA
can be conveniently determined by hybridizing a probe for
the inserted DNA to a Southern transfer of DNA from the
follicle cells of transformed strains following digestion with
an appropriate restriction endonuclease.

Figure 2 illustrates that sequences capable of inducing
specific replication can be isolated using this approach.
Two Sal I fragments from the chorion gene clusters were
ligated into the Xho I site of the P element transposon p6.1
(see Figure) and the resulting transposons were introduced
into germline chromosomes. Figure 2A-C shows DNA from
premaplification (stages 1-8) and postamplification (stage
9-14) follicle cells from a strain transformed with a transposon containing a 3.7 kb Sal I fragment from the third
chromosome gene cluster. DNAs were digested with Eco RI,
separated on a .7% agarose gel, blotted, and hybridized to a
probe for the 3.7 kb chorion region sequences. A 7.7 kb
band represents the gene in its normal location, while four
additional Eco RI fragments, labeled a-d, derive from four
independent sites of integration of the transposon. Comparison of the specific intensity of hybridization of each
fragment (the amount of DNA transferred is shown by the EtBr
stain in Figure 2C) shows that amplification of all four
fragments occurs during oogenesis. The ability to induce
such amplification is a specific property of the 3.7 kb
fragment, since a 0.8 kb Sal I fragment from the X chromosome
chorion gene cluster did not induce amplification at two autosomal sites   (a and b in Figure 2D) where it was introduced
(compare Figure 2D and 2E). While out studies using this
approach are still preliminary, it is already clear that they
provide a powerful means for analyzing the regulation of
chorion gene programming.

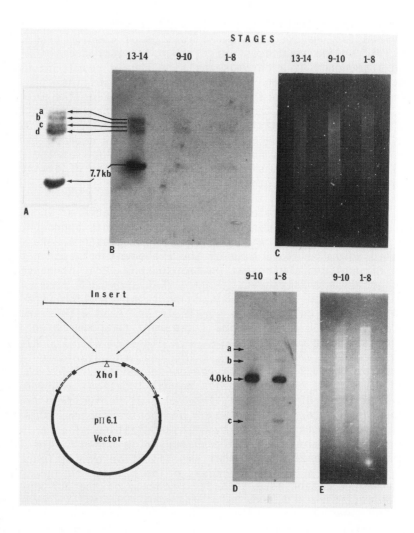

FIGURE 2. Induction of specific amplification by chorion
region sequences introduced at multiple chromosomal sites
by P element-mediated transformation. See text for details.

REFERENCES

Engels, W.R. (1981). Cold Spring Harbor Symp. Quant. Biol. 45, 1025.
Griffin-Shea, R., Thireos, G., and Kafatos, F. (1982). Dev. Biol. 91, 325.
Mahowald, A.P., and Kambysellis, M. (1980). In "The Genetics and Biology of Drosophila", Vol. 1 (M. Ashburner and T. Wright, eds.), p. 141. Academic Press, New York.
Rubin, G., and Spradling, A. (1982). Science, in press.
Rubin, G., Kidwell, M., and Bingham, P. (1982). Cell 29, 987.
Shortle, D., DiMaio, D., and Nathans, D. (1981). Ann. Rev. Genet. 15, 265.
Spradling, A.C., and Mahowald, A.P. (1980). Proc. Natl. Acad. Sci. USA 77, 1096.
Spradling, A.C. (1981). Cell 27, 193.
Spradling, A.C., and Mahowald, A.P. (1981). Cell 27, 203.
Spradling, A.C., Levine, J., Parks, S., and Wakimoto, B. (1981). Carnegie Inst. of Wash. Year Book 80, 185.
Spradling, A.C., and Rubin, G. (1982). Science, in press.

21
Studies on the Compostion and Conformation of Native Salivary Chromosomes of *Drosophila melanogaster*

R. J. Hill

F. Watt

M. R. Mott

A. Underwood

CSIRO Molecular and Cellular Biology Unit, Sydney

S. M. Abmayr

Ky Lowenhaupt

S. C. R. Elgin[1]

Department of Biology
Washington University, St. Louis

Eileen M. Lafer

B. David Stollar[2]

Department of Biochemistry and Pharmacology
Tufts University, Boston

Since the early 1930's *Drosophila* salivary chromosomes have been prepared for the cytological examination of gene loci by squashing the salivary gland in 45% acetic acid (Painter, 1934). This treatment, which involves exposure to pH 1.6, may extract histones, alter protein and nucleic acid antigenicity, disrupt higher orders of chromatin structure and even depurinate the DNA itself. Over the past seven years our group at Sydney has been developing a micro-surgical procedure for the isolation of *Drosophila* salivary chromosomes at neutral pH and physiological ionic strength

[1,2]*Supported by NIH Grants GM30273 and AM27232*

MANIPULATION AND EXPRESSION
OF GENES IN EUKARYOTES
ISBN 0 12 513780 x

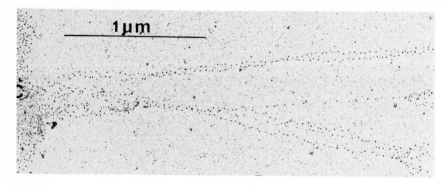

FIGURE 1.   Portion of native salivary chromosomes spread by the Miller procedure showing nucleosome bearing filaments emanating from the mass of chromosome C.

(Hill and Watt, 1977). Since the salivary nucleus is relatively small for microsurgery, since the chromosomes are
quite fragile and since for the greater portion of larval
life they are attached at many points to the internal surface
of the nuclear membrane, this has not been an easy task.
However, there is a brief window towards the end of third
instar when the chromosomes detach from the nuclear membrane.
Taking advantage of this we have been able to operate on the
nucleus with very fine needles and withdraw the entire chromosome complement. The preservation of ultra-structure at the
level of individual gene loci of chromosomes isolated in this
way is better than that of classical preparations (Mott *et
al.*, 1980).
     Whilst nucleosomes have been observed in a wide variety
of organisms, tissues and isolated mitotic chromosomes, there
has been a paucity of demonstrations of these structures in
Drosophila salivary chromosomes. This probably results from
the difficulty in isolating these chromosomes in a native
state. When salivary chromosomes are dissected out in the
physiological buffer A of Hewish and Burgoyne (1973) and
placed directly onto a carbon coated electron microscope grid
they may be spread by a slight variation of the procedure of
Miller and Beatty (1969) to reveal nucleosome bearing filaments emanating from the mass of the chromosomes (Fig. 1).
     In order to biochemically confirm the presence of nucleo-
somes in the polytene chromosomes of *Drosophila* salivary
glands, nuclei were isolated and digested with micrococcal
nuclease. The pattern of DNA fragments produced from total
chromatin showed a series of bands on electrophoresis corres-

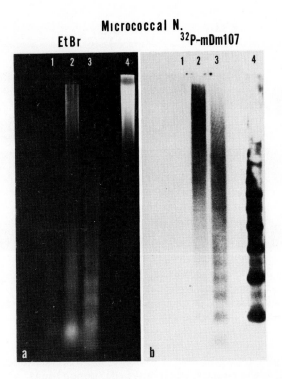

FIGURE 2. Nuclease digestion pattern of salivary chromosomes. (a) ethidium bromide stained electropherogram of digests obtained with increasing concentrations of micrococcal nuclease (lanes 1 to 3). Lane 4 is an incomplete Hae III digest. (b) southern blot of the same gel probed with the 1.688 satellite probe pDm107.

ponding to multiples of approximately 200 base pairs (Fig. 2a). Such a pattern is diagnostic for the presence of nucleosomes. In addition the specific DNA sequence of a simple satellite $\rho = 1.688$ g/cm^3) has been shown to be liberated in a similar family of fragments (Fig. 2b). The above results firmly establish the presence of nucleosomes in native *Drosophila* salivary chromosomes.

When acid squashed chromosomes are treated with antibodies against histone H1 the results obtained are dependent on the details of the preparative procedure. In particular there is a tendency for lowered levels of H1 to be observed at active loci (Bustin *et al.*, 1978; Jamrich *et al.*, 1978).

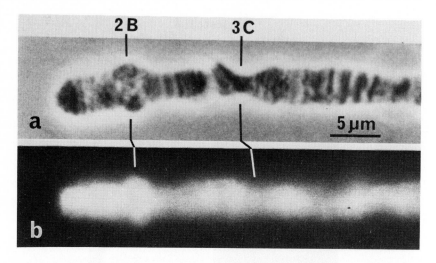

FIGURE 3.   Phase contrast (a) and fluorescence micrographs (b) of distal region of the X-chromosome prepared by micromanipulation.   In (b) the chromosome was treated with a monoclonal antibody against H1.

However, when chromosomes isolated by microdissection at neutral pH are treated with anti-H1 monoclonal antibodies no lowering of H1 content of active puffed loci such as 2B is observed (Fig. 3b).   It is likely that acetic acid extracts H1 even from formaldehyde fixed chromosomes.
    There is currently considerable interest in the existence and significance of alternative conformations of DNA to the right-handed B-form described originally by Watson and Crick in 1953.   The indirect immunofluorescence observations by Nordheim et al. (1981) that antibodies against left-handed Z-DNA bind to the interbands of Drosophila polytene chromosomes have thus assumed considerable significance.   In these observations a number of controls were performed.   However, it was not possible for these workers to test chromosomes that had not been exposed to 45% acetic acid.
    When we tested native chromosomes that had never been exposed to low pH we obtained a strikingly different result. Antiserum against Z-DNA gave negligible fluorescence over the chromosome (Fig. 4b).   However if the same chromosome was exposed to 45% acetic acid and then retreated with antibody at neutral pH a striking increase in fluorescence occurred. A short exposure to acid lead to fluorescence over interbands

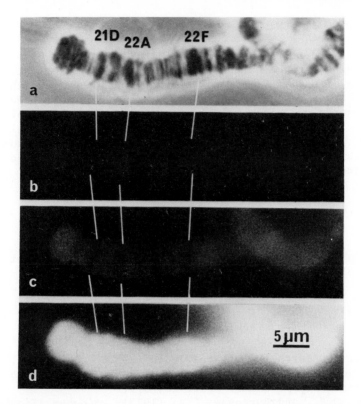

FIGURE 4.   Phase contrast (a) and anti-Z-antibody stained fluorescence micrographs (b)-(d) of the distal region of
2L.   The chromosome was initially isolated by the procedure
of Hill and Watt (1978) without exposure to 45% acetic acid
and treated directly with antibody (b).   In (c) and (d) it
was exposed to 45% acetic acid for 5 and 25 seconds, respectively, before antibody treatment.

(Fig. 4c).   Longer exposure lead to intense fluorescence over
the entire chromosome (Fig. 4d).   In view of these findings
the relevance to events and structures occurring *in vivo* of
the binding of anti-Z-DNA antibodies to polytene chromosomes
prepared by classical squashing in 45% acetic acid should be
very carefully assessed.   It may be that acetic acid is
simply unmasking an extensive amount of Z-DNA occurring *in
vivo*.   However, the possibility that the action of acid on
chromatin is causing a conversion of B- to Z-DNA is also
worthy of consideration.

CONCLUDING STATEMENT

A procedure has been developed for the cytological preparation of native *Drosophila* salivary chromosomes which
avoids exposure to low pH. These chromosomes have been shown
to preserve a nucleosomal organization. They should prove
useful for a number of applications including the localization of proteins, the study of higher orders of chromosome
structure and the microcloning of chemically unmodified DNA
by microdissection of particular loci.

ACKNOWLEDGMENTS

The skilled technical assistance of D. MacPherson is
gratefully acknowledged.

REFERENCES

Bustin, M., Kurth, P.D., Moudrianakis, E.N., Goldblatt, D.,
    Sperling, R., and Rizzo, W.B. (1978). *Cold Spring
    Harbor Symp. Quant. Biol. 42*, 379.
Hewish, D.R., and Burgoyne, L.A. (1973). *Biochem. Biophys.
    Res. Commun. 52*, 504.
Hill, R.J., and Watt, F. (1978). *Cold Spring Harbor Symp.
    Quant. Biol. 42*, 859.
Jamrich, M., Greenleaf, A.L., Bautz, F.A., and Bautz, E.K.F.
    (1978). *Cold Spring Harbor Symp. Quant. Biol. 42*, 389.
Miller, O.L., Jr., and Beatty, B.R. (1969). *Science (Wash-
    ington D.C.) 164*, 955.
Mott, M.R., Burnett, E.J., and Hill, R.J. (1980). *J. Cell
    Sci. 45*, 15.
Nordheim, A., Pardue, M.L., Lafer, E.M., Moller, A., Stollar,
    B.D., and Rich, A. (1981). *Nature 294*, 417.
Painter, T.S. (1934). *Genetics 19*, 175.
Watson, J.D., and Crick. F.H.C. (1953). *Nature 171*, 737.

22
Gene Amplification in *Rhynchosciara*

Francisco J. S. Lara

Department of Biochemistry
Institute of Chemistry
University of São Paulo
São Paulo, Brazil

CYTOLOGICAL EVIDENCE

The cytological evidence for gene amplification in the salivary gland chromosomes of
Rhynchosciara and those of larvae of other flies
belonging to the family Sciaridae consists in the
demonstration of an unusual accumulation of DNA in
certain chromosomal bands when they are involved
in a process of puff formation. This accumulation
is indicated by: a) microscopic observations
(Breuer and Pavan, 1955); b) ^{3}H-thymidine
incorporation (Ficq and Pavan, 1957; Gabrusewycz-
-Garcia, 1964; Sauaia *et al.*, 1971) and c)
microspectrophotometric measurements (Rudkin and
Corlette, 1957; Crouse and Keyl, 1968). These novel
puffs have been called "DNA puffs" by several
authors, a practical if improper designation.
    The measurements of Crouse and Keyl, 1968, car-
ried out on *Sciara coprophila* chromosomes indicate
rather conclusively that in a DNA puff site
studied there occurred two extra DNA replication
cycles in relation to that observed in a
neighbouring site in which no puff developed.
These results constituted, until recently, the
most conclusive evidence for gene amplification in
Sciara and by implication in *Rhynchosciara*.

113

MANIPULATION AND EXPRESSION
OF GENES IN EUKARYOTES
ISBN 0 12 513780 x

BIOCHEMICAL AND MOLECULAR STUDIES

Most of these studies have been carried out with *Rhynchosciara americana*. The large size of the larvae of this species and the biological peculiarities of this organism facilitate the collection of sufficient amounts of material from specialized organs, such as the salivary gland, which are necessary for the analysis.

The main object of these studies was to determine the biological function of the DNA puffs. Experimental approaches to this problem were facilitated by the fact that these puffs develop only in the last 10 days of larval life and it is possible to single out periods in which they are entirely absent or abundant. These are respectively developmental periods III and IV-VI in late forth instar (see Terra *et al.*, 1977) for a definition of these periods). Significant progress was made in the search for changes in the patterns of poly (A)$^+$RNA and of protein synthesis at these characteristic periods. It was found that the appearance of DNA puffs was accompanied by the appearance of new species of poly (A)$^+$RNA (Okretic *et al.*, 1979) Bonaldo *et al.*, (1979) and of new polypeptides (Winter *et al.*, 1977 and Winter *et al.*, 1980). It has been possible to identify by "*in situ*" hybridization and "*in vitro*" translation the messengers and the polypeptides which are synthesized by two of the major DNA puffs of *Rhynchosciara* (de Toledo, 1981). These data are summarized in Table I.

TABLE I. Characteristic of the messengers and polypeptides coded by sequences of two major Rhynchosciara DNA puffs

DNA puff	Messenger RNA Kb	Polypeptide d
B-2	0.67	20,300
C-3	1.25	28,200

The proteins transcribed from the RNA puff sequences are important components of the salivary gland secretion at the time in which these puffs appear. Calculations indicate that the amount of one of these proteins (that coded by puff C-3) synthesized in a period of 60 hours is equivalent to the weight of the gland itself (100μg protein) (Winter *et al.*, 1977).

CLONING OF DNA PUFF SEQUENCES

Advantage was taken of the fact that the DNA puffs are involved in messenger RNA synthesis in order to clone the sequences which they contain. Using reverse transcriptase, cDNA's where made from a preparation of glandular poly (A)$^+$RNA, obtained at the time in which the DNA puffs are present. These were then used to construct a cDNA library of recombinant plasmids by inserting them into the pAT153, by GC tailing. A total of 600 recombinants was obtained and after appropriate screening 5 clones were selected for further study (Glover *et al.*, 1982). Four of these cor-respond to DNA puff sequences and one does not. Their characteristics are given in Table II.

TABLE II. Characteristics of different recombinant plasmids from a cDNA bank, containing Rhynchosciara sequences.

Recom-binant plasmid	cRNA in salivary gland (KB)	"in situ" hybridization to chromo-somal locus	Period of tran-scription III V		Amplifi-cation
pRa 3.46	0.6	–	–	+	–
pRa 3.59	1.25	C3	–	+	+
pRa 6.20	1.25	C3	–	+	+
pRa 3.65	1.25	C3	–	+	+
pRa 3.81	1.95	C8	–	+	+

QUANTITATION OF THE AMPLIFICATION OF THE C-3 AND
C-8 DNA PUFF   SEQUENCES

The number of chromosomal sequences complementary to *Rhynchosciara* sequences contained in
the recombinants pRa. 3.65 (C-3) and pRa 3.81 (C-8)
were measured using the technique of quantitative
Southern hybridization with EcoRl digested salivary
gland DNA both of period III (no DNA puffs present)
and period VI larvae (when the DNA puffs have
regressed and DNA accumulation, at the chromosomal
site is evident). In addition the sequences complementary to pRa 3.46, which transcription, like
those of the DNA puffs is developmentally
regulated, were also measured at these two periods
(Glover *et al.*, 1982). The results indicated that:
a) the sequences complementary to pRa 3.65 and
pRa 3.81 are 16 fold more abundant in period VI
than in period III salivary gland DNA; b) the
sequences complementary to pRa 3.46 are equally
abundant in both periods studied; c) the
hybridization patterns of EcoRl digested glandular
DNA with either of these plasmids do not change
with the developmental period studied. These
measurements, since they involve cloned material,
are the most precise measurements of amplification
in this system obtained to date.

MECHANISM OF AMPLIFICATION

Any model to explain the formation and organization of the amplified DNA in the chromosomes of
Rhynchosciara has to account for the fact that
this DNA does not leave the chromosome and is
associated with high molecular weight DNA. This
indicates that this type of amplification differs
substantially from that which is found in the
oocytes of many organisms (reviewed by Tobler
1975) and that which is found for ribosomal RNA
cistrons in *Tetrahymena* (Yao and Gall, 1977).

It is significant that gene amplification in
Rhynchosciara does not lead to the formation of
new restriction fragments. This finding rules out
the organization of amplified DNA in tandem

arrays, such as that found in the case of the development of resistance to metabolic inhibitors (see Schimke, 1982, for a review.

The type of amplification which takes place in *Rhynchosciara* seems to have many parallels to that of the *Drosophila* chorion genes discovered recently (Spradling and Mahowald, 1980). The best model to explain this type of amplification is that of differential replication at the DNA regions involved in amplification, leading to the formation of a multiforked, and multistranded structure. Several predictions from this model have been experimentally verified by (Spradling 1981; Spradling and Mahowald, 1981).

It is known that different degrees of strandness in polytene chromosomes is achieved in many cases (Laird *et al.*, 1974). In general the euchromatic bands appear to replicate regularly, but there is a marked under-replication on the heterochromatic regions. At the present, the mechanisms which lead to this are unknown. In the case of gene amplification in *Rhynchosciara* it is known that this phenomenon is a late effect of the increase in ecdysone levels,which take place in larval hemolymph in late fourth instar larvae (Stocker and Pavan, 1974; Berendes and Lara, 1975). Currently these problems are being re-investigated in my laboratory.

CHARACTERIZATION OF THE DNA PUFFS REPLICATION ORIGINS

Attempts are being made both in the laboratories of David M. Glover and my own to characterize the origins of replication in the DNA puffs. To this end the cDNA clones were used to isolate chromosomal DNA which contain these sequences from a library of EcoRl fragments cloned on λ phage. We reasoned that in both the genomic sequences containing the C-3 and the C-8 genes there must occur specialized replication origins which permit them to undergo additional rounds of DNA replication in contrast to other sequences in which such additional replication cycles are not allowed. As a first step in characterizing these sequences, we tested whether they permit autonomous

replication in yeast. EcoRl fragments of recombinant phages containing either the C-3 or C-8
genes or their flanking regions were subcloned
into YIp5 (a plasmid which is capable of transforming Ura⁻ yeast at very low frequencies, only
by chromosomal integration, because it is incapable of autonomous replication in yeast cells).
(Struhl *et al.*, 1979). We found that the efficiency of transformation by the different subcloned
fragments differs by several orders of magnitude.
Further subcloning and sequencing should lead to
identification of these origins. Ultimately it
should be possible to assay for the properties and
function of these specialized DNA regions in
insect cells.

REFERENCES

Berendes, H. D. and Lara, F. J. S. (1975). *Chromo-
   soma 50*, 259.
Breuer, M. D. and Pavan, C. (1955). *Chromosoma 7*,
   371.
Bonaldo, M. F., Santelli, R. V., and Lara, F. J.
   S. (1979). *Cell 17*, 827.
Crouse, H. V. and Keyl, H. C. (1968). *Chromosona 25*
   357.
de Toledo, S. M., (1981). Ph. D. Thesis, University
   of São Paulo.
Ficq, A. and Pavan, C. (1957). *Nature 180*, 983.
Gabrusewycz-Garcia, N. (1964). *Chromosoma 15*, 312
Glover, D. M., Zaha, A., Stocker, A. J., Santelli,
   R. V., Pueyo, M. T., de Toledo, S. M. and
   Lara, F. J. S. (1982). *Proc. Natl. Acad. Sci.
   USA. 79*, 2947.
Graesmann, A., Graesmann, M. and Lara, F. J. S.
   (1974). *in* "Molecular Cytogenetics" (B. A.
   Hamkalo and J. Papaconstantinou, eds.) p. 209
   Plenum Press, New York.
Laird, C. D., Chooi, W. Y. Cohen, E. H. Dickson,
   Okretic, M. C., Penoni, J. S. and Lara, F. J.
   S. (1977). *Arch. Biochem. Biophys. 178*, 158.
Rudkin, G. T. and Corlette, S. L. (1957). *Proc.
   Natl. Acad. Sci. USA. 43*, 964.

Sauaia, H., Laicine, E. M. and Alves, M. A. R.
    (1971). *Chromosoma 34*, 129.
Schimke, R. T. (1982). *in* "Gene Amplification"
    (R. J. Schimke, ed.) p.1. Cold Spring Harbour
    Laboratory, New York.
Spradling, A. C. (1981). *Cell 27*, 193.
Spradling, A. C. and Mahowald, A. P. (1980). *Proc.*
    Natl. Acad. Sci. USA. 77.
Spradling, A. C. and Mahowald, A. P. (1981). *Cell.*
    27, 203.
Stocker, A. J. and Pavan, C. (1974). *Chromosoma*
    45, 295.
Struhl, K., Stinchcomb, D. T., Scherer, S., and
    Davis, R. W. (1979). *Proc. Natl. Acad. Sci.*
    USA. 76, 1035.
Terra, W. R., de Bianchi, A. G., Gambarini, A. G.
    and Lara, F. J. S., (1973). *J. Insect Physiol.*
    19. 2097.
Tobler, H. (1975). *in* "Biochemistry of Animal
    Development" p. 91. Academic Press, New York.
Winter, C. T., de Bianchi, A. G., Terra, W. R.
    and Lara, F. J. S. (1977). *Chromosoma 61*, 193.
Winter, C. E., de Bianchi, A. G., Terra, W. R.
    and Lara, F. J. S. (1980). *Devel. Biol. 75*, 1.
Yao, M. C. and Gall, J. G. (1977). *Cell 12*, 121.

Part III
Simple Eukaryotes

Introduction

The current emphases of molecular genetics are on: (1) the structure of genes, (2) regulation of gene expression in response to environmental conditions, and (3) regulation of gene expression during development. While the simple eukaryotes offer little opportunity to study development, some are eminently suitable for the study of structure and regulation in response to environmental stimuli.

In several simple eukaryotes, their complex genetic regulatory systems are now being researched at the molecular level. Chapter 28 is on the acetamidase gene of *Aspergillus nidulans*; this gene is known to be subject to several regulatory genes. This approach should lead to the sequencing of a *cis*-acting regulatory region. Chapter 29 describes the alcohol utilisation system in *A. nidulans*; here one may find the sequence of the gene for a *trans*-acting regulatory protein and the structural genes under its control.

The study of molecular genetics of *Saccharomyces cerevisiae* is more advanced than in other lower eukaryotes, partly because of the extensive application of plasmid vectors. Structural studies of *S. cerevisiae* nuclear genes and their products are well represented (Chapters 24, 25 and 26). In each case the nuclear genes were isolated through complementation of specific nuclear mutations. The appropriate mutant yeast strain was transformed with a pool of yeast/*E.coli* shuttle plasmids that carries a collection of yeast nuclear DNA fragments. Chapter 23 describes an innovative system that uses a plasmid vector carrying a *HIS*3 gene with its promoter deleted. This composite vector is introduced into a *his*3 mutant of *S. cerevisiae*. The selection and analysis of his$^+$ revertants in this vector indicate that many sequences can possess promoter activity. Interestingly, it may prove more useful to establish sequence characteristics that do not have promoter activity rather than those that do.

Other chapters show how molecular and recombinant DNA techniques provide fresh approaches to a variety of biological problems. These approaches include attempts to clone liver fluke genes coding for functional antigens (Chapter 32), the analysis of genomic differences between avirulent and virulent strains of the protozoan *Babesia*, which causes disease in cattle (Chapter 31), the study of gene expression after fusion of cells from different yeast species (Chapter 27) and the characterisation of a plasmid in *Dictyostelium* that may provide a gene transfer system for use with this organism (Chapter 30).

23

In vivo and *in vitro* Reversion of a Promoter Deletion in Yeast

Ronald W. Davis, Stewart Scherer,[1] Marjorie Thomas, Carl Mann
and Mark Johnston

Department of Biochemistry
Stanford University School of Medicine
Stanford, California

REVERTANTS OF THE HIS3-Δ4 PROMOTER DELETION

Revertants obtained in this study were derived using
the same plasmid vector, 2961 (Fig. 1). This plasmid
contains a functional centromere, an ARS, and additional
selectable markers. The promoter mutant his3-Δ4, deletes
to within a few nucleotides of the start of HIS3 transcription in wild type strains and produces no detectable HIS3
mRNA. Two host strains were used, YNN209, a rad52 strain
that contains the his3-Δ1 allele (Scherer and Davis, 1980)
and YNN211, that contains a total deletion of the his3
gene, his3-Δ200 (M. Fasullo, unpublished strain). In both
strains, the frequency of spontaneous His$^+$ revertants was
less than one in 10^9 viable cells.
  His$^+$ revertants of yeast strains YNN209 and YNN211
containing plasmid 2961 fall into three broad classes:
insertions of the transposable element Ty1, rearrangements
of the plasmid DNA, and apparent chromosomal mutations. The
revertants were first characterized by isolation of the

[1]Present address: California Institute of Technology,
Pasadena, California.

MANIPULATION AND EXPRESSION
OF GENES IN EUKARYOTES
ISBN 0 12 513780 x

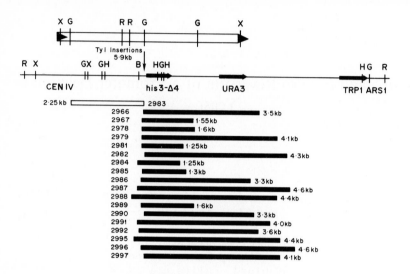

FIGURE 1. Plasmid rearrangements associated with reversion of the his3 promoter deletion. A linearized map of plasmid 2961 with the positions of certain restriction endonuclease cleavage sites is shown. The total length of the plasmid is 11.1 kb. The symbols are: R, EcoRI; B, BamHI; H, HinDIII; G, BglII; and X, XhoI. Above the map of the plasmid is a map of the prototype Ty1 element (Cameron et al., 1979). The terminal repeats of Ty1, the δ sequences, are indicated as filled triangles. One of the Ty1 elements which inserted into 2961 lacked the XhoI sites in the δ sequences. The Ty1 elements all inserted into the vector in the orientation shown. Transcription of Ty1 is right to left as drawn. The single open box represents the deletion contained in one of the revertants. The solid bars indicate the regions duplicated in the other revertants. The sizes of the deletion and the tandem duplications are shown. The sizes are ±50 bp for the smaller rearrangements and ±100 bp for the largest ones.

plasmid DNA from the revertant cells. E. coli was then transformed with each DNA preparation. Certain revertants yielded plasmids that appeared identical to the starting plasmid. All of these strains grew poorly on selective media and the plasmids obtained from them would not transform yeast his3 mutants to His⁺. They are apparently chromosomal mutations. The plasmid-encoded mutations are described in the following sections.

REVERSION BY INSERTION OF THE TRANSPOSABLE ELEMENT Ty1

Of fourteen revertants isolated using the rad52/centromere-plasmid system described above, six were insertions of the transposable element Ty1. This was determined by comparison with restriction site maps of other cloned Ty1 elements (Cameron et al., 1979) and confirmed by hybridization of ^{32}P-labelled Ty1 sequences with the plasmid DNA recovered from the revertants (not shown). Restriction site mapping placed the insertion of the element near the 5' end of the his3 coding sequences. The orientation of the element, as diagrammed in Fig. 1, was the same relative to the gene as observed at the cytochrome-c (Errede et al., 1980) and alcohol dehydrogenase (Williamson and Young, 1981) loci. The Ty1 elements inserted into at least four distinguishable sites. This was determined by digestion of heteroduplexes of revertant DNA and end-labelled parental DNA with S1 nuclease and measurement of the distance from the labeled end to the novel joint in the revertant DNA by agarose gel electrophoresis.

All six plasmids containing Ty1 insertions transformed his3 mutants to His$^+$. The insertions which were closer to the start of the the the HIS3 coding sequences expressed HIS3 at a somewhat higher level as judged by growth in the presence of 10mM 3-amino-1,2,4-triazole, an inhibitor of the HIS3 gene product, IGP-dehydratase. Attempts to correlate the level of HIS3 expression with insertion site were complicated by the fact that the Ty1 elements that had inserted were not identical. Digestion with two restriction endonucleases, SalI and XhoI, revealed that at least three different Ty1 elements were present. Many polymorphisms of Ty1 have been characterized (Cameron et al., 1979; Kingsman et al., 1981) and it is possible that the six insertions were six different Ty1 elements. These studies have demonstrated that Ty1 elements can transpose to a non-promoter sequence (λ DNA) near a gene and activate that gene. Also, Ty1 transposition can occur in a rad52 strain that blocks gene conversions.

Of twenty-one revertants of his3-Δ4 obtained in a strain containing a total deletion of the chromosomal his3 gene, none were due to the insertion of Ty1 elements. The frequency of transposition relative to the efficiency of other mechanisms available for reversion of the promoter deletion appears to be different for different strains. The source of the difference between these two strains is unknown.

REVERSION BY TANDEM DUPLICATION

Plasmids recovered from fifteen revertants contained tandem duplications of a portion of the starting plasmid. Plasmids were screened for duplications by cleavage of the plasmid DNA with BglII endonuclease, which cleaves once in the HIS3 gene and has no sites for approximately 6kb downstream from the gene (Fig. 1). As will be shown below, all the duplications had one breakpoint near the 5' end of HIS3 and one breakpoint well beyond the 3' of the gene in the vector sequences. Cleavage of the revertant DNA with BglII produced a pattern of fragments that contained all the fragments of the vector plus one additional fragment which is the size of the duplicated region. Cleavage of the plasmid DNA from a revertant with any other enzyme which cleaves once in the duplicated segment will produce all the vector fragments normally observed for that enzyme and an additional fragment whose size is the same for all enzymes of this type.

The endpoint of the duplicated segment near the 5' end of the HIS3 gene was localized by an S1 nuclease mapping procedure similar to that used for the Ty1 insertions. The results of these mapping experiments are shown in Fig. 2. The rearrangement breakpoints were spread over a wide region beginning near the start of the wild-type HIS3 mRNA and extending back as far as 500 base-pairs from that point.

Given the size of the duplication and one of its endpoints, the other endpoint is determined. The regions that were duplicated in each revertant are shown in Fig. 1. There are no obvious hotspots for the duplication breakpoints. There is no obvious correlation between the duplications that break in a given region of the vector DNA and a particular distance from the 5' end of the HIS3 gene for the other breakpoint. It is interesting that no duplications were found which fuse the his3 gene to URA3 sequences. The URA3 gene is constitutively expressed and transcribed in the appropriate direction. About half of the revertants were selected under conditions where the URA3 gene on the vector was also being selected. Under these conditions, URA3 is transcribed at a higher level (Bach et al., 1979).

Seven of the duplication-containing plasmids, including the closest and the two most distant breakpoints relative to the HIS3 gene, were tested for the ability of the plasmid DNA to transform his3 mutant strains to His$^+$. All seven

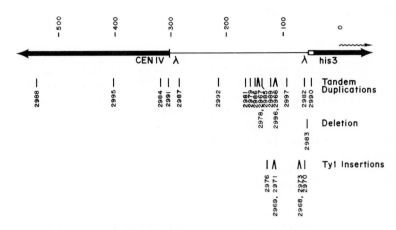

FIGURE 2.   Rearrangements breakpoints near HIS3. The
position of the breakpoint for the tandem duplications,
deletion, and Ty1 insertions are shown. The positions are
±10bp for the closest breakpoints and ±20 base pairs for
the more distant ones.   Distances from the start of the
HIS3 coding sequences (ATG) are given in base pairs. The
yeast DNA sequences in 2961 (his3-Δ4 and CEN4) are indicated as thick lines. The HIS3 structural gene sequences
are indicated by the wavy line. The small region of λ DNA,
to the right of the att site, attached to his3-Δ4 is shown
as a thin line. The λ att site, which has partial homology
to the yeast DNA that it replaced, is represented as an
open box. All revertants shown were indepentently isolated
except 2966, 2967, 2968, and 2969 which form one group. The
strains containing the rearranged plasmids are: pNN163
(2966), pNN164 (2967), pNN165 (2967), pNN166 (2969), pNN167
(2970), pNN168 (2971), pNN169 (2973), pNN170 (2976), pNN171
(2978), pNN172 (2979), pNN173 (2981), pNN174 (2982), pNN175
(2983), pNN176 (2984), pNN177 (2985), pNN178 (2986), pNN179
(2987), pNN180 (2988), pNN181 (2989), pNN182 (2990), pNN183
(2991), pNN184 (2992), pNN186 (2995), pNN187 (2996), pNN207
(2997).

plasmids yielded His$^+$ transformants. All of the transformants except those generated using 2988 DNA were resistant
to 10mM 2-amino-1,2,4-triazole and therefore must express
HIS3 at near wild-type levels.   The duplication was responsible for the HIS3 expression since removal of the duplication by homologous recombination resulted in reversion of
the His$^+$ phenotype to His$^-$.

MISCELLANEOUS REVERTANTS

One revertant was a simple deletion of the starting plasmid. One breakpoint of the deletion was near the 5' end of the HIS3 gene (Fig. 2). The size of the deletion was ~2.25kb, which placed the other breakpoint in a region of yeast DNA normally transcribed into polyA-containing RNA (Stinchcomb et al., 1982). The deletion-containing plasmid retained at least partial CEN4 function during mitotic growth.

Several revertants appeared to be duplications of most of the starting plasmid, excluding the CEN4 region (not shown). The size of these duplications resulted in the production of a large number of coincidentally sized restriction fragments and made their characterization less certain.

Two plasmids were found to contain multiple rearrangements of the starting plasmid DNA. One was a triplication or two similar duplications. The other appeared to contain two different duplications (not shown).

Plasmids isolated from the revertants described in this section transformed his3 mutants to His$^+$ and presumably represent the event in those cells that restored HIS3 function.

IN VITRO REVERSION BY DNA FUSION

A synthetic EcoRI restriction site was introduced at two positions near the native start of transcription for the His3 gene. One site was at + and the other at -8 from the native transcription start site. Random fragments can be fused to this promoterless gene in a centromere vector and the resulting fusion assayed for function in yeast. Numerous fragments from E. coli, yeast, Drosophila and corn were found to have promoter activity by this assay.

Sequences to the 5' side of the GAL1 gene from yeast were fused to the promoterless his3 gene. In these fusions his3 expression was under both galactose regulation and catabolite repression. These GAL1 sequences should serve as the bases for the construction of highly regulated expression vectors for foreign genes in yeast.

CREATION OF PROMOTERS BY DNA REARRANGEMENT

In almost all of the HIS3 revertants described above, a new promoter was created by in vivo or in vitro DNA rearrangement near the 5' end of the gene. As can be seen with the tandem duplications, the distance from the start of the HIS3 protein to the rearrangement break point can be quite large, up to 500 base pairs.

The Ty1 insertions are clustered more tightly than the duplications; however, the events that have been detected may reflect both the target specificity of the transposon, and the spectrum of sites from which it can activate a gene.

Activation of cryptic genes or enhanced expression of wild-type functions by insertion of transposable elements is found in many systems. For example, in E. coli, the major mechanism for activation of the genes for the metabolism of β-glucosides is the insertion of IS1 and IS5 (Reynolds et al., 1981). Similarly the c-myc gene, cellular homology of the transforming virus MC29, is overproduced after nearby integration of a retrovirus (Hayward et al., 1981).

Clearly, the orientation of the Ty1 insertions is significant as all the insertions found here and all the insertions seen at cytochrome-c (selecting overproduction) and at alcohol dehydrogenase (selecting altered regulation) are the same relative to the gene. This orientation is the reverse of that of transcription of the Ty1 element itself (Elder et al., 1980). One might imagine bidirectional positive control (promoter) sequences at one end of the element. Some portion of this information must reside outside the identical end repeat sequences of Ty1 (Scherer and Davis, 1980).

The duplications, which represent changes only in the arrangement of the plasmid DNA sequences, are more difficult to interpret. Most of the rearrangements involve only bacterial DNA sequences (λ and pBR322). Presumably several sequences which will pass for all or part of a promoter are to be found in the vector DNA; although they do not function in their current positions. The in vitro fusion indicates, however, that many sequences can act as a promoter in yeast. These results suggest that it may be as important to define what is not a promoter sequence as it is to define a promoter sequence.

Duplication is an important mechanism in the evolution of chromosomes. It allows for new regulation of genes without loss of either the original copy of the gene or the source of the new regulation. One well characterized example of this process is found in the mammalian globin genes. A series of related structural genes has evolved with different regulation for the various members of the family. Interspersed among these are several nonfunctional copies (Proudfoot and Maniatis, 1980). The new regulation developed after duplication need not be at the level of transcription. With the immunoglobulin heavy chain constant region genes it appears to reside in additional genomic rearrangements (Sakano et al., 1979).

The experiments described here indicate that DNA rearrangements can create new promoters. As studies on the various aspects of gene expression proceed it will become increasingly important to understand the mechanics of the chromosomes on which they reside.

REFERENCES

Bach, M.L., Lacroute, F., and Botstein, D. (1979). Proc. Natl. Acad. Sci. 76, 386.

Cameron, J.R., Loh, E.Y., and Davis, R.W. (1979). Cell 16, 739.

Elder, R.T., St. John, T.P., Stinchcomb, D.T., and Davis, R.W. (1980). Cold Spring Harbor Symp. Quant. Biol. 45, 581.

Errede, B., Cardillo, T.S., and Sherman, F. (1980). Cell 22, 427.

Hayward, W.S., Neel, B.G., and Astrin, S.M. (1981). Nature 290, 475.

Kingsman, A.J., Glimlich, R.L., Clarke, L., Chinnault, A.C., and Carbon, J. (1981). J. Mol. Biol., 145, 619.

Proudfoot, N., and Maniatis, T., (1980). Cell 21, 537.

Reynolds, A.E., Felton, J., and Wright, A. (1981). Nature 293, 625.

Sakano, H., Hüppi, K., Heinrich, G., and Tonegawa, S. (1979). Nature 280, 288.

Scherer, S. and Davis, R.W., (1980). Science 209, 1380.

Stinchcomb, D.T., Mann, C., and Davis, R.W. (1982). J. Mol. Biol., 158, 157.

Williamson, V.M., and Young, E.T. (1981). Cell 23, 605.

This investigation was supported by grant AM CAN NP 286B from the American Cancer Society.

24
Yeast Fatty Acid Synthetase: Structure–Function and Molecular Cloning of Subunits

Salih J. Wakil and Michael A. Kuziora

Department of Biochemistry
Baylor College of Medicine
Houston, Texas

Fatty acid synthetases of animal tissues and yeast are complexes of multifunctional proteins. Homogenous preparations of the synthetase were isolated from these sources and shown to catalyze the synthesis of long chain fatty acids from acetyl-CoA and malonyl-CoA in the presence of NADPH. The overall reactions catalyzed by the animal and yeast synthetases are basically the same. However, there are some subtle differences both in the structural organization of the two enzymes and in some of the partial reactions they catalyze. The animal synthetase is a homodimer of two polypeptides (M_r = 263,000). Each of the subunits contains the acyl carrier protein (ACP) site and the seven different catalytic domains required for the synthesis of palmitic acid. Recent investigations of the structure-function relationships of the animal synthetase showed that the dimer subunits are arranged in a head-to-tail configuration so that the active cysteine-SH of the β-ketoacyl synthetase (condensing activity), where the acyl-CoA is bound, is juxtapositioned within 5 Å of the pantetheine-SH of the ACP domain, where the malonyl group is attached, so that a condensation of the acyl group with the malonyl group occurs with simultaneous formation of the β-ketoacyl derivative and CO_2.

MANIPULATION AND EXPRESSION
OF GENES IN EUKARYOTES
ISBN 0 12 513780 x

Two such sites of chain elongation have been predicted in the head-to-tail model proposed by our laboratory (1).

The yeast synthetase has a different structure and minor variations in component enzymes from the animal synthetase. An example of this variation is the requirement for FMN as a co-factor for enoyl reductase reaction. Palmitoyl-CoA is the product of yeast fatty acid synthetase, the result of the exchange of thioesterase activity for palmitoyl transferase, which catalyzes the transfer of palmitoyl or stearoyl products from ACP to acceptor CoA-SH. Structurally, the yeast synthetase is completely different from the animal enzyme. The native enzyme ($M_r = 2.4 \times 10^6$) contains equal amounts of two polypeptide subunits: α (M_r = 213,000) and β (M_r = 203,000) (2), thus the active yeast fatty acid synthetase is an $\alpha_6\beta_6$ complex (4).

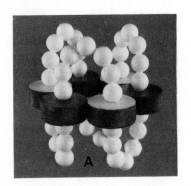 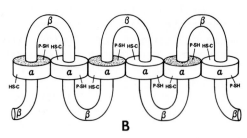

Fig. 1.   Schematics of a model of the yeast fatty acid synthetase complex. (A) The ovate structure with arches (β) and plates (α). The black and white faces of the plates depict the alternate arrangement of α subunits. (B) A linear drawing of the same model showing the six sites of palmitate synthesis and the complementary arrangements of the 4'-phosphopantetheine-SH (P-SH) and the active cysteine-SH (HS-C) at the β-ketoacyl synthetase centers in each site.

Electron microscopic studies of negatively stained yeast synthetase (4) led us to propose a model for the enzyme: an ovate structure containing on its short axis plate-like protein structures to which six arch-like proteins are equally distributed on either side (Fig. 1). The structural organization of the enzyme became apparent when stereoscopic images of particles were studied. These particles appear to contain three "arches" on each side of the plate. An arch begins on one side of the plate and terminates on the opposite side of an adjacent "plate" subunit. This conclusion was later supported by the results obtained from our studies of the mechanism of action of the β-ketoacyl synthetase component of the fatty acid synthetase (3).

The α and β subunits of the yeast synthetase have been isolated (4,5). The α subunits was found to have the β-ketoacyl synthetase domain, the ACP site and the β-ketoacyl reductase activity. The β subunit had the remaining five activities: enoyl reductase, dehydratase, and the transacylases for acetyl, malonyl, and palmitoyl groups (Fig. 2). This distribution of activities reconfirmed earlier genetic analyses which showed (6,7) that all partial activities of the fatty acid synthetase are coded by only two gene loci, designated as fas1 and fas2. Each of these genes (Fig. 2) codes for one single multifunctional protein. The fas1 gene codes for the β subunit and is located on chromosome XI. The fas2 gene codes for the α subunit and its chromosomal location has not been determined as yet.

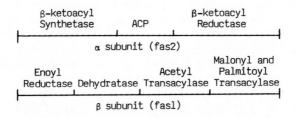

Fig. 2. Distribution of partial activities on the α and β subunits

We have recently investigated the role of thiols in the mechanism of fatty acid synthesis and have gained insight into the nature of the β-ketoacyl synthetase site. The bifunctional reagent 1,3-dibromo-2-propanone inhibits the yeast synthetase by reacting rapidly with two juxtapositioned active sulfhydryl groups. Sodium dodecyl sulfate-polyacryl-amide gel electrophoresis of the dibromopropanone-inhibited synthetase shows that the β subunit is intact and the α sub-unit is cross-linked into oligomers (M_r = 0.4 -1.2 X 10^6). This coupling of the α subunits is independent of protein concentration and requires the integral structure of the complex $\alpha_6 \beta_6$. Because the reactive centers of the 1,3-dibromo-2-propanone are about 5 Å apart, it is concluded that the α subunits are within 5 Å of each other. In our model of Fig. 1, the "plate" subunits are the only struc-tures that are spatially arranged so as to meet this require-ment for cross-linking. Consequently, we concluded that the "plate" structures are the α subunits and the "arches" are the β subunits. This assignment is consistent with that previously based on electron microscope studies (5).

The inhibition studies by dibromopropanone yielded impor-tant information concerning the structure-function relation-ship of the yeast fatty acid synthetase. As stated above, the α subunits or "plate" contained the domains of the β-ketoacyl synthetase with its active cysteine-SH, the acyl carrier protein ACP with its 4'-phosphopantetheine-SH, and the β-ketoacyl reductase (Fig. 2). Assay of the partial reactions of the synthetase shows that the dibromopropanone inhibits the β-ketoacyl synthetase reaction, but none of the other six partial reactions, indicating that the site of the bifunctional reagent is the condensing reaction. These observations and others (3) led us to propose that the site of action of the dibromopropanone was the active cysteine-SH of the β-ketoacyl synthetase of one α subunit and the pante-theine-SH of the ACP moiety of an adjacent α subunit. Thus, the active center of the β-ketoacyl synthetase consists of an acyl·group attached to the cysteine-SH of one α subunit (plate) and a malonyl group attached to the pantetheine-SH of an adjacent α subunit (Fig. 1). This arrangement appears to be necessary in order for the coupling of the acyl group and β-carbon of the malonyl group to occur, yielding CO_2 and the β-ketoacyl-S-pantetheinyl derivative. The β-keto-acyl group is then reduced by NADPH to the β-hydroxy homolog at the β-keto reductase site of an α subunit. Dehydration of the β-hydroxyacyl derivative by the dehydratase of the "arch" β subunit yields the α-β unsaturated acyl homolog, which is then reduced by NADPH through the FMN of the enoyl reductase of the β subunit to the saturated acyl deriva- tive. The latter is then transferred from the pantetheine- SH to the active cysteine-SH of the β-ketoacyl synthetase of the α subunit, where the acyl group was bound prior to con- densation. The free pantetheine-SH is then charged with another malonyl group and the sequence of reactions commen- ces again. The sequential reactions are repeated many times until the acyl chain is elongated to 16 or 18 carbons which are then transferred to CoA-SH by the palmitoyl transferase located on the "arch" β subunit.

The essence of this mechanism is fatty acid synthesis in- volves the participation of two half-α subunits and one β subunit; condensation occurs by engaging the acyl-S-cysteine of one α subunit and the malonyl-S-pantetheine of the second α subunit; and chain elongation occurs by transferring the acyl group back and forth between the pantetheine-SH and the cysteine-SH of the two complementary halves of the α sub- units. In an $\alpha_6\beta_6$ structure, therefore, six sites exist for fatty acid synthesis, all of which may function simul- taneously, Fig. 1B. This arrangement is a novel feature of our mechanism for the condensation reaction and for the syn- thesis of fatty acids by the yeast fatty acid synthetase.

As stated earlier, the α and β subunits of yeast synthe- tase are coded by the two multicistronic genes, fas2 and fas1, respectively. We have used the yeast transformation technique (8) to isolate DNA clones of the genes coding for the α and β subunits of the synthetase. Fatty acid auxo- trophs of S. cerevisiae were used to select clones from a bank of yeast DNA sequences which complement mutations in the genes coding for the α and β subunits. The stable leu2 muta- tion was introduced into the fas1 and fas2 yeast strains to use it as a marker in mitotic segregation experiments.

A bank of yeast genomic DNA sequences was prepared from DNA sequences obtained from a partial Sau 3A digestion of yeast DNA and ligated into YEp13, a yeast - E. coli shuttle vector (10). The YEp13 plasmid contains the entire pBR322

genes, the yeast 2 μcircles (B form), and the entire leu2
gene of yeast (11). E. coli was transformed by the ligated
plasmids and amp^R tet^S colonies were selected, pooled and
amplified. The plasmid DNA from the latter pool was then
used in two separate experiments to transform strain fas190
leu2 can^R and strain fas2, leu2 Can^R as described (8,9).
Cells transformed with a plasmid containing yeast DNA sequences that complemented the fas1 or fas2 mutation were selected. One colony from each transformation was obtained. The
complementary plasmid from each colony was then isolated and
named YEpFAS1 and YEpFAS2. It has been shown (11) that yeast
cells transformed with YEp13 and grown in nonselective media
yield mitotic segregates lacking YEp13 and any inserted DNA
at a frequency of about 1% per generation. Transformed
Fas$^+$ Leu$^+$ yeast strains were grown in media supplemented
with leucine and fatty acids for several generations. After
appropriate dilution, the cells were plated on supplemented
media. After allowing two days growth, the colonies were
replicated onto leucine or fatty acid lacking plates. As
expected, the mitotic cosegregation of the Leu$^+$ Fas$^+$
phenotype is about 100% (12), indicating that the selected

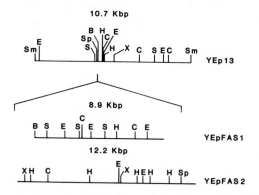

Fig. 3.   Restriction maps of YepFAS1 and YEpFAS2. Plasmid DNA was digested with various restriction endonucleases
and lengths of fragments were determined on agarose gels to
within 0.3 kilobase pairs (kbp). Abbreviations are: Sm,
SmaI; E, EcoRI; H, HindIII; X, XbaI; S, SalI; C, ClaI; Sp,
SphI; and B, Bam HI.

phenotypes Leu⁺ FAS⁺ are plasmid encoded and that the DNA sequences cloned into the vectors follow an identical segregation pattern as cells grown under nonselective conditions.

Restriction endonuclease maps of YEpFASl amd YEpFAS2 are presented in Fig. 3. The size of yeast DNA inserts are 8.9 and 12.1 kilobase pairs in YEpFASl and YEpFAS2, respectively. The yeast DNA inserts shown in Fig. 3 are well above the number of bases required to code for the α and β subunits: 5.5 and 5.3 kilobase pairs, respectively. The inserts shown in Fig. 3 are large enough to encode an entire α or β subunit gene as well as 6.6 and 3.6 kilobase pairs, respectively, that may be needed for flanking sequences and/or noncoding intervening sequences if present.

The maxicell technique (11) was used to characterize plasmid encoded polypeptides. E. coli was transformed with YEpl3, YEpFASl and YEpFAS2. Plasmid containing cells were irradiated briefly with UV light and incubated in the presence of [^{35}S]methionine. The radioactive proteins, including those encoded by the plasmids, were detected by radioautography of cellular proteins obtained after lysis of cells and electrophoresis on sodium dodecyl sulfate-polyacrylamide gels. As shown in Fig. 4A, no predominantly labeled proteins were observed in untransformed cells (lane 2), suggesting that nearly all protein synthesis coded for by E. coli chromosomal genes were lost by UV irradiation. Cells transformed with YEpl3 synthesized three polypeptides (lane 3) with molecular weights of 44,000, 35,000, and 29,000, which correspond to gene products of leu 2, tet and amp genes, respectively. Cells transformed with YEpFASl yielded three unique polypeptides (Fig. 4A, lane 4) with moleculr weights of 48,000, 46,500 and 21,500. On the other hand, the maxicells harboring YEpFAS2 (Fig. 4A, lane 5) produced four polypeptides with molecular weights of 160,000, 120,000, 110,000 and 100,000. The peptide ($M_r = 36,000$), corresponding to the tet gene product, is absent in lanes 4 and 5 since fas complementing sequences were ligated into the Bam Hl site of YEpl3, which lies within the tet gene.

The antigenicity of the polypeptides synthesized by the maxicells was tested with the anti-yeast fatty acid synthetase antibodies. Unlabeled polypeptides were obtained from maxicells prepared by above procedure, except nonradioactive methionine was used. Cellular proteins were separated on denaturing gel as before and transferred to nitrocellulose paper by electrophoresis. The proteins were probed with affinity purified anti-yeast synthetase antibodies raised in goat. Antigen-antibody complexes were then detected by incubating the paper in ^{125}I-anti-goat IgG raised in rabbit and radioautography of the dried paper. As shown in Fig. 4B, there was some nonspecific binding of the antibodies to E. coli proteins obtained from nontransformed E. coli CSR603 (lane 2) which appear to be in all other tracks. Transformation of CSR603 with YEp13 did not produce any other antigenetically reactive polypeptide (cf. Fig. 4B, lane 3). The three polypeptides produced by the maxicells transformed by YEpFAS1 did not react with the anti-yeast fatty acid synthetase either (cf. Fig. 4, lanes 3). However, the polypeptides synthesized in maxicells harboring YEpFAS2 showed positive immunologic reaction with anti-yeast fatty acid synthetase (Fig. 4, lanes 5). These results showed that although full length α or β subunits of fatty acid synthetase were not observed in E. coli maxicells transformed with YEpFAS1 or YEpFAS2, several polypeptides of lower molecular weights were detected in each case (Fig. 4). These peptides could possibly be products of protein degradation and/or may result from the inability of E. coli to completely and accurately express large eucaryotic genes. The fact that E. coli maxicells transformed by YEpFAS2 produced proteins of large molecular weights (100,000 to 160,000) is by itself remarkable. That these proteins carry antigenic determinants as the yeast synthetase would identify them as being of the α subunit of yeast synthetase.

The peptides produced by maxicells containing YEpFAS1 are relatively small (21,500 to 48,000) and are not antigenetically reactive (Fig. 4). It is possible that these peptides are truncated β subunits which do not contain antigenetic determinants due to incomplete synthesis. Alternatively, these labeled bands may represent nonsense peptides resulting from an E. coli promoter region out of phase with any synthetase coding sequences. That the plasmids YEpFAS1 and YEpFAS2 contain DNA sequences coding, respectively, for the

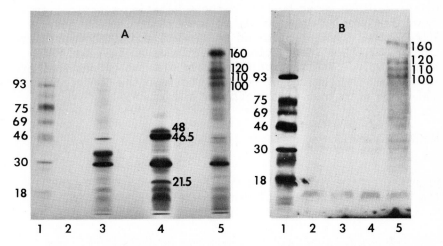

Fig. 4.   Detection and immunological reactivity of plasmid encoded peptides.   E. coli strains C5R603 was transformed with no plasmid (lane 2), YEp12 (lane 3), YEpFAS1 (lane 4)
and YEpFAS2 (lane 5) and subjected to maxicell procedure
(11).   Lane 1 contains standard proteins with the indicated
molecular weights X 10^{-3}.   (A) Autoradiograph of [^{35}S]-
methionine labeled peptides after electrophoresis on denaturing gel (9) and (B) Same as A except non-labeled methionine was used and the polypeptides were transferred to
nitrocellulose paper and probed with anti-yeast synthetase
raised in goat and then with ^{125}I-labeled anti-goat IgG.

α and β subunits of yeast synthetase was further reaffirmed
by their hybridization with plasmids 33F1 and 102B5, respectively, isolated from two colonies of E. coli selected from
a yeast genome bank by Chalmers in our laboratory.   By employing the antibody selection method, he screened approximately 5,000 clones containing randomly sheared yeast DNA
inserted into ColE1 vector and successfully identified two
clones (33F1 and 102B5) that express fatty acid synthetase
related antigen.   The sizes of the inserts were 8.2 kilobase
pairs for 33F1 and 8.0 kilobase pairs for 102B5.   Fragments
generated from plasmid 33F1 and 102B5, consisting of yeast
DNA and free from vector sequences, were transferred to

nitrocellulose paper by the Southern blotting technique and hybridized to ^{32}P-labeled nicked translated YepFAS1 and YEpFAS2 DNA. 33F1 hybridized to YEpFAS1 and 102B5 hybridize(to YEpFAS2, thus identifying these clones as FAS1 and FAS2 respectively. Restriction mapping of the two clones further identified the regions of homology between the DNA of 33F1 and 102B5 with that of the YEpFAS1 and YEpFAS2.

ACKNOWLEDGMENTS

Grants from the National Science Foundation (PCM 77-00969), the National Institutes of Health (GM 19091, AM 21286, and HL 17269) supported in part this investigation.

REFERENCES

1. Stoops, J., and Wakil, S. (1982) *J. Biol. Chem.* <u>257</u>, 3230.
2. Stoops, J., Awad, E., Arslanian, M., Gunsberg, S., Wakil, S., and Oliver, R. (1978) *J. Biol. Chem.* <u>253</u>, 4464.
3. Stoops, J., and Wakil, S. (1981) *J. Biol. Chem.* <u>256</u>, 8364.
4. Stoops, J., and Wakil S. (1978) *Biochem. Biophys. Res. Comm.* <u>84</u>, 225-231.
5. Wieland, F., Siess, E., Renner, L., Verfürth, C., and Lynen, F. (1978) *Proc. Natl. Acad. Sci. U.S.A.* <u>75</u>, 5792.
6. Schweizer, E., Dietlein, G., Gimmler, G., Knobling, A., Tahedl, H., Schwietz, H.,and Schweizer, M. (1975) *Proc. FEBS Meeting* 40, 85.
7 Henry, S. and Fogel, S. (1971) *Molec. Gen.* <u>113</u>, 1-19.
8. Beggs, J. (1978) *Nature* <u>275</u>, 104.
9. Kuziora, M., Douglas, M., and Wakil, S. *Proc. Natl. Acad. Sci, USA*, submitted.
10. Williamson, V., Bennetzen, J., Young, E., Nasmyth, K. and Hall, B. (1980) *Nature* <u>283</u>, 214.
11. Broach, J., Strathern, J., and Hick, J. (1979) Gene <u>8</u>, 121.

25

Yeast Nuclear Genes Necessary for Expression of Cytochrome *b*

Alexander Tzagoloff, Patricia McGraw and Carol L. Dieckmann

Department of Biological Sciences
Columbia University
New York, New York

INTRODUCTION

The advent of rapid methods for DNA sequencing has facilitated the analysis of mitochondrial DNA (mtDNA)from widely divergent phylogenetic sources. Our own efforts have been aimed at defining the composition and structure of genes in yeast mtDNA (1-5). Although the genome of yeast mitochondria is some five times larger than that of mammalian mitochondria, its coding capacity is probably more limited. The larger size of yeast mtDNA is accounted for by the presence of introns in some genes and of long intergenic regions consisting for the most part of AT-rich sequences. At least three genes have been shown to contain intervening sequences; these code for the 21S rRNA (6,7), the subunit 1 of cytochrome oxidase (4,8) and cytochrome b (5,9).

The cytochrome b gene has become the focus of intensive studies in a number of laboratories (5,8-13). Earlier genetic and physical studies of the gene indicated that it is interrupted by intervening sequences (10,14,15). This has been confirmed by more recent data on the sequences of the gene in several different strains of yeast (5,9). In addition, some unique features of the introns have led to imaginative proposals for the mechanism of processing of the primary transcript (9,16).

Despite a wealth of new information on the genetic function of mitochondrial DNA, there remains a gap in our knowledge of the manner in which mitochondrial genes are regulated. The cytochrome b gene is an excellent system for studying this

141

MANIPULATION AND EXPRESSION
OF GENES IN EUKARYOTES
ISBN 0 12 513780 x

problem. To help clarify the nuclear contribution towards the synthesis of cytochrome b we have screened for Mendelian mutants specifically defective in this respiratory component. The initial aim has been to identify nuclear genes necessary for the production of cytochrome b message. We anticipated such genes might code for factors essential in either transcription or processing of the pre-mRNA.

In this paper we report the properties of mutants lacking only the message for cytochrome b. The mutants define two different genes, CBP1 and CBP2. Both genes have been cloned and sequenced. CBP1 has been shown to be required either for correct transcription or processing of the pre-mRNA. The product of the CBP2 gene, on the other hand, is essential for the excision of the second intron from the pre-mRNA.

PROCESSING OF THE CYTOCHROME B TRANSCRIPT

The studies reported in this paper have been done with S. cerevisiae D273-10B, a haploid strain previously shown to have a cytochrome b gene with two intervening sequences (5). Most other laboratory stocks of yeast, however, have a longer gene with three additional introns (9-13). The extra or optional introns occur in the first exon (B₁) of the short gene (Fig. 1). In other respects the two genes are identical; thus the first and second introns of the short gene correspond to the fourth and fifth introns of the long gene, respectively.

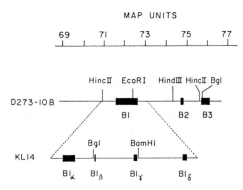

Fig. 1. Structures of the short and long cytochrome b genes in in S. cerevisiae. The three exons (B₁,B₂,B₃) of the short gene are denoted by the solid bars. The first exon (B₁) is fragmented into four shorter exons in the long gene of strain KL14. Transcription is from left to right.

The exon-intron junctions of the short gene have been determined by reverse transcription and sequencing of the fully
spliced mRNA (17) and in the case of the long gene by sequence
comparison (9). The consensus nucleotides normally found at
the intron boundaries of eucaryotic split genes are absent.
The yeast genes also do not reveal any other obvious sequences
at the boundaries that might be important for splicing.

In $\underline{S}$. $\underline{cerevisiae}$ D273-10B, the first intron of the gene
contains an open reading frame capable of coding for a protein
with splicing function ("maturase") (5,9). Translation of this
sequence appears to be necessary for the excision of the first
intron. This is supported by the observation that ρ^- mutants
defective in mitochondrial protein synthesis fail to process
the first intron (17). Processing of the second intron,
however, does occur in ρ^- mutants and therefore must be catalyzed entirely by nuclearly encoded products (17).

Four different cytochrome $\underline{b}$ transcripts are detected in
mitochondrial RNA extracts of wild type yeast (Fig. 2).

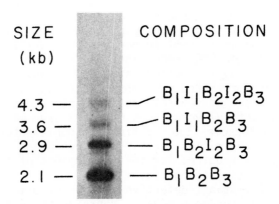

SIZE
(kb)

COMPOSITION

4.3 — $B_1I_1B_2I_2B_3$
3.6 — $B_1I_1B_2B_3$
2.9 — $B_1B_2I_2B_3$
2.1 — $B_1B_2B_3$

Fig. 2. Cytochrome b *transcripts in* S. *cerevisiae D273-10B.
Total mitochondrial RNA was separated on 1% agarose, transferred to DBM paper and probed with a 5'-end labeled restriction
fragment from exon B₁. The sizes and compositions of the RNA
species are indicated in the margins.*

Their sequence compositions have been ascertained by a combination of S_1 mapping and Northern hybridizations (17). They
all have a 940 nucleotide long 5' non-translated leader and a
110 nucleotide 3' extension. The longest and least abundant
RNA with an estimated size of 4.3 kb contains both introns and

therefore could be the primary transcript. Two other transcripts have been identified as partially spliced intermediates. These are also of low abundance in wild type yeast; their sizes are 3.6 kb ($B_1I_1B_2B_3$) and 2.9 kb ($B_1B_2I_2B_3$). The dominant transcript detected on Northern blots is 2.1 kb long. This RNA hybridizes only to DNA probes with exon sequences and therefore corresponds to the fully spliced cytochrome b mRNA. The fact that transcripts lacking either the first or second intron are present in mitochondria suggests a non-obligatory order of excision of these sequences. This is also evident from the transcripts observed in strains with mutations in the introns. For example, certain mitochondrial mutants are incapable of processing the first intron due to mutations in the intron region. Such strains accumulate the 3.6 kb intermediate indicating that excision of the second intron does not depend on a prior excision of the first intron. The converse situation prevails in mutants blocked in the excision of the second intron. The latter strains accumulate the 2.9 kb intermediate (17).

ISOLATION AND CHARACTERIZATION OF NUCLEAR MUTANTS

The expression of mitochondrial genes can be regulated either at the level of transcription or of translation. The presence of introns in certain genes offers still another means of control through RNA processing. Since there are no direct means of screening for mutants affected at any of these three stages, we have resorted to a more laborious method of mutant isolation. Upward of several thousand independent respiratory deficient mutants of yeast (pet mutants) have been tested and assigned to some 130   different complementation groups. As a first approximation these should represent an equivalent number of genes required for the development of fully respiratory competent mitochondria.

To identify mutants defective in transcription or processing of the cytochrome b mRNA, a representative mutant from each complementation group was analyzed for the presence of cytochrome b. Mutants lacking cytochrome b both by spectral and protein gel criteria were studied further. The mitochondrial RNA from these mutants was analyzed for cytochrome b specific transcripts by Northern hybridizations. In addition we have looked for the presence of other mitochondrial RNAs, e.g. rRNA, tRNA and mRNA. Based on the results of these studies, five complementation groups were found to be deficient in cytochrome b message. The gross phenotypes of mutants from each complementation group are presented in Table I. Mutants from complementation groups G36 and G60 transcribe and process

TABLE I.   Phenotypes of nuclear cytochrome b deficient mutants.

Complementation group	Message		
	cytochrome b	sb.1 cyt. ox.	other
G26	−	−	+
G69	−	−	+
G85	−	−	+
G36	−	+	+
G60	−	+	+

all the known mitochondrial genes except cytochrome b. Mutants from complementation groups G26, G69 and G85 are deficient in both cytochrome b and cytochrome oxidase. Since the characterization of the pleiotropic mutants is still in progress, the remainder of this discussion will deal with the first two groups for which we have the most data at present.

PROPERTIES OF MUTANTS IN COMPLEMENTATION GROUPS G36 and G60

At least twenty different mutants have been assigned to complementation group G36. These mutants have an unambiguous phenotype indicative of a lesion in a component necessary for the excision of the second intervening sequence from the cytochrome b pre-mRNA. The mutants do not grow on a non-fermentable carbon source due to a deficiency in cytochrome b. All the mutants in this complementation group fail to form mature cytochrome b message. Northern hybridizations of their mitochondrial RNAs indicate the accumulation of a transcript

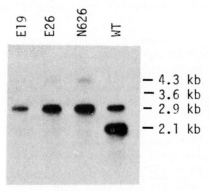

Fig. 3. Northern hybridizations of G36 mutants. Mitochondrial RNA from wild type (WT) and G36 mutants (E19, E26, N626) was hybridized to a probe from the first exon of the cytochrome b gene.

with a size identical to the partially spliced intermediate
$B_1B_2I_2B_3$ (Fig. 3). This transcript hybridizes to probes from
the exon and second intron regions but does not hybridize to
probes from the first intron. G36 mutants are therefore able
to correctly transcribe the gene and to process the first
intron. They are, however, blocked in the excision of the
second intron. The gene defined by this complementation group
has been designated CBP2.

   Mutants from complementation group G60 are also specifically deficient in cytochrome b. The function of this gene
(CBP1) is still not clear. Mutations in CBP1 are expressed in
a low level of cytochrome b transcripts, although other mitochondrial RNAs are present at normal concentrations (18). The
transcripts detected with cytochrome b probes do not correspond to those seen in wild type mitochondria. A similar phenotype is observed when a cbp1 allele is introduced into a
strain with a long cytochrome b gene. In this genetic background, however, the small amount of cytochrome b transcripts
do migrate identically to the mRNA and the precursors seen in
wild type. A reasonable interpretation of these results is
that CBP1 codes for a factor that promotes transcription of
the gene. Alternatively, the CBP1 product may be necessary
for correct processing of the pre-mRNA.

CLONING OF CBP1 and CBP2

   The wild type CBP1 and CBP2 genes have been cloned by transformation of the appropriate mutants with a recombinant plasmid library of yeast nuclear DNA. The library consisted of
partial Sau3A fragments of wild type nuclear DNA inserted at
the BamHI site of the yeast/E.coli hybrid vector CV13 (YEp13)
(19).

   To clone the CBP1 gene, a strain carrying mutations in CBP1
and LEU2 (the latter marker was introduced to permit double
selection for respiratory competency and prototrophy) was
transformed with the plasmid pool. Transformants capable of
growth on glycerol in the absence of leucine were isolated.
It was established that the clones harbored freely replicating plasmids with overlapping fragments of yeast nuclear DNA.
One of the plasmids (pG60/T10) had a DNA insert of 6.7 kbp(18).
To obtain the gene on a smaller fragment of DNA, the insert
was digested with Sau3A to yield partial fragments averaging
2-3 kbp. These were ligated into the CV13 vector and the new
pool was used to transform a cbp1 mutant. A number of new
respiratory competent clones were obtained and their plasmids
were analyzed with restriction endonucleases. The smallest

plasmid capable of complementing G60 mutants (pG60/T31) had
an insert of 2.5 kbp (Fig. 4). A similar approach was used
to clone the CBP2 gene. This gene was isolated in a plasmid
(pG36/T5) with a nuclear DNA insert of 2.5 kbp.

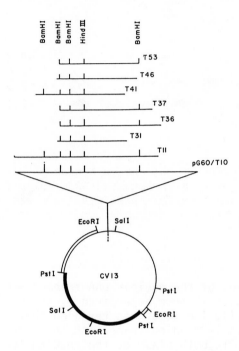

*Fig. 4. Restriction maps of pG60/T10 and derivative plasmids.
The CV13 vector is shown as a circle and the cloned fragments
of nuclear DNA as lines on the upper part of the figure. Only
some of the restriction sites have been marked.*

Several criteria were used to confirm that pG60/T31 and
pG36/T5 contain the CBP1 and CBP2 genes, respectively. Each
plasmid confer respiratory competency to all members of the
complementation group. Stable transformants acquire the
ability to produce cytochrome b message at wild type levels
(Fig. 5). Lastly, the cloned fragments of nuclear DNA in each
plasmid were found to be physically linked to the mutational
sites after induction of homologous recombination between the
plasmid and the chromosome (20). The last finding constitutes
strong evidence that complementation is due to the presence
in the plasmid of the wild type copy of the mutated gene
rather than to suppression by some other gene.

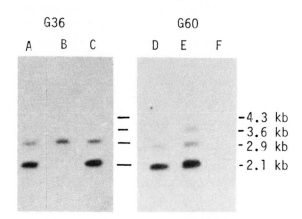

Fig. 5. *Restoration of cytochrome b message with the cloned CBP1 and CBP2 genes. Mitochondrial RNA from wild type (A,D), mutants (B,F) and transformants (C,E) were separated on 1% agarose, transferred to DBM paper and hybridized to a 5'-end labeled restriction fragment from the first exon of the cytochrome b gene. The cbp1 mutant (F) was transformed with pG60/T31 and the cbp2 mutant with pG36/T5.*

The sequences of the nuclear DNA fragments cloned in pG60/T31 and pG36/T5 were determined by the method of Maxam and Gilbert (21). The sequences revealed that each plasmid contained a single reading frame with the potential to code for a protein. In both plasmids the reading frame occupies almost the entire length of the insert. We have provisionally identified these coding sequences to be the CBP1 and CBP2 genes. The properties of the proteins encoded by the genes are summarized in Table II. Both have a high molecular weight and based on their amino acid sequences, they carry a net positive charge

TABLE II.   *Properties of CBP1 and CBP2 gene products.*

	G36	G60
Complementation group	G36	G60
Gene	CBP2	CBP1
Molecular weight	75,000	76,000
Number amino acids	630	654
Basic amino acids	89	97
Acidic amino acids	75	63

CONCLUDING COMMENTS

There are two important areas of mitochondrial biogenesis still poorly understood. The first deals with the mechanism of import of externally synthesized proteins into the organelle. The factors governing the segregation of proteins among the different compartments will ultimately explain how the structure and topography of mitochondria are constantly maintained during growth. The second issue deals with those regulatory phenomena insuring a temporal order of production of nuclearly and mitochondrially encoded constituents. There exists reasonable evidence that while the expression of mitochondrial genes is under nuclear control, the converse is not the case. To understand the nature of this regulation we have adopted a genetic approach designed to saturate by mutation the set of nuclear genes directly involved in mitochondrial biogenesis. Such mutants should be useful in future studies not only of the structural but of the regulatory genes as well. The results of the present study point to the feasibility of this approach. Even though we are still short of having saturated the genome, the mutants at hand have already provided us with valuable tools for probing two important aspects of cytochrome b synthesis. Our evidence clearly indicates that the CBP2 gene product functions at some early step in the excision of one of the cytochrome b introns. The CBP1 gene may be of even more interest since it is a possible candidate for a transcriptional effector.

ACKNOWLEDGEMENTS

This research was supported by Research Grant HL22174 and a Research Service Award GM07619 (to C.L.D) from the National Institutes of Health.

REFERENCES

1. Coruzzi, G. and Tzagoloff, A. (1979) J. Biol. Chem. 254, 9324-9330.
2. Macino, G. and Tzagoloff, A. (1980) Cell 20, 507-517.
3. Thalenfeld, B.E. and Tzagoloff, A. (1980) J. Biol. Chem. 255, 6173-6180.
4. Bonitz, G.S., Coruzzi, G., Thalenfeld, B.E., Tzagoloff, A. and Macino, G. (1980) J. Biol. Chem. 255, 11927-11941.
5. Nobrega, F.G. and Tzagoloff, A. (1980) J. Biol. Chem. 255, 9828-9837,
6. Bos, J.L., Osinga, K.A., Van der Horst, G., Hecht, N.B., Tabak, H.F., Van Ommen, G.J.B. and Borst, P. (1980) Cell 20, 207-214.
7. Dujon, B. (1980) Cell 20, 185-197.

8. Grivell, L.A., Arnberg, A.C., Hensgens, L.A.M., Roosendaal, E., Van Ommen, G.J.B., Van Bruggen, E.F.J. (1980) in "The organization and expression of the mitochondrial genome" (C. Saccone and A.M. Kroon, eds) North Holland, Amsterdam, pp. 37-49.
9. Lazowska, J., Jacq, C. and Slonimski, P.P. (1980) Cell 22, 333-348.
10. Grivell, L.A., Arnberg, A.C., Boer, P.H., Borst, P., Bos, J.L., Van Bruggen, E.F.J., Groot, G.S.P., Hecht, N.B., Hensgens, L.A.M., Van Ommen, G.J.B. and Tabak, H.F. (1979) ICN-UCLA Symp. Mol. Cell Biol. 15, 305-324.
11. Hanson, D.K., Miller, D.H., Mahler, H., Alexander, N.J. and Perlman, P.S. (1979) J. Biol. Chem. 254, 2480-2490.
12. Solioz, M. and Schatz, G. (1979) J. Biol. Chem. 254, 9331-0334.
13. Bechmann, H., Haid, A., Schweyen, R.J., Matthews, S., and Kaudewitz, F. (1981) J. Biol. Chem. 256, 3525-3531.
14. Slonimski, P.P., Claisse, M.L., Foucher, M., Jacq, C., Kochko, A., Lamouroux, A., Pajot, P., Perrodin, G., Spyridakis, A., Wambier-Kluppel, M.L. (1978) in "Biochemistry and Genetics of Yeast: Pure and applied aspects" (M. Bacila et al eds) Academic Press, N.Y. pp. 391-399.
15. Foury, F. and Tzagoloff, A. (1978) J. Biol. Chem. 253, 3792-3797.
16. Church, G.M. and Gilbert, W. (1980) in "Mobilization and reassembly of genetic information" (D.R. Joseph et al eds) Academic Press, N.Y. pp. 379-385.
17. Bonitz, S.G., Homison, G., Thalenfeld, B.E., Tzagoloff, A. and Nobrega, F.G. (1982) J. Biol. Chem. 257, 6268-6274.
18. Dieckmann, C.L., Pape, L.K. and Tzagoloff, A. (1982) Proc. Natl. Acad. Sci. U.S.A. 79, 1805-1809.
19. Broach, J.R., Strathern, J.N. and Hicks, J.B. (1979) Gene 8, 121-133.
20. Hinnen, A., Hicks, J.B. and Fink, G.B. (1978) Proc. Natl. Acad. Sci. U.S.A. 75,1929-1933.
21. Maxam, A.M. and Gilbert, W. (1977) Proc. Natl. Acad. Sci. U.S.A. 74, 560-564.

26
Isolation of Genes Coding Yeast F_1 ATPase Subunits

Michael G. Douglas, Jo Saltzgaber, Satya Kunapuli
and Marc Boutry

Department of Biochemistry
University of Texas Health Science Center
San Antonio, Texas

I. INTRODUCTION

The yeast mitochondrial ATPase complex is assembled from the products of two mitochondrial and seven nuclear genes (1,2). As many as 6 of the nuclear gene products are made as higher molecular weight precursors which contain additional amino terminal sequences of several thousand daltons (3,4). The specific post translational segregation and processing events which direct assembly of the ATPase complex (5,6,7) suggest that there may be common structural features among the various F_1-ATPase precursors. To test this notion we have initiated studies to take advantage of the molecular genetic approach available in yeast. By genetic complementation of defined pet mutants, the nuclear genes coding different F_1-ATPase subunits may be selected. Analysis of these genes would allow a primary sequence comparison of the subunits. Further, important genetic tools would be available for a detailed examination of the mitochondrial import-assembly pathway.

MANIPULATION AND EXPRESSION
OF GENES IN EUKARYOTES
ISBN 0 12 513780 x

151

II.   RESULTS AND DISCUSSION

Yeast nuclear mutants which had been shown to lack detectable F_1-ATPase activity (8) were utilized in this study.
Strain constructions to combine the non-reverting leu 2 (9),
and pet alleles were performed using standard methods (10).
The specific F_1-ATPase activities of the leu⁻pet mutants
(Table I) are similar to those described previously. Spheroplasts of each strain were transformed with 20 μg of a pool
of partial Sau 3A DNA fragments in YEp 13 (13). Selection of
Leu⁺ Pet co-transformants and genetic analysis to confirm
the plasmid origin of the transforming genes was as previously described (14). Transformants from each complementation
group contained 35–100% the oligomycin sensitive ATPase
activity of the parental strain (Table II).

TABLE I. F_1-ATPase Activities* of pet-leu 2⁻ Mutants

Complementation Group	Parental Mutant	Mutant pet leu2	Mitochondrial		Post Ribosomal	
			no inhibitor	efrapeptin	no inhibitor	efrapeptin
I	N9-84	XJ12	0.23	0.18	0.02	0.02
II	N9-168	XJ2	0.15	0.14	0.73	0.10
III	E2-126	XJ14	0.07	0.09	0.05	0.06
	DC5	–	3.97	0.24	0.45	0.09

*μmole/min/mg All strains were grown on YPGalactose (10)
Mitochondria and post ribosomal supernatant preparations
(11) and ATPase activities, Efrapeptin, 0.1 μg/ml (12)
were as described.

TABLE II. F_1-ATPase Activities* of Pet Leu⁺ Strains

Complementation Group	Transformant	Mitochondrial		Post Ribosomal	
		no inhibitor	oligomycin	no inhibitor	efrapeptin
I	TJ12-5	2.76	0.46	0.61	0.27
II	TJ2-1	5.56	0.37	0.37	0.20
III	TJ14-1	1.82	0.26	0.18	0.11
Wild-type	DC5	5.24	0.76	0.36	0.23

*μmole/min/mg Strains grown on YPGE were analyzed as
above; oligomycin 8 μg/ml.

Plasmid Encoded Protein

The complementing plasmid (pJ14-1) which transformed mutant XJ14 to TJ14 was selected as amp^R in E coli strain RR1 (15). pJ14-1 DNA propagated in E coli was used for further analysis. When poly A RNA prepared from DC-5 harboring YEp13 was hybridized with plasmid YEp13 or pJ14-1

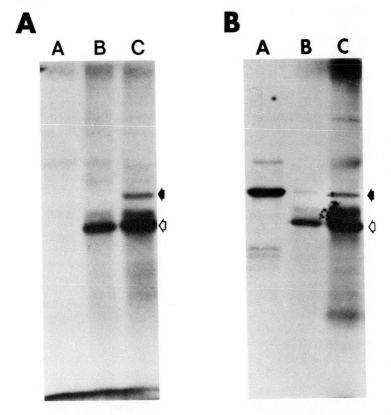

FIGURE 1. Plasmid pJ14-1 selects F$_1$-ATPase β subunit mRNA

Panel A. Fifteen μg of yeast poly A RNA from DC5 (YEp13) was hybridized with 2 μg of A. pBR322, B. YEp13, and C. pJ14-1 fixed to nitrocellulose filters. After removing excess RNA, hybridizable RNA was eluted, translated, and the products resolved on gels (14). Panel B, lanes B and C are translations with pJ14-1 hybridizable RNA which were pooled and immunoprecipitated with antiserum against F$_1$-ATPase β subunit, lane A.

immobilized on nitrocellulose, released then translated in a
reticulocyte lysate, gel analysis revealed the products
shown in Figure 1, panel A. The product with an apparent
molecular weight of 45,000 (open arrow) is the Leu2 gene
product identified in earlier studies (14,16). The component with an apparent molecular weight of 55,000 (closed
arrow) comigrated with the precursor of the F_1-ATPase β
subunit and was immunoprecipitated by specific antiserum to
this subunit (Figure 1, panel B).

Expression in Yeast

Preliminary analysis of the expression of pJ14-1 gene
product in S. cerevisae revealed that the F_1-ATPase β subunit was overproduced and accumulated in mitochondria (not
shown). Mitochondria prepared from transformant TJ12-5
exhibited a specific accumulation of the F_1-ATPase α subunit. "Western blot" analysis with subunit specific antiserum against the α and β subunits was used to confirm the
identity of these accumulated proteins which were present in
normal or reduced amounts in the respective pet mutants.
Quantitative immunological analysis of mitochondria and post-
mitochondrial supernatants revealed an 8-10 fold increase of
the specific subunit in the respective transformants over
the wild-type level. The gene dose of pJ14-1 in TJ14
measured by hybridization techniques relative to URA3 was
approximately 10 per haploid genome. Thus, the steady
state level of the ATPase β subunit in the cell under derepressed conditions may be gene dose dependent.
Additional support for the assignment of pJ14-1 as the
β subunit structural gene was provided by the observation
that pJ14-1 can transform the nuclear pet mutant of Schizo-
saccharomyces phombe (B59-1) lacking the F_1-ATPase β subunit (17) to restore growth on a non-fermentable carbon
source. By taking advantage of the difference in mobility
and immunological reactivity of the β subunit from each
source it could be shown that F_1-ATPase in the S. phombe
B59-1 transformed (18) with pJ14-1 contained the β subunit
of S. cerevisiae (Figure 2C). In addition to confirming the
gene product identification of pJ14-1 this result demonstrates that the processing, localization, and assembly of
this multi-subunit complex in these two yeasts is very
similar if not identical.

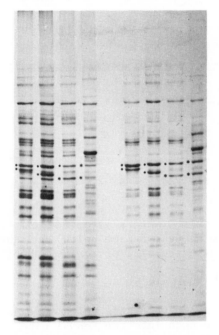

FIGURE 2.   Expression of pJ14-1 in S. phombe

Panel 1 mitochondrial preparations, 40 μg and Panel 2
chloroform solubilized crude F_1-ATPase prepared and
analyzed as described (17) a. S. phombe F25-28 with
polymorphic β subunit (17); b. S. phombe 972h wild
type; c. S. phombe B59-1 (pJ14-1) transformant; d.
S. cerevisiae D273-10B; (•) denote S. phombe and (*)
denote S. cerevisiae subunits.

Preliminary Characterization of pJ14-1

Restriction analysis of pJ14 (Figure 3) indicated the
size of pJ14-1 as approximately 21kb with an estimated insert size of 9.3kb. The unique Bam H1 site in YEp13 used
for cloning the Sau 3A pool of yeast genomic DNA was lost
at both ends of the pJ14-1 insertion.

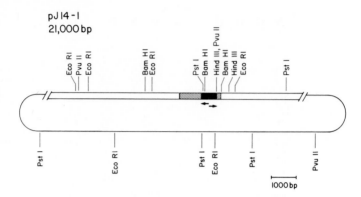

FIGURE 3.   Restriction Map of pJ14-1

The double line represents inserted DNA.   The black area
is the region sequenced (see below) and the shaded region is the expected size of the β subunit coding
region.

The Eco R1 restriction fragments which were not vector fragments were ligated each into the Eco R1 site of RF M13 mp7.
The RF of M13 with each fragment were then prepared (19) and
used to hybridize yeast poly A RNA which was translated in
a reticulocyte lysate.   This hybridization translation
analysis revealed that Eco R1 fragment 3 (E3) 3.3kb contained the yeast F_1 β gene.   M13 plaques containing fragments of E3 were picked and used to prepare single strand
templates (19) which were sequenced by the primer extension
dideoxy chain termination method (20).   Comparison of protein sequence derived from DNA sequence analysis confirmed
the location of the β subunit gene on E3.   The S. cerevisiae
protein sequences (Figure 4) were derived from a portion of
DNA sequences (arrows in Figure 3) which map in E3 as shown.
Additional analysis (not shown) strongly suggests that the
yeast ε subunit is also contained within E3 near its 3'
end.   Thus, the gene organization of the F_1-ATPase β and
ε genes in yeast may retain features of that documented in
E. coli (23,24) and chloroplast (21).

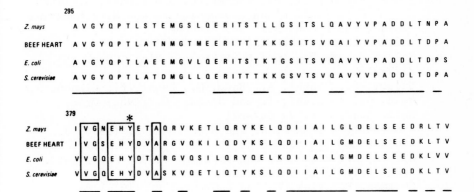

FIGURE 4. β subunit protein sequence comparison

The numbers refer to amino acid residues from the beginning of the Z. mays β subunit (21). Beef heart (22) and E coli (23,24) are as published. The (*) denotes the p-fluorosulfonyl benzoyl adenosine modified tyrosine in the nucleotide binding site of the beef heart (25) or S. cerevisiae (26) β subunit.

ACKNOWLEDGMENTS

This study was supported by NIH Grant GM25648 and Robert A. Welch Grant AQ-814. Thanks to Janet Kendall for typing this manuscript.

REFERENCES

1. Schatz, G. and Mason, T. (1974) Ann. Rev. Biochem. 43 51-87
2. Tzagoloff, A. and Meagher, P. (1971) J. Biol. Chem. 246 7328-7336
3. Maccechini, M., Rudin, Y., Blobel, G. and Schatz, G. (1979) Proc. Natl. Acad. Sci. 76 343-347
4. Neupert, W. and Schatz, G. (1981) Trends in Biochem. Sci. 6 1-4

5. Bohni, P., Gasser, S., Leaver, C. and Schatz, G. (1980) in The Organization and Expression of the Mitochondrial Genome, eds Kroon, A.M. and Saccone, C. (Elsevier/ North Holland, Amsterdam) pp. 423-433
6. McAda, P. and Douglas, M. (1982) J. Biol. Chem. 257 3177-3182
7. Gasser, S., Ohashi, A., Daum, G., Bohni, P., Gibson, J., Reid, G., Yonetani, T. and Schatz, G. (1982) Proc. Natl. Acad. Sci. 79 267-271
8. Tzagoloff, A., Akai, A. and Needleman, R. (1975) J. Biol. Chem. 250 8228-8235
9. Hinnen, A., Hicks, J. and Fink, G. (1978) Proc. Natl. Acad. Sci. 75 1929-1933
10. Sherman, F., Fink, G. and Lawrence, C. (1974) Methods in Yeast Genetics, Cold Spring Harbor Laboratory, New York
11. Woodrow, G. and Schatz, G. (1979) J. Biol. Chem. 254 6088-6093
12. Todd, R., Griesenbeck, T. and Douglas, M. (1980) J. Biol. Chem. 255 5461-5467
13. Williamson, V.M., Bennetzen, J., Young, E., Nasmyth, K. and Hall, B. (1980) Nature 283 214-216
14. O'Malley, K., Pratt, P., Robertson, J., Lilly, M. and Douglas, M. (1982) J. Biol. Chem. 257 2097-2103
15. Broach, J., Strathern, J. and Hicks, J. (1979) Gene 8 121-133
16. Hsu, Y. and Kohlhaw, G. (1982) J. Biol. Chem. 257 34-41
17. Boutry, M. and Goffeau, A. (1982) Eur. J. Biochem., in press
18. Beach, D. and Nurse, P. (1981) Nature 290 140-142
19. Winter, G., Fields, S. and Gait, M. (1981) Nucleic Acids Res. 9 237-245
20. Sanger, F., Coulson, A., Barrell, B., Smith, A. and Roe, B. (1980) J. Mol. Biol. 143 161-178
21. Krebbers, E., Larrinua, I., McIntosh, L. and Bogorad, L. (1982) Nucleic Acids Res., in press
22. Walker, J., personal communication
23. Kanazawa, H., Kayano, T., Mabuchi, K. and Futi (1981) Biochem. Biophys. Res. Commun. 103 604-612
24. Saraste, M., Gay, N., Eberle, A., Runswick, M. and Walker, J. (1981) Nucleic Acids Res. 9 5287-5296
25. Esch, F. and Allison, W. (1978) J. Biol. Chem. 253 6100-6106
26. Bitar, K. (1982) Fed. Proc. 41 a 1173

27

Changes of Gene Expression in Fusion Products between *Saccharomyces cerevisiae* and *Kluyveromyces lactis*

Cesira L. Galeotti[1] and G. D. Clark-Walker

Department of Genetics
Research School of Biological Sciences
Australian National University
Canberra, Australia

In terms of the petite colony mutation, whereby large deletions are formed in mitochondrial DNA, yeasts can be divided into those that undergo this mutation, termed petite-positive yeasts and those that do not, called petite-negative species. Three possible causes have been advanced to account for the petite negative character, either the structure of mtDNA or the enzymes responsible for deletion in these species are such that large deletions do not occur or else the physiological properties of the yeast (lack of fermentative growth) precludes survival of petite mutants (Clark-Walker and Miklos, 1974). To gain insight into this problem we decided to examine whether mtDNA from a petite negative yeast could undergo large deletions in the nuclear background of a petite positive species. Our reasoning for this approach is that we could exclude mtDNA structure from contention if deletions could subsequently be obtained, thereby leaving a decision to be made between physiological background and lack of appropiate enzymes.

Since there are no reports of petite positive yeasts interbreeding with petite negative species, we decided to

[1]*Present address: Department de Biologie
Moléculaire, Université de Genève, Geneve,
Switzerland.*

MANIPULATION AND EXPRESSION
OF GENES IN EUKARYOTES
ISBN 0 12 513780 x

try protoplast fusion in an attempt to combine the nuclear
genome of the former type with mitochondria of the latter.
For this experiment we chose double auxotrophically marked
strains of *Saccharomyces cerevisiae* and *Kluyveromyces lactis*
as the petite positive and negative yeasts respectively with
the additional constraint being that the *S. cerevisiae*
strains lack mitochondrial DNA. Our approach has been to
select prototrophic colonies on minimal medium after
inducing fusion of protoplasts by polyethylene glycol
(Ferenczy & Maraz, 1977), full details of these techniques
have been described elsewhere (Galeotti *et al.*, 1981).

I. FUSION BETWEEN *K. LACTIS* (K25) *a lys, ura* AND
    S. CEREVISIAE (D6) α *arg8, met*

A. *Isolation of Fusion Products*

     In the first fusion experiment using *S. cerevisiae*
strain D6eρ⁻ α *arg8, met* and *K. lactis* K25 a *lys, ura,*
fourteen colonies were observed on minimal medium plates
after two weeks incubation at 25°C.   This number represents
about 3 colonies per 10^7 regenerated protoplasts.   When all
fourteen colonies were subcultured again onto minimal medium
none survived, whereas on glucose/yeast extract/peptone
(GYP) medium two types of colony morphology, typical of the
parental strains, were seen from each isolate.   Testing of
each colony morph from the fourteen isolates showed only
four exceptions to the D6 or K25 phenotypes and these
colonies, KD3, 11, 13 and 14, which had *K. lactis* like
morphology, now required arginine in addition to lysine and
uracil.   This unexpected result was reproduced on repeating
the fusion experiment.   Of the eighteen colonies that grew
initially on minimal medium, fourteen produced segregants
showing the *arg, lys, ura* phenotype.   Subsequently we used
the four initial isolates for further analysis.

B. *Characterization of Fusion Products*

     All physiological parameters tested such as maltose
assimilation, cycloheximide resistance hydrolysis of β
glucosides, pulcherrimin production and absence of anaerobic
growth, indicated that the fusion products resembled *K.
lactis* rather than *S. cerevisiae*.   Estimation of DNA content
per cell showed that levels in fusion products were not significantly different from the K25 and D6 strains (Table 1).

TABLE 1. DNA Content of KD Fusion Products

Strain	fg DNA/cell
K25	7.25 ± 0.7
D6eρ⁻	8.22 ± 1.6
KD3	7.34 ± 0.7
KD11	8.21 ± 1.4
KD13	8.59 ± 1.5
KD14	8.33 ± 0.8

Taken together these results suggested that the *arg⁻* phenotype of the fusion products may have resulted from the presence of only a small amount of *S. cerevisiae* genetic material such as a single chromosome or a chromosomal fragment carrying the *arg8* locus and that in the *K. lactis* background the *arg⁻* phenotype was dominant. We tested for the presence of *S. cerevisiae* chromosome 15, which contains the *arg8* locus at the extremity of the left arm, by using a recombinant DNA Yep6 (Struhl *et al.*, 1979), containing the *his3* gene from the right arm of this chromosome, as a probe in a DNA hybridization experiment following gel blotting. No hybridization to the *K. lactis* genome or any of the fusion products could be detected (not illustrated). This experiment demonstrates firstly, that the fusion products do not contain a complete *S. cerevisiae* chromosome 15, secondly, that 2 μm DNA is absent from these strains as the Yep6 probe contains a fragment of this plasmid, and finally that insufficient homology is present between the *his3* gene of *S. cerevisiae* and the equivalent gene in *K. lactis* to allow detectable hybridization. As this result suggested that only a portion of a *S. cerevisiae* chromosome may be present which may lack a centromere, we were interested to learn if anomalous *arg8* distribution in meiosis could be detected by tetrad analysis.

Preliminary experiments established that all fusion products could mate with *K. lactis* WM52 α *ade1*, *his7* and subsequently we chose diploids from KD14 mating for sporulation and ascus dissection. From 67 four spored asci only 26 gave rise to 4 viable spores, 29 asci had 3 viable spores and the remaining asci contained fewer viable spores. By contrast, dissection of asci from the control cross K25 x WM52 showed normal levels of 4:0 viable spores indicating that the drop in viability was not due to abnormalities in the parental strains.

When segregation of auxotrophic markers was examined from asci containing four viable spores the *arg* phenotype showed 2:2 distribution along with other markers. Further data from marker segregation analysis also show that the *arg* mutation is apparently linked to the *his7* locus as the parental ditype class of *arg/his* segregants exceeds the nonparental ditype (Table 2).

TABLE 2. *Analysis of linkage between markers in asci from cross WM52 x KD14*

	PD*	NPD*	TT*	Total no. of tetrads
arg/his	15	0	10	25
arg/lys	4	9	12	25
arg/ura	6	12	6	24
lys/ura	12	0	14	26
his/lys	2	11	12	25
his/ura	5	8	11	24

* *PD, Parental ditype; NPD, nonparental ditype and TT, tetratype asci.*

In turn *arg* and *ura* also appear to be centromere linked as the tetratype segregants in the *arg/ura* analysis is exceeded by the parental and nonparental ditypes. Taken together these results suggest that a fragment of *S. cerevisiae* chromosome 15, containing the *arg8* locus, has become integrated into a *K. lactis* chromosome and can undergo normal 2:2 segregation in the absence of chiasma formation. Integration may have occurred in the vicinity of the *arg8* segment leading to inactivation of the *K. lactis* gene and disruption of the chromosome in this region so that chiasmas are not resolved to yield four viable products. Molecular analysis using a cloned *arg8* gene from *S. cerevisiae* should help to resolve these issues.

II. FUSION BETWEEN *K. LACTIS* (K25) AND *S. CEREVISIAE* (Feρ⁻) α *ade1, arg 4-16*

A. *Characterization of Fusion Products*

At the same time as the first fusion experiments were being conducted we also undertook similar experiments with a different *S. cerevisiae* (Fe ρ⁻) replacing the D6 strain. Unlike the previously described experiments, three of the six initial colonies appearing on minimal medium plates retained their prototrophic character on subsequent testing, the other three colonies were inviable. Examination of cell morphology under the light microscope revealed that many cells in all three isolates were larger than the parental forms and had unusual morphology. This irregularity of cell form, characterized by the presence of elongated and multiple budding cells, precluded accurate cell counting and therefore DNA estimation was not attempted. Testing of physiological properties showed in all cases that the colonies behaved like *K. lactis* as noted above for fusion products from the D6 strain, with the exception that growth at 30° was poor for KF3 and 4 and absent for KF6.

Survival to X-ray irradiation was ascertained for each isolate to try and distinguish the degree of ploidy, diploids, and some aneuploids being more resistant to killing than haploids. In each case, however, the fusion products, while showing linear dose-response curves, proved to be more sensitive to X-rays than K25 but more resistant than Feρ⁻ (not illustrated). Although this result is inconclusive regarding the possible aneuploid nature of the fusion products we did notice the occurrence of smaller colonies amongst the survivors to X-ray irradiation. Auxotrophic segregants could be identified amongst the smaller colonies, as shown in Table 3.

In addition to *lys, ura* or *lys, ura* phenotypes we found an *ade, lys* requiring colony that appeared as a red sector on a prototropic colony. The *ade, lys* colony phenotype was again isolated when the X-ray irradiation was repeated with KF4. Since this phenotype, containing simultaneously adenine and lysine requirements, was isolated on two occasions amongst only a few hundred survivors to X-rays it can be discounted that the strains showing these requirements have arisen from conventional mutational events. These isolates termed KF4.1r and KF4.2r were subjected to further tests.

TABLE 3. *Mitotic segregation in "small colonies" produced after X-ray exposure of fusion products between K25 and Feρ⁻*

Fusion products	Total No. of small colonies	Proto- trophic colonies	Auxo- trophic colonies	lys ura	lys ade ura lys	
KF3	7	4	3	2	1	-	-
KF4	25	21	4	2	1	-	1
KF6	17	16	1	-	-	1	-

B. *Characterization of KF4.1r and KF4.2r*

   Analysis of physiological characteristics of the two isolates showed them to be unchanged from KF4 apart from the adenine and lysine requirements.   Moreover DNA hybridization using purified ribosomal DNA or Yep6 containing the *his3* and 2 μm DNA sequences did not reveal the presence of these *S. cerevisiae* genome fragments.   Nevertheless our working hypothesis for the formation of prototropic colonies on fusion of K25 and Feρ⁻ is that the auxotrophic requirements of each parent are covered by genes from the opposite one.   Hence in the present case we postulate that X-ray irradiation has caused the *ade* and *lys* genes, one from each parent, to be blocked.   "Blockage" for the *lys* gene may be due to loss of a *S. cerevisiae* contributed fragment by mitotic segregation, as is often observed after X-ray treatment of aneuploids, but this cannot be the explanation for the production of *ade⁻*.   An indication that a complex process is operating in the KF4.1r and KF4.2r strains comes from studies on petite colony formation induced by ethidium bromide.

   Both sets of fusion products from the two experiments (KD and KF isolates) proved to be petite negative when treated with ethidium bromide at a concentration of 10 mg/ml.   By contrast when the treatment was repeated with KF4.1r or KF4.2r we recovered colonies that could not grow on non-fermentable carbon sources.   To test if the respiratory-deficient colonies were cytoplasmic petites, we isolated whole cell DNA, digested it with EcoRl and, after gel electrophoresis and transfer to nitrocellulose, we hybridized the filter with mitochondrial DNA isolated from *K. lactis* (Fig. 1).

a b c   d e f g   h

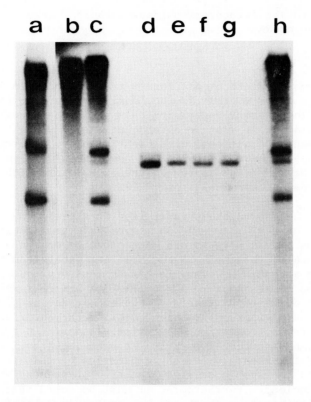

FIGURE 1. *Hybridization patterns of whole cell DNA from*
E.Br.-induced petites of KF4.1r and KF4.2r to ^{32}P *labelled*
K. lactis mtDNA. All samples except (b) were cleaved with
EcoRI. In (a) purified K. lactis mtDNA was used as a
control. (a) K25 mtDNA; (b) and (c) KF4.2r; (d) KF4.2r7;
(e) KF4.2r5; (f) KF4.2r1; (g) KF4.1r5 and (h) K25.

It can be seen from the figure that all four respiratory
deficient isolates lack *K. lactis* mtDNA, that is they are
petites of the $\rho°$ type (compare lane a with lanes d-g). In
addition to this interesting result, another unexpected
observation is revealed by the experiment whereby a DNA
fragment of approximately 5.5 kb hybridized to the mtDNA in
all digested samples apart from the purified mtDNA (lane
a). This discrete hybridizable fragment is produced by
EcoR1 digestion of nuclear DNA as it is absent from
untreated whole cell DNA (lane b). Subsequent studies (not
illustrated), using a variety of mtDNA probes, has shown
that *K. lactis* nuclear DNA contains a sequence showing

homology to the cytochrome oxidase subunit 2 gene of *S. cerevisiae* mtDNA.

III. CONCLUSIONS

A common feature of all fusion products between *K. lactis* and *S. cerevisiae* is the "dominance" of the former genome. The reasons underlying this phenomenon remain unexplained but it must be noted that the *S. cerevisiae* strains lacked functional mitochondria and to ascertain the importance of these organelles in the "dominance" phenomenon it would be important to repeat the experiments with respiratory competent strains.

Nevertheless in the present case it appears that only fragments of the *S. cerevisiae* genome have been retained in the fusion products as it has been shown by molecular hybridization that ribosomal DNA and *his3* sequences from this yeast are absent. However, in the KF fusion products there must be at least enough *S. cerevisiae* genome to complement the *lys* and *ura* requirements of K25 and in KF4.1r and KF4.2r there is sufficient information to allow fermentative growth as indicated by formation of ρ° type petites. In the latter case, it appears that a mutation brought about by X-ray treatment is necessary for the expression of this trait but the reason for the "recessive" nature of fermentative growth remains unclear. Furthermore, it must be noted that mtDNA of a petite negative yeast can be completely deleted without affecting viability (presumably only in the presence of *S. cerevisiae* gene(s)), which suggests but does not prove, that the underlying cause behind the failure of some yeasts to form these mutants under normal circumstances rests on the lack of fermentative growth.

REFERENCES

Clark-Walker, G.D. and Miklos, G.L.G. (1974).
   Genet. Res. *24*, 43.
Ferenczy, L. and Maraz, M. (1977). Nature *268*,
   524.
Galeotti, C.L., Sriprakash, K.S., Batum, C.M. and
   Clark-Walker, G.D. (1981). Mutation Res *81*,
   155.
Struhl, K., Stinchcomb, D.T., Scherer, S., and
   Davis, R.W. (1979). Proc. Natl. Acad. Sci.
   U.S.A. *76*, 1035.

28
Molecular Analysis of the Structure and Regulation of the *amd*S Gene of *Aspergillus nidulans*

Michael J. Hynes,[1] Catherine M. Corrick and Julie A. King

Department of Genetics
University of Melbourne
Parkville, Victoria

I. INTRODUCTION

Isolation and analysis of mutations affecting gene regulation *in vivo* coupled with molecular analysis provide the most powerful means of understanding mechanisms of the control of gene activity. Only in lower eukaryotes with well developed genetic systems do these methods approach the level of analysis realized in prokaryote systems.

We have been analyzing one system in *Aspergillus nidulans* in detail. The *amd*S gene codes for an acetamidase enzyme which hydrolyzes acetamide to acetate and ammonium. Synthesis of this enzyme is subject to multiple control mechanisms which have been elucidated by physiological and genetic studies (Hynes, 1978a). Each control mechanism is characterized by a *trans*-acting regulatory gene and in most cases a *cis*-acting site of action adjacent to the structural gene. Genetic evidence indicates that the product of each regulatory gene acts positively to control gene expression and the control mechanisms are independent of each other.

[1]*Research supported by Australian Research Grants Committee*

MANIPULATION AND EXPRESSION
OF GENES IN EUKARYOTES
ISBN 0 12 513780 x

II. CLONING AND CHARACTERIZATION OF THE *AMDS* GENE

Lambda genomic clones containing the *amdS* gene have been isolated using ^{32}P-labelled cDNA made from RNA from induced wild-type and *amdS* deletion strains as probes. Whole genome DNA Southern blots using DNA isolated from the wide range of *amdS* mutations isolated and analyzed by fine structure mapping (Hynes, 1979) has been carried out. Some results are summarized in Table 1 and Figure IA and IB. The analysis of deletions has allowed the *amdS* region to be localized within a 5kb EcoRI-SalI fragment. Furthermore, the *cis*-dominant controlling site mutations *amdI*18 (Hynes, 1978b), *amdI*9 (Hynes, 1975), *amdI*66 (Hynes, 1982) and *amdI*93 (Hynes, 1980) have all been shown to be due to point mutations (or small rearrangements of less than 100 base pairs). This indicates that the regulatory alterations due to these mutations are not due to the coupling of *amdS* to new promoters.

The mutation, *amd*-205, which is inseparable genetically from the *amdI*18 "up-promoter" site (Hynes, 1979) has been found to be a point mutation. Several other closely linked point mutations in this region show reductions in the level of mRNA produced. The analysis of the small deletions, *amd*-406 and *amd*-407 indicates that a region of at least 500 base pairs 5' to the coding region is necessary for transcription. In addition, from genetic analysis, the *amdI*66 mutation is 5' to this region (Table 1 and Figure IA, IB). A relatively large promoter-controlling region is therefore indicated.

TABLE I.   Characterization of some amdS *mutations*

Mutation	Nature of mutation	amdS polypeptide	amdS mRNA
amd*S135*	*missense*	+	+
amd*S1005*	*nonsense*	−	+
amd*S224*	*missense*	+	+
amd-*205*	*point-"promoter"*	−	±
amd*I18*	*point-"up-promoter"*	++	++
amd-*406*	*small deletion*	−	−
amd-*407*	*small deletion*	−	−
amd*I66*	*point*	+++	+++

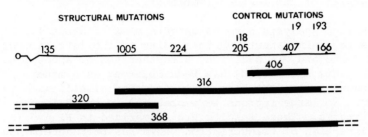

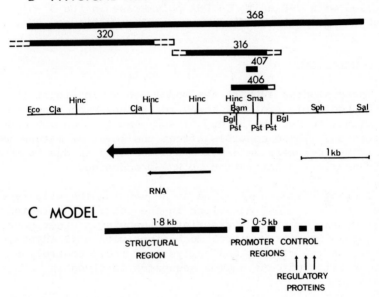

FIGURE 1. *Correlation of physical and genetic maps of* amdS.

III. TRANSCRIPTIONAL STUDIES

RNA blot analysis has shown that the major *amdS* mRNA is 1.6-1.8 kilobases, correlating well with the *amdS* polypeptide encoded (55-60,000). Analysis of wild-type and regulatory mutant strains has shown that regulation by all control mechanisms is at the mRNA level. Furthermore, the nature of *amd*S mutations has been correlated with their production of *amd*S polypeptide and mRNA and the physical and genetic properties of the lesions (Table 1).

R-loop analysis has indicated a mRNA of about 1.6 kilobases beginning about 200 base pairs to the left of the *Bam*Hl site and, in addition, a further RNA of about 1 kilobase. This smaller RNA has been observed as a minor component in the RNA blot analysis. Its significance is unclear. No large introns were observed in the R-loop experiments. These results are summarized in Figure IB. The direction of transcription shown has been determined by DNA blots using incomplete oligo dT-primed cDNA probes, by probing single stranded BamHl-EcoRl fragments cloned in both orientations in Ml3 vectors with cDNA and by using DNA labelled with ^{32}P using T4 DNA polymerase in the 3'-5' direction from the BamHl site to probe immobilized RNA.

IV. CONCLUSIONS

These studies support the model, shown in Figure IC, in which regulatory proteins coded for by unlinked genes interact with sites at the 5' end of the gene to modulate gene expression. These controls affect the level of mature mRNA but it has not been determined whether this is due to effects on transcription initiation or RNA processing.

Current research is aimed at sequencing the wild-type gene and the various mutations in the controlling region as well as more accurate transcription mapping. Other genes in this organism share control mechanisms with *amd*S (Hynes, 1978a) and the cloning and analysis of their controlling regions will indicate common sequences involved in regulation.

REFERENCES

Hynes M.J. (1975). *Nature* 253, 210-212.
Hynes M.J. (1978a). *Mol. Gen. Genet.* 161, 59-65.
Hynes M.J. (1978b). *Mol. Gen. Genet.* 166, 31-36.
Hynes M.J. (1979). *Genetics 91*, 381-392.
Hynes M.J. (1980). *J. Bacteriol.* 142, 400-406.
Hynes M.J. (1982). *Genetics*, in press.

29
Genes for Alcohol Utilization in the Lower Eucaryote *Aspergillus nidulans*

J. A. Pateman, C. H. Doy, Jane Olsen and Heather Kane

Department of Genetics
Research School of Biological Sciences
Australian National University
Canberra, Australia

Alcohol dehydrogenase (ADH) is responsible for the oxidation and synthesis of alcohols in many organisms. It has been described in Yeast, Maize, *Drosophila*, Mice, Humans, etc., and in all cases the amount and type of ADH expressed is dependent upon physiological conditions and/or the type of tissue or organ. Aldehyde dehydrogenase (AldDH) oxidases aldehydes, notably acetaldehyde to acetic acid. It is usually the second enzyme in the pathway for ethanol oxidation but in comparison with ADH little is known about its genetic determination and regulation. Since ADH in *A. nidulans* was known to be carbon repressed it was decided to undertake an analysis of the structure and expression of the genes responsible for alcohol utilization in a lower eucaryote which is particularly favourable for genetic and biochemical studies. This article describes some of our progress.

Induction and Repression of ADH and AldDH

Carbon catabolite repression in *A. nidulans* affects a number of enzyme and uptake systems, including ADH. Wild-type cells were grown on a variety of carbon sources and assayed for both ADH and AldDH. There was no ADH or AldDH activity in cells grown on 50mM glucose or acetate which are strong carbon repressors, and low activity in cells grown on 50mM glycerol and arabinose, which are weak carbon repressors. There were intermediate levels of both enzymes

MANIPULATION AND EXPRESSION
OF GENES IN EUKARYOTES
ISBN 0 12 513780 x

when cells were grown on 2mM glucose. If either ethanol or
threonine were the carbon sources there were high activities
of both enzymes. A simple explanation is that both ADH and
AldDH are subject to carbon repression by glucose or some
derived product and both enzymes are induced by ethanol and
threonine or some derived product. Similar experiments with
mutant cells lacking ADH or AldDH activity strongly indicate
that both ethanol and acetaldehyde are inducers of both ADH
and AldDH. Acetaldehyde is a product of the breakdown of
both ethanol and threonine.

Genetics of the alcA, alcR and aldA genes

Allylalcohol is not toxic to *A. nidulans*, but it is a
substrate for ADH and the resultant product acrolein is
toxic. It is possible in *A. nidulans* as in other organisms,
to obtain mutants which lack ADH activity by selection for
resistance to allylalcohol toxicity. Allylalcohol resistant
mutants were made using gamma rays and diepoxyoctane as
mutagens. About a hundred independently derived
allylalcohol resistant mutants were crossed and the frequency of wild type recombinants estimated. The frequencies
of wild type recombinants (RF) obtained indicated clearly
that there were two closely linked classes of mutants which
were located on chromosome VII. One class was named alcA
(the ADH structural gene) and the other alcR (an alcohol
utilization regulatory gene). The RF data indicated that
the probable size of the alcA gene is 1 to 2 cM and 0.7 to
1.0 cM for the alcR gene. The RF values for crosses between
alcA and alcR alleles were in the range 0.7 to 2.0 cM and a
deletion mutant extends over the majority of both the alcA
and alcR genes. This shows that the alcA and alcR genes are
very closely linked and probably adjacent. Six sizable
deletion mutants in the alcA gene are known so far and a
fine structure deletion map will be used to locate the
structural and control regions of the alcA gene. The
alcR125 mutant (formerly alcA125) was isolated by M. Page
(1971) and thought to be in the ADH structural gene, but it
is now known on genetical and biochemical criteria to be an
alcR allele. Since alcR125 (as alcA125) was shown by
Roberts *et al.* (1979) to be a nonsense mutant, this proves
that the product of the alcR regulatory gene is a protein.
Three allelic mutants which are unable to utilise ethanol or
threonine are located on chromosome VIII. They define a
gene which has been named aldA since it is probably the
AldDH structural gene.

Enzyme activities determined by the alcA, alcR, and aldA genes

The allylalcohol resistant and other relevant mutants were grown for 20h with 2 mM glucose, 100 mM threonine as the carbon sources and assayed for ADH and AldDH activities. Under these conditions the cells are carbon derepressed and induced and the mutants grow to a similar degree to the wild type. The results are summarized in Table 1. Forty seven mutant alleles form a well defined group, they all had subnormal ADH and essentially normal AldDH levels relative to the wild type. This group corresponds exactly with one of the genetic class of mutants and on the basis of their genetic and biochemical characteristics, this group defines the *alcA* gene.

TABLE 1. Enzyme activities of wildtype and mutants

Genotype	No. strains tested	ADH	AldDH
Wildtype	2	213.380	2.119
alcA	42	<0.020	2.860
aldA	2	10.810	<0.020
alcR	35	<0.020	0.050

All strains independently assayed at least 3 times. Enzyme units μmole, min^{-1}, mg^{-1} protein.

After assay for ADH and AldDH cell extracts of the wild-type and *alcA* mutants were given an SDS treatment and run on SDS polyacrylamide gels (SDS PAGE). SDS PAGE of extracts of carbon derepressed, induced wild type cells showed a band of approximate rmw 41,000 which sometimes resolved into two or three bands. There was also a band of approximate rmw 57,000 which sometimes resolved into two bands. These bands in the rmw 41,000 and 57,000 regions were all absent from extracts of carbon repressed non-induced wild-type. The bands in the rmw 41,000 region were absent from extracts of thirty nine *alcA* mutants. Five *alcA* mutants produced rmw 41,000 bands as strong as the wild type although all five lacked detectable ADH activity. The characteristics of the *alcA* mutants identify the rmw 41,000 bands as the probable polypeptide sub-units of ADH and *alcA* as the ADH structural gene.

The rest of the allylalcohol resistant mutants, a total of thirty eight, all had subnormal ADH and subnormal AldDH activities. This group corresponds exactly with the second genetic class of mutants which maps adjacent to the *alcA* gene and defines the *alcR* gene. SDS PAGE of extracts from the thirty eight *alcR* mutants showed that all lacked the bands in both the rmw 41,000 and 57,000 regions. The *alcR125* mutant, which is known to be a nonsense mutant, does not give bands in the rmw 41,000 and 57,000 regions.

A heterozygous diploid which carried the *alcA51* and *alcA*$^+$ alleles had similar ADH and AldDH activities to the haploid wild type *alcA*$^+$. So an allele of the ADH structural gene is recessive. A heterozygous diploid which carried the *alcR1* and *alcR*$^+$ alleles had similar ADH and AldDH activities to the haploid *alcR*$^+$. So *alcR1*, an allele of a regulatory gene controlling both ADH and AldDH enzyme activities, is completely recessive in function to the wild-type allele. This indicates that the protein product of the *alcR* gene has a positive function in the regulation of the ADH and AldDH enzyme activities.

The three known alleles of the *aldA* gene had normal ADH, but lacked AldDH activity relative to the wild type. SDS PAGE of extracts of all three *aldA* mutants gave similar bands to the wild type in the rmw 41,000 ADH region. Two of the *aldA* mutants lacked the band in the rmw 57,000 AldDH region. The third mutant *aldA15* had a similar band in the rmw 57,000 region to the wild type although *aldA15* lacked AldDH activity. The characteristics of the three *aldA* alleles indicate that *aldA* is the structural gene for AldDH.

In vitro translation of mRNA for ADH and AldDH

Messenger RNA preparations were made from the wild type, *alcA* and *alcR* mutants grown under carbon repressed, non-induced and carbon derepressed, induced conditions. Messenger RNA from the carbon derepressed, induced wild-type directed the synthesis in a rabbit reticulocyte translation system of protein bands in the rmw 41,000 (ADH) and rmw 57,000 (AldDH) regions as shown by SDS PAGE. Figs. 1 and 2. The mRNA from the *alcA51* and *alcA52* mutants directed the synthesis of the rmw 57,000 band, but not the rmw 41,000 band. The mRNA from the *alcR3* and *alcR22* mutants did not direct the synthesis of bands in the rmw 41,000 or, 57,000 regions. Thus the situation with respect to functional mRNA

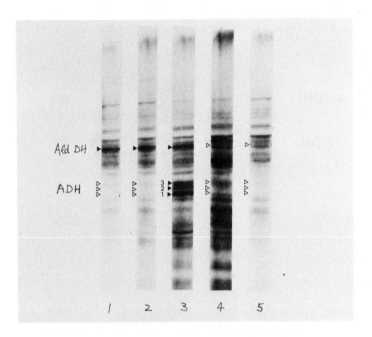

FIGURE 1. SDS PAGE of protein products of in vitro translation of A. nidulans mRNA. ▶ *band present* ▷ *band absent. Channel 1. alcA51. 2. alcA52. 3. wild type, three bands numbered 1 to 3 in the rmw 41,000 ADH region. 4. deletion mutant alcA55 alcR3. 5. alcR22.*

for ADH and AldDH corresponds to that for the ADH and AldDH peptide subunits in cell extracts of the *alcA* and *alcR* mutants.

The mRNA from carbon repressed, non-induced wild-type cells, did not direct the synthesis of either the rmw 41,000 or the rmw 57,000 proteins, in contrast to the mRNA from carbon derepressed induced cells. This indicates that the carbon catabolite repression and induction regulation of ADH and AldDH is at the level of transcription or post-transcriptional modification of the mRNA.

On some occasions the major ADH band in the rmw 41,000

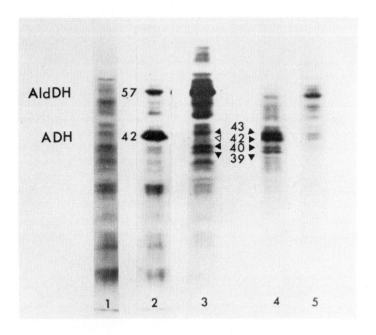

FIGURE 2.   SDS PAGE of protein products of in vitro translation of A. nidulans mRNA.   Numbers give rmw x 10⁻³ of bands ▶ *band present.* ▷ *band absent.   Channel 1. Deletion mutant alcA55 alcR. 2. Wild type. 3. alcA83. 4. Sucrose gradient fraction enriched for ADH mRNA. 5. Sucrose gradient fraction enriched for AldDH mRNA.*

region was resolved into two or three closely associated bands which differed by about rmw 1,000. It is difficult to be certain that all of the multiple bands were missing after translation of mRNA from the *alcA* and *alcR* mutants because of other fainter bands in the rmw 41,000 region, but it is probable that all the bands are lacking in both classes of mutants. The most likely explanation for the multiple ADH bands is that post-transcriptional processing produced two or more forms of ADH mRNA.

Regulation of ADH and AldDH

     Our current hypothesis concerning the regulation and

gene-enzyme relationships of ADH and AldDH is as follows:
The structural genes *alcA* on chromosome VII and *aldA* on
chromosomes VIII are subject to dual control; induction by
ethanol and acetaldehyde together with carbon catabolite
repression. A regulatory gene, *alcR*, is adjacent to *alcA*.
The gene product of *alcR* is a regulatory protein
specifically concerned with the induction control of
transcription of the *alcA* and *aldA* genes. The *alcR*
regulatory protein, together with the probable effectors
ethanol and acetaldehyde, has a function which is necessary
for the transcription of the *alcA* and *aldA* genes to
proceed. The *creA* gene - a general carbon catabolite
regulator gene, Arst *et al.* (1977) - product is concerned
with the carbon repression control of *alcA*, *aldA* and other
carbon catabolite repressed genes. There are at least two
different polypeptide sub-units for ADH and two for AldDH,
but in each case only one structural gene. Post-transcriptional processing produces at least two forms of mRNA
coded by *alcA* which are translated to give at least two
similar, but not identical ADH polypeptides. These
different peptides may allow the production of several types
of ADH enzyme with different characteristics and functions
with regard to the utilisation and detoxification of
alcohols and aldehydes.

Cloning and fractionation of ADH and AldDH mRNA

    Two kinds of *A. nidulans* gene banks have been made: (1)
Double stranded cDNA cloned in the pBR322 Bam site.   (2) *A.
nidulans* DNA partially digested with Sau3A and the 9K to 18K
base pair fragments cloned in the λ derivatives Charon 28
and Charon 30.

    The pBR322 derivatives were probed with single stranded
cDNA made from carbon derepressed induced wild type mRNA. A
number of pBR322 clones of potential interest were obtained
and single stranded cDNA from these is being used to probe
genomic blots of wildtype and deletion mutant DNA. In one
case, so far, there is a difference in the restriction
enzyme fragments indicating that the pBR322 clone may
contain a part or the whole of the *alcA* region.

    Many clones in Charon 28 showed strong hybridization to
a single-stranded cDNA probe made from carbon derepressed
induced wild type. None of these clones on further testing
showed differential hybridization with the cDNA from the

wild type and an *alcA alcR* deletion mutant. The most
probable explanation for this result is that this Charon 28
bank is incomplete and does not contain the *alcA* gene. The
cDNA from most of the Charon 28 clones gave patterns of very
rapid hybridization to wildtype fragments which may represent a commonly occurring class of highly or moderately
repeated sequences of *A. nidulans* DNA. The screening of two
Charon 30 genomic banks (one obtained from M. Hynes) is also
being done using single stranded cDNA made from mRNA preparations separately enriched for ADH and AldDH mRNA by
fractionation on a sucrose gradient. A number of Charon 30
clones have been obtained which possibly contain part or all
of the *alcA* region and some possible clones of the *aldA*
region and these are now being analysed.

CONCLUSIONS

The gene-enzyme system for alcohol utilization in *A.
nidulans* is a favourable one for the molecular analysis of
gene structure and regulation in a lower eucaryote. It is
expected that both the ADH and AldDH structural genes will
soon be cloned and sequenced together with adjoining control
regions. This will allow a direct comparison of the control
regions of two structural genes *alcA* and *aldA* which are
under the same induction and repression regulation mediated
at least in part by two regulator genes *alcR* and *creA*. The
regulator gene *alcR* is adjacent to the structural gene
alcA. Thus it will easily be cloned along with *alcA* from
one of the gene banks. Once *alcR* has been cloned it may be
possible to get expression of the gene in *E. coli* or
Yeast. This would provide the positive acting regulatory
protein along with the DNA adjacent to the coding regions of
alcA and *aldA* for direct studies of the regulatory
mechanism. Lastly, the cloning of the *alcA* gene will allow
a direct test for the existence of more than one form of ADH
mRNA and their possible role in the production of functionally differentiated ADH proteins.

REFERENCES

Arst, H.N., and Bailey, C.R. (1977). In *Genetics and
     Physiology of Aspergillus* (ed. J.E. Smith and
     J.A. Pateman) pp 131-146. London: Academic Press.
Roberts, T., Martinelli, S., and Scazzocchio, C.
     (1979). *Molec. gen. Genet. 177*, 57-64.
Page, M.M. (1971). PhD Thesis, University of Cambridge.

30
Endogenous Plasmid in *Dictyostelium discoideum*

Birgit A. Metz, Thomas E. Ward,[1] Dennis L. Welker and
Keith L. Williams

Max-Planck-Institut für Biochemie
D-8033 Martinsried bei München F.R.G.

Cobalt resistance in Dictyostelium discoideum

Cobalt resistance has been discovered in the simple eukaryote D.discoideum, and a number of alleles mapped to a
single locus cobA on linkage group VII (Welker & Williams
1980). Anomalous results with one "mutation" (cob-354) led us
to propose that this "mutation" was extrachromosomal. This
"mutation" has been transferred extensively into different
genetic backgrounds using the parasexual cycle. It is clear
that it does not map to linkage group I, II, III, IV, VI or
in particular VII, the linkage group containing the cobA
locus (Table 1).

Discovery of plasmid Ddp1 in D.discoideum

The failure to locate the cob-354 "mutation" to a chromosome, coupled with the fact that the cobalt resistance is unstable under non-selective conditions and dominant to wild
type (Williams, 1978), led to the hypothesis that this "mutation" may reside on a plasmid. A plasmid, Ddp1, has been iso-

[1]Visiting scientist. Permanent address: University of
Massachusetts, Worcester, U.S.A.

MANIPULATION AND EXPRESSION
OF GENES IN EUKARYOTES
ISBN 0 12 513780 x

179

Table 1 Exchange of chromosomes in strains carrying the cob–354 "mutation"

Strain	Parents	Linkage group					
		I	II	III	IV	VI	VII
HU32	NP81	+	axeA1 axeC1 oaaA1	axeB1 tsgA1	bwnA1 ebrA1	+	+
HU1159	DU1569 (HU32/HU892)	cycA1	acrA1 tsgD12 whiA1	bsgA5	bwnA1	+	couA351 tsgK21 frtB353
HU1184	DU1716 (HU1159/AX3)	+	axeA1 axeC1 oaaA1	bsgA5	bwnA1	+	couA351 tsgK21 frtB353
HU1315	DU1865 (HU1184/HU485)	+	axeA1 axeC1 oaaA1	axeB1	whiC351	+	couA351 tsgK21 frtB353
HU1491	DU2143 (HU1315/HU526)	cycA1	axeA1 axeC1 oaaA1	bsgA5	whiC351	manA2	+

All strains carry the unstable cob–354 "mutation"

lated from strains carrying the cob-354 "mutation" using an adaption of the method of Birnboim and Doly (1979), which is a rapid technique for isolating plasmid DNA from bacteria based on the use of selective alkaline denaturation of chromosomal DNA. Subsequent purification of ethanol precipitated plasmid DNA was based on spermine precipitation (Hoopes & McClure, 1981). In initial experiments cured (cobalt sensitive) strains derived from strains carrying cob-354 were found not to carry plasmid Ddp1.

At first we thought that plasmid Ddp1 arose in strain HU32, which was isolated as a spontaneous cobalt resistant mutant (Williams, 1978). Using improved techniques for plasmid isolation, we have shown that a number of our standard laboratory strains carry a plasmid that is similar, if not identical to, plasmid Ddp1 on the basis of the restriction enzyme digestion pattern using ten different restriction enzymes. The apparent copy number of plasmid in these strains is lower than that in strains bearing the cob-354 "mutation". A number of strains, including axenic strains used by most biochemists and molecular biologists, do not carry plasmid detectable by our methods. As a result of these findings it is not yet clear exactly where the plasmid comes from or how it potentiates cobalt resistance.

Characterization of plasmid Ddp1

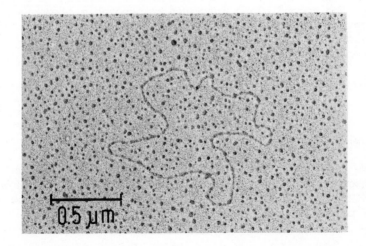

Fig. 1. Electron micrograph of D.discoideum Ddp1 plasmid

The following features of plasmid Ddp1 are established:

(1) It is a covalent closed circular DNA duplex of
    ~12.9 kilobases (Fig. 1).
(2) It is located in the nucleus.
(3) The restriction pattern is different from that of
    Dictyostelium mitochondrial DNA and ribosomal DNA.
(4) Preliminary estimates suggest that there are ~ 200
    copies of the plasmid in each cell.
(5) The plasmid is not lost during the developmental
    cycle.

Discussion

Few naturally occurring plasmids have been discovered in eukaryotes and none to our knowledge are associated with selective markers. The plasmid described here in D.discoideum is associated with a heavy metal (cobalt) resistance. Many bacterial plasmids code for heavy metal resistance (Chakrabarty, 1976).

We are currently investigating the relationship, if any, between Ddp1 DNA and chromosomal DNA of Dictyostelium. In other eukaryotes two classes of extrachromosomal (but probably nuclear) closed circular duplex DNA have been discovered. The first class, exemplified by the 2μm circle in the yeast Saccharomyces cerevisiae, does not integrate into chromosomal DNA under normal conditions (Cameron et al., 1977). The second class involves transposable elements such as copia in Drosophila (Flavell & Ish-Horowicz, 1981) and DNA associated with Alu repeated sequences in man (Calabretta et al., 1982).

The association with a selectable marker makes plasmid Ddp1 a candidate for a transformation vector in D.discoideum.

Acknowledgements

We thank Dr. Michael Claviez for help with electron microscopy and grants from the Max-Planck-Gesellschaft and the NIH (No. GM27757 to Dr. A. Jacobson) for making the visit of TW possible.

References

Birnboim, H.C., and Doly, J. (1979) Nucleic Acids Res. 7, 1513–1523.
Calabretta, B., Robberson, D.L., Barrera-Saldana, H.A., Lambrou, T.P., and Saunders, G.F. (1982) Nature 296, 219–225.
Cameron, J.R., Philippsen, P., and Davis, R.W. (1977) Nucleic Acids Res. 4, 1429–1448.
Chakrabarty, A.M. (1976) Ann. Rev. Genet. 10, 7–30.
Flavell, A.J., and Ish-Horowicz, D. (1981) Nature 292, 591–595.
Hoopes, B.C., and McClure, W.R. (1981) Nucleic Acids Res. 9, 5493–5504.
Welker, D.L., and Williams, K.L. (1980) J. Gen. Microbiol. 120, 149–159.
Williams, K.L. (1978) Genetics 90, 37–47.

31

cDNA Clones from Poly (A)$^+$ RNAs which Differ between Virulent and Avirulent Strains of the Protozoan Parasite *Babesia bovis*

Alan F. Cowman

David J. Kemp

The Walter and Eliza Hall Institute of Medical Research
Melbourne, Australia

Peter Timms

Bob Dalgleish

Tick Fever Research Centre
Brisbane, Australia

Babesia bovis, an intraerythrocytic protozoan cattle parasite transmitted by the tick Boophilus microplus, causes the economically significant disease babesiosis. Attenuated Babesia bovis strains can be used successfully as live vaccines - for example, the K-avirulent strain (K_A) derived from the virulent K geographical isolate (K_V) by ~26 serial passages through splenectomized calves (Callow and Mellors, 1966). While attenuation can be reversed either by passage through the tick vector or through intact animals (Callow et al., 1979) its molecular basis is not clear. Two dimensional gel-electrophoretic studies revealed that the K_A and K_V strains differ in only a limited number of polypeptides (Kahl et al. in press). Assuming that these polypeptide differences are relevant to avirulence, and that they are reflected at the RNA level, we have approached the isolation of their genes by differential colony hybridization.

MANIPULATION AND EXPRESSION
OF GENES IN EUKARYOTES
ISBN 0 12 513780 x

A   B

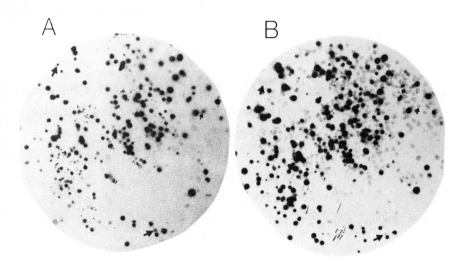

Fig. 1   Differential hybridization of a random array of K
avirulent cDNA clones probed with in panel A, ^{32}P cDNA of K
avirulent, and panel B, ^{32}P cDNA of Kvirulent isolate.

cDNA Clones from Differentially Expressed RNAs

Poly $(A)^+$ RNA from K_A was isolated by extraction from
washed, white cell-depleted, 8-20% parasitized bovine
erythrocytes with phenol/chloroform/sarkosyl (Alwine
et al., 1977) followed by oligo-dT cellulose chromatography. The RNA was copied into double-stranded cDNA which
was inserted into the Pst site of plasmid pBR322 by the
dG:dC tailing procedure (Gough et al., 1980). Transformed
cells were plated directly onto nitrocellulose filters
(Hanahan and Meselson, 1980) on tetracycline plates.
Replica filters were hybridized with ^{32}P-cDNA probes synthesized from poly $(A)^+$ RNA of K_A and K_V. In Fig. 1,
hybridization of the K_A probe to a random array of K_A
clones is shown. As might be expected, the colonies
hybridized to different extents, reflecting the abundance
of the relevant RNAs. While most colonies hybridized
equally to both probes, ~0.1% of them hybridized strongly
with K_A probe but not detectably with the K_V probe - for
example, the arrowed colonies (Fig. 1) hybridized in this
manner. A few colonies hybridized strongly with the K_V
probe but weakly with the K_A probe. Presumably, these
sequences are more abundant in K_V than in K_A - they may
correspond to proteins which are necessary for virulence
and lost during attenuation.

K$_a$ K$_v$ C$_a$ C$_v$ S$_v$ B

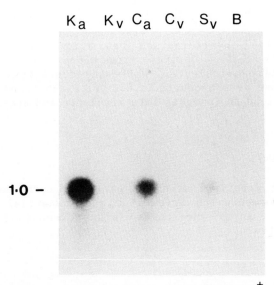

1·0 –

Fig. 2   Northern hybridization of poly (A)$^+$ RNA from
different isolates of Babesia bovis with the Pst insert from
the differential cDNA clone K5. The tracks contain 1μg of
poly (A)$^+$ RNA of; K$_A$, Kavirulent; K$_V$, Kvirulent; C$_A$,
Cavirulent; C$_V$, Cvirulent; S$_V$, Svirulent and B, Bovine bone
marrow.

Increased Levels of an RNA Molecule Correlate with
Avirulence

     To confirm that the cDNA plasmids selected above repre-
sent RNAs of differing abundance, we hybridized probes from
the cDNA inserts to RNA from K$_A$ and K$_V$ by the "Northern"
procedure (Alwine et al., 1977). One such clone, designated
K5, hybridized to a poly (A)$^+$ RNA of ~1.0kb that is present
in K$_A$ but not detectable in K$_V$ (Fig. 2). This result of
course does not necessarily implicate the K5 gene product
directly in avirulence and so we examined other B.bovis
strains. The distinct C virulent geographical isolate (C$_V$)
has also been attenuated by serial passage, generating the
C$_A$ isolate. The 1.0kb RNA species was present in RNA from
C$_A$ but not C$_V$, a virulent "revertant" of C$_A$ obtained by a
series of intact calf passages (Fig. 2). Further, only very
low levels of the RNA species were present in the virulent
S geographical isolate (Fig. 2). Uninfected bone marrow
did not contain the RNA, indicating its origin from Babesia
bovis.

CONCLUSION

Differential hybridization has allowed us to select B.bovis cDNA clones derived from mRNAs where abundance correlates with avirulence. While the functional signifi-cance of these gene products is not yet established, further studies may lead to insights into virulence and attenuation.

ACKNOWLEDGMENTS

We thank Kirstin Easton for excellent technical assis-tance and Commonwealth Scientific and Industrial Research Organisation, Animal Health Division, Queensland for the virulent S strain.

REFERENCES

Alwine, J.C., Kemp, D.J., and Stark, G. (1977). Proc.Natl. Acad.Sci.USA 72, 3962.
Callow, L.L. and Mellors, L.T. (1966). Aust.Vet.J. 42, 464.
Callow, L.L., Mellors, L.T. and McGregor, W. (1979). Int.J.Parasitol. 9, 333.
Gough, N.M., Webb, E.A., Cory, S. and Adams, J.M. (1980). Biochem. 19, 2702.
Hanahan, D, and Meselson, M. (1980). Gene 10, 63.
Kahl, L.P., Anders, R.F., Rodwell, B.J., Timms, P. and Mitchell, G.F. J.Immunol. (in press).

32
Cloning of Genes Coding for the Functional Antigens of the Liver Fluke, *Fasciola hepatica*

D. O. Irving, K. C. Reed* and M. J. Howell

Dept of Zoology and Dept of Biochemistry*
Australian National University
Canberra, Australia

The liver fluke, *Fasciola hepatica* is a parasitic hel-
minth which causes substantial losses to the wool and cattle
industries throughout the world. To date, control of *F.
hepatica* has been through the use of chemical drenching
regimes, pasture rotation and control of the snail intermediate host. All of these methods are expensive in terms of
capital outlay and are labour intensive. Moreover, the
continued prevalence of *F. hepatica* infestations indicates
that present methods are of limited utility. The development
of alternative methods of control would therefore seem desirable.
       Judging from the success of vaccination campaigns against
diseases of viral and bacterial origin, immunoprophylactic
treatment would appear to be an attractive method for the
control of *F. hepatica*. That this may be possible is indicated by the observation that rats have been successfully
vaccinated against challenge infections of *F. hepatica* using
excretory-secretory (ES) antigens derived from flukes maintained in culture (Rajasekariah *et al*. 1979). One major
drawback to the further investigation of ES antigens is that
they are produced in only minute quantities when flukes are
cultured *in vitro*. In order to obtain larger quantities of

*This work was supported by a grant from the Wool Research
Trust Fund on the recommendation of the Australian Wool
Corporation.*

MANIPULATION AND EXPRESSION
OF GENES IN EUKARYOTES
ISBN 0 12 513780 x

ES antigens it was decided to use recombinant DNA technology. The gene(s) coding for *F. hepatica* functional antigens could then be identified, and potentially limitless quantities of ES antigens could be synthesised.

A genomic DNA library was constructed from the total DNA of *F. hepatica*. DNA was extracted from whole flukes, using phenol/chloroform. The isolated DNA was then digested partially with Sau 3A into random fragments ranging in size from several hundred bases to 20 kb and then treated with alkaline phosphatase. Fragments in the size range of 9.6 - 19 kb were eluted from a 0.8% preparative agarose gel and ligated to the λ phage cloning vector, Charon 30. The vector was prepared by ligating its cohesive ends with T4 DNA ligase, cutting with Bam HI and separating the internal 'stuffer' fragments from the ligated arms by preparative gel electrophoresis. The partially digested *F. hepatica* DNA was then ligated to the vector using T4 DNA ligase. Recombinant molecules were subsequently packaged *in vitro* and amplified to establish a permanent library. Approximately 10^6 pfu/µg insert DNA were obtained after packaging.

In order to screen the genomic DNA library for functional antigen coding sequences, specific cDNA probes are being prepared. These are being isolated from a cDNA library, using immunological detection methods. In order to establish the cDNA library, total RNA was isolated from *F. hepatica* using guanidinium thiocyanate extraction followed by centrifugation through a 5.7 M CsCl cushion. Polyadenylated RNA was then separated from the total RNA by a single passage through an oligo (dT)-cellulose column. About 25% of the total RNA bound to and was eluted from the column. This material was subsequently used to construct cDNA (using reverse transcriptase) which was cloned into pBR 322. To determine whether intact mRNA coding for functional antigens was being isolated using this procedure, the polyadenylated RNA was tested in a rabbit reticulocyte *in vitro* translation system. Indeed functional antigens were identified in the translation products by immunoprecipitation with specific antisera. Several polypeptides in the molecular weight range of 10,000 - 31,000 were positively identied. Screening of the genomic DNA library is now in progress. The results will be reported subsequently.

REFERENCE

Rajasekariah, G.R., Mitchell, G.F., Chapman, C.B., and
    Montague, P.E. (1979). *Parasitology 79,* 393.

Part IV
Plants

Introduction

The study of plant molecular biology is at an exciting stage and is catching up quickly to the level of knowledge of gene expression that exists for some animal systems. Plant genes have essentially the same organization as animal genes (e.g. Chapter 35). There are intervening sequences that have the same GT–AG border-recognition signals as those in animal genes. The various transcription-control signals have similar sequences and positions: the CCAAT and TATAA boxes are positioned above the transcription start point at about the same distances, and the cases analysed so far fit the consensus sequences derived from animal genes. It is not clear whether all plant genes have an AATAAA polyadenylation signal, and so far nothing is known about transcription termination sequences for plant genes.

Another feature that genomes of higher plants and animals have in common is that many genes exist as families of closely related genes. The chapters in this part show this to be true both for storage protein genes (Chapter 34) and for genes that code for enzymes (Chapter 37). Already there is evidence that different members of the multi-gene families have specific transcriptional patterns, which suggests they may have distinct functions either at different times or in different cell types.

There is a considerable promise that gene analysis will help us understand the intricacies of the photosynthetic system in higher plants. These studies will eventually encompass an understanding of the interactions that occur in a plant cell between the nuclear genome and the genome of the chloroplasts, both of which code for proteins that are directly involved in photosynthesis.

Of great interest is the development of gene transfer systems in plants. Plant cell culture has progressed considerably in recent years, and in many plant species it is possible to establish single cell, and even protoplast, cultures that can be manipulated to regenerate whole plants. Therefore, we would expect to be able to introduce genes into single plant cells and recover a plant with the desired genetic modification. So far only the crown gall organism, *Agrobacterium*, has yielded a workable transformation system, but this is restricted to some dicotyledons. Other techniques are being researched, including one in which a chloroplast DNA replicating sequence is involved in the construction of possible gene vehicles (Chapter 40). Another chapter describes the earliest DNA sequence information on mobile genetic elements in plants (Chapter 36). It is conceivable that mobile gene segments, such as controlling elements in maize, could be used as vectors for the introduction of genes into host chromosomes. The *Ds/Ac* controlling element system of maize may be comparable to the P element system in *Drosophila*.

191

Already plants are providing valuable leads on the control of gene expression during development. Several groups are taking advantage of the tissue-specific expression of genes coding for a series of storage proteins in seeds of legumes or cereals (e.g. Chapter 33). These studies may lead to controlled mutations that provide correction of amino acid imbalances of these proteins when used for human or animal consumption.

Another exciting area of plant molecular biology which is just beginning to be explored is the relationship between host and pathogen in disease (Chapter 38). An understanding of the genetic basis of host plant resistance to disease will be of great benefit to the stability and efficiency of crops and pastures.

33
Regulation of Soybean Seed Protein Gene Expression

Robert B. Goldberg, Martha L. Crouch[1] and Linda Walling

Department of Biology
University of California
Los Angeles, California

I. INTRODUCTION

Soybean seed protein genes provide on excellent system for investigating the molecular events which regulate gene expression during plant development. These genes are expressed at specific periods in the life cycle, and encode highly prevalent products which can be idendified readily at both the mRNA and protein levels (Goldberg et al., 1981a, 1981b; Vodkin, 1981; Sengupta et al., 1981; Tumer et al., 1981; Beachy et al., 1981; Barton et al., 1982). In addition, soybean seed proteins are of considerable nutritional and economic importance to man, since they comprise over 40% of the seed mass (Larkins, 1981).

In this paper we review briefly the research which has been carried out in our laboratory on seed protein gene expression during soybean development (Goldberg et al., 1981a, 1981b). Seed protein genes are organized into several, small, nonhomologous gene families which are not dispersed among each other in soybean chromosomes (Goldberg et al., 1981a; Fischer and Goldberg, 1982). The genes we have investigated encode the storage proteins glycinin and β-conglycinin, the Kunitz trypsin inhibitor, seed lectin, and a 15,500 dalton protein whose identity is not yet known. The storage proteins are rapidly degraded after seed germination and serve as a "food" supply for the rapidly growing seedling (Larkins, 1981). The physiological functions of the Kunitz trypsin inhibitor

---

[1] *Present address: Indiana University, Bloomington, Indiana.*

MANIPULATION AND EXPRESSION
OF GENES IN EUKARYOTES
ISBN 0 12 513780 x

and seed lectin are still unidentified, but they have been implicated in several cellular processes (Ryan, 1981; Lis and Sharon, 1981). The results will show that seed protein gene expression is under strict developmental control, and that the primary control level appears to be transcriptional.

II. CLONING OF SEED PROTEIN mRNAs

To study the expression of specific seed protein genes, a cDNA library of mid-maturation stage embryo mRNA was constructed in the plasmid pBR322 (Goldberg et al., 1981a). Figure 1 shows that seed protein mRNAs are highly

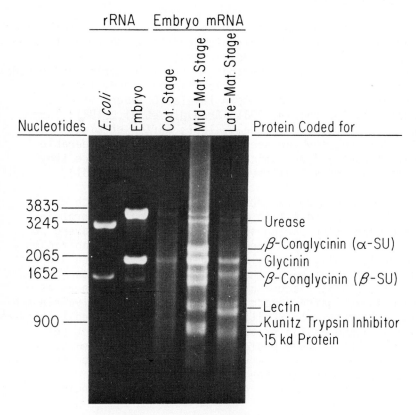

Fig. 1. *Presence of superprevalent seed protein mRNAs in embryo mRNA populations.   Adapted from Goldberg et al. (1981a, 1981b).*

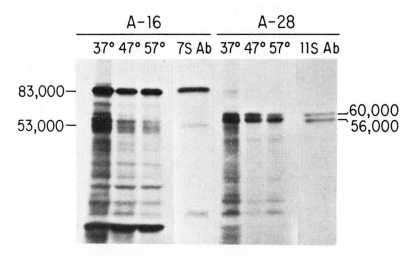

Fig. 2. Translation products of mRNAs selected with the A-16 and A-28 seed protein plasmids. 7S and 11S Ab refer to β-conglycinin and glycinin antibodies, respectively. M.L. Crouch and R.B. Goldberg, unpublished data.

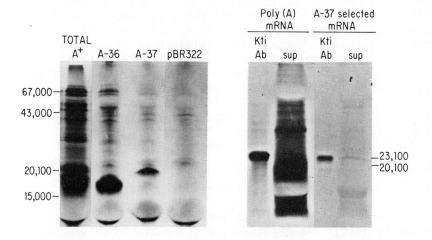

Fig. 3. Translation products of mRNAs selected with the A-36 and A-37 and seed protein plasmids. Kti Ab refers to Kunitz trypsin inhibitor antibody. Sup refers to supernatant after immunoprecipitation of protein with antibody. M.L. Crouch and R.B. Goldberg, unpublished data.

prevalent at this developmental stage, representing
approximately half of the mRNA mass (Goldberg et al.,
1981b). An abundant cDNA fraction was used to select seed
protein cDNA clones, and both mRNA gel blot studies
(Goldberg et al., 1981a) and hybrid- selected translation
experiments (M.L. Crouch and R.B. Goldberg, unpublished
data) were used to correlate these clones with specific
seed protein mRNAs. Figures 2 and 3 show representative
translation results with plasmid-selected mRNAs. It can
be seen that plasmids A-16 and A-28 represent the
β-conglycinin (7S) and glycinin (11S) storage protein
mRNAs, respectively. On the other hand, plasmid A-37
represents the Kunitz trypsin inhibitor mRNA, while A-36
represents an unidentified message which directs the
synthesis of a 15,500 dalton protein. Similar procedures
were used to show that plasmid L-9 represents lectin mRNA
(R.B. Goldberg, G. Hoschek, and L.O. Vodkin, unpublished
results).

III. SEED PROTEIN GENE EXPRESSION DURING DEVELOPMENT

A. Expression During Embryogenesis

To describe the temporal program of seed protein gene
expression during embryogenesis, mRNAs were isolated from
different embryonic stages, and the prevalence of each
seed protein message was measured by solution cDNA excess
DNA/RNA titration experiments (Goldberg et al., 1981a).
The results, summarized in Figure 4, show that there is
striking temporal control of seed protein gene expression
at the mRNA level. Each seed protein mRNA class
(including lectin, data not shown) accumulates during
early development, and then decays during late development
when seed dehydration occurs. At dormancy, and in
post-germination cotyledons, seed protein mRNAs have
diminished to the level of only a few molecules per cell.

B. Expression in Different Embryo Cell Types

Soybean embryos possess two, developmentally distinct
organ systems. The terminally differentiated cotyledons
are responsible for food reserve synthesis during
embryogenesis, while the axis develops into the mature
plant after seed germination. The inset to Figure 4
compares seed protein mRNA prevalences in axis and
cotyledon cells at the mid-maturation developmental
stage. Striking differences are evident. While glycinin,

REGULATION OF SEED PROTEIN GENE EXPRESSION

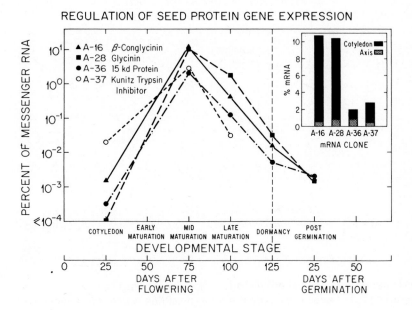

Fig. 4.   Accumulation and decay of seed protein mRNAs during embryogenesis.   Adapted from Goldberg et al. (1981a).

β-conglycinin, Kunitz trypsin inhibitor, and 15 Kd protein mRNAs comprise approximately 25% of the cotyledon mRNA mass, they constitute less than 3% of the axis message population (Goldberg et al., 1981a). Thus, not only is seed protein gene expression regulated temporally in development, but it is regulated with respect to embryo cell type as well.

C. Expression in Organ Systems of the Mature Plant

Are seed protein genes expressed in the cells of the mature plant? To answer this question, leaf, root, and stem mRNAs were isolated, and the presence of seed protein gene transcripts was measured by (1) hybridization of labeled cDNA clones to RNA gel blots (Goldberg et al., 1981a), (2) solution titration analyses (Goldberg et al., 1981a), and (3) reaction of labeled, random-primed cDNAs with DNA gel blots containing seed protein genomic clones (Fischer and Goldberg, 1982; J.K. Okamuro and R.B.

TITRATION OF cDNA CLONES WITH LEAF
AND EMBRYO NUCLEAR RNAs

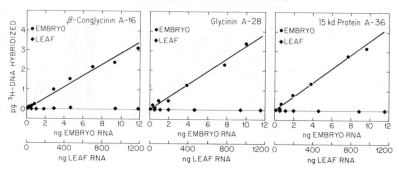

Fig. 5.   Absence of seed protein gene transcripts in leaf steady-state nuclear RNA.   Adapted from Goldberg et al. (1981a).

Goldberg, unpublished data). No physiologically meaningful levels of seed protein mRNAs were observed in any organ system, indicating that seed protein genes are not expressed at the mRNA level in the cells of the mature plant.

To determine whether seed protein genes are expressed at the nuclear RNA level, leaf nuclear RNA was isolated, and solution titration studies were used to detect the presence of seed protein gene transcripts (Goldberg et al., 1981a). The results, presented in Figure 5, indicate that the leaf *steady-state* nuclear RNA population contains no detectable seed protein gene sequences.

IV. TRANSCRIPTIONAL CONTROL OF SEED PROTEIN GENE EXPRESSION

The experimental findings outlined above strongly suggest that seed protein genes are activated during early embryogenesis, are repressed during late embryogenesis, and remain inactive during the post-germination phase of the life cycle. To test this transcriptional control model, studies were carried out with labeled "runoff" RNA synthesized *in vitro* by isolated nuclei (L. Walling and R.B. Goldberg, unpublished results). Leaf nuclei were isolated, and preinitiated RNA chains were allowed to "runoff" in the presence of ^{32}P-UTP (McKnight and Palmiter, 1979; Luthe and Quatrano, 1980a, 1980b). The

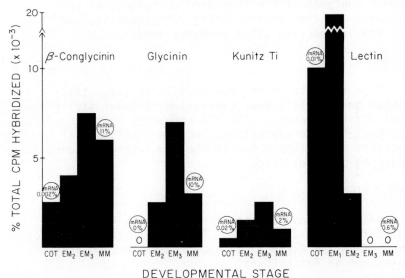

RELATIVE SEED PROTEIN GENE TRANSCRIPTION RATES

Fig. 6. Relative seed protein mRNA synthesis rates during embryogenesis. L. Walling and R.B. Goldberg, unpublished data.

labeled leaf nuclear "runoff" RNA was isolated, and then hybridized with filter-bound cDNA plasmids and/or genomic blots containing seed protein genomic clones. No significant hybridization signals were observed with any seed protein gene sequence (L. Walling and R.B. Goldberg, unpublished data). This finding, and that shown in Figure 5 with leaf steady-state nuclear RNA, strongly suggest that seed protein genes are "off" in leaf cells, and that seed protein gene expression is controlled primarily at the transcriptional level.

To determine the developmental periods in which seed protein genes are activated and repressed, labeled nuclear "runoff" RNAs from embryos at different embryonic periods were hybridized with filter-bound plasmid DNAs. Preliminary findings, shown in Figure 6, indicate that the relative seed protein RNA synthesis rates increase during early development, and then decrease during late development (L. Walling and R.B. Goldberg, unpublished data). Assuming that *in vitro* nuclear "runoff" transcription studies reflect actual *in vivo* events, these findings strongly support the proposition that seed protein genes are activated and inactivated during embryogenesis, and that they remain inactive after dormancy ends.

V. CONCLUSION

From the data presented above it is clear that soybean seed protein gene expression is under strict developmental control. The DNA sequences which regulate seed protein gene expression, and the physiological processes which interact with the regulatory regions, remain to be identified.

ACKNOWLEDGMENTS

We express our appreciation to Gisela Hoschek for expert technical assistance. This work was supported by grants from the National Science Foundation and the U.S. Department of Agriculture.

REFERENCES

Barton, K.A., Thompson, J.F., Madison, J.T., Rosenthal, R., Jarvis, N.P., and Beachy, R.N. (1982). J. Biol. Chem 257, 6089-6101.
Beachy, R.N., Jarvis, N.P., and Barton, K.A. (1981). J. Mol. Appl. Genet. 1, 19-27.
Fischer, R.L. and Goldberg, R.B. (1982). Cell 29, 651-660.
Goldberg, R.B., Hoschek, G., Ditta, G.S., and Briedenbach, R.W. (1981a). Dev. Biol. 83, 218-231.
Goldberg, R.B., Hoschek, G., Tam, S.H., Ditta, G.S., and Briedenbach, R.W. (1981b). Dev. Biol. 83, 201-217.
Larkins, B.A. (1981). Biochemistry of Plants 6, 449-489. Academic Press, New York.
Luthe, D.S. and Quatrano, R.S. (1980a). Plant Physiol. 65, 305-308.
Luthe, D.S. and Quatrano, R.S. (1980b). Plant Physiol. 65, 309-313.
Lis, H. and Sharon, N. (1981). Biochemistry of Plants 6, 371-433. Academic Press, New York.
McKnight, G.S. and Palmiter, R.D. (1979). J. Biol. Chem. 254, 9050-9058.
Ryan, C.A. (1981). Biochemistry of Plants 6, 351-369. Academic Press, New York.

Sengupta, G., Deluca, V., Bailey, D., and Verma, D.P.S. (1981). Plant Molec. Biol. 1, 19-34.
Tumer, W.E., Thanh, V.K., and Nielsen, N.C. (1981). J. Biol. Chem. 256, 8756-8760.
Vodkin, L.O. (1981). Plant Physiol. 68, 766-771.

34
French Bean Storage Protein Gene Family: Organisation, Nucleotide Sequence and Expression

Jerry L. Slightom, Michael J. Adang, Duncan R. Ersland,
Leslie M. Hoffman, Michael J. Murray and Timothy C. Hall

Agrigenetics Research Park
5649 East Buckeye Road
Madison, WI   53716

INTRODUCTION

The major seed storage protein in French bean is phaseolin, because it represents about 50% of the dry seed's nutritional protein (Ma and Bliss, 1978). One-dimensional SDS–polyacrylamide gel electrophoresis of purified phaseolin isolated from the cultivar 'tendergreen' resolves three polypeptides, α, β and γ, which have apparent molecular weights of 53, 47, and 43 kilo daltons (kd), respectively (McLeester et al., 1973). Each of these polypeptides is glycosylated (Hall et al., 1977). Peptide mapping of these proteins following proteolytic and chemical cleavages reveals that all three proteins are highly homologous (Ma et al., 1980), which suggests that they may be encoded in a homologous multigene family.

ANALYSIS OF THE PHASEOLIN MULTIGENE FAMILY

Analysis of the phaseolin gene copy number by reassociation kinetics and genomic blots indicates that

MANIPULATION AND EXPRESSION
OF GENES IN EUKARYOTES
ISBN 0 12 513780 x

there are about 10-14 homologous phaseolin genes (Ersland *et al*., manuscript in preparation). Hybridization of ^{32}P-nick-translated phaseolin subclone pPVPh 3.0 (see Fig. 3) onto *P. vulgaris* genomic DNA digested with different enzymes is shown in Fig. 1. These digests reveal three distinct

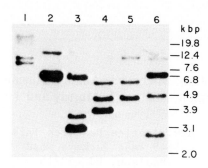

FIGURE 1. *Analysis of the phaseolin gene family, after digestion with different restriction enzyme followed by Southern (1975) blot hybridization to phaseolin gene sequences. Restriction digest are, lane: (1) Bam HI, (2) Eco RI, (3) Bam HI and Eco RI, (4) Bam HI and Bgl II, (5) Bgl II, and (6) Bgl II and Eco RI.*

classes of phaseolin gene fragments, *Class I*, consisting of 2 to 3 gene copies, is represented as a 12 kb Eco RI fragment which is cut to 3.0 kb in a *Bam* HI–*Eco* RI double digest; *Class II*, consisting of 5-7 gene copies, is represented on a 7.0 kb *Eco* RI fragment which is cut to 3.0 kb in a *Bam* HI–*Eco* RI double digest and is not cut with *Bgl* II; and *Class III*, consisting of 3-4 gene copies, is represented on a 6.7 kb *Eco* RI fragment which is cut to 2.8, and 4.8 kb fragments in a *Bgl* II–*Eco* RI double digest, see Fig. 1.

Cloning the Phaseolin Gene Family

We previously reported the construction of a Charon 24A phage library and the isolation of the phaseolin genomic clone λ177.4, see Fig. 3 (Sun et al., 1981). Since the construction of this Ch 24A library, several new lambda cloning vectors have been developed, which are more versatile. We selected the lambda vector, λ1059, developed

by Karn et al. (1980), which is capable of accepting *Mbo* I
(GATC) generated DNA fragments up to 24 kb long into its *Bam*
HI site. We have isolated 29 phaseolin containing λ1059
clones and analyzed about 15 of these cloned by detailed
restriction enzyme site mapping and blot hybridization to
known phaseolin sequences. Restriction endonuclease maps of
several of these clones are shown in Fig. 2, and they reveal

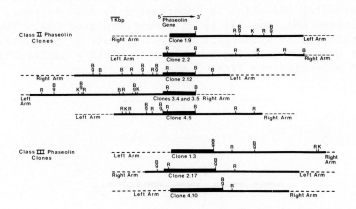

FIGURE 2. Restriction enzyme site mapping of λ1059
*phaseolin clones shows non-allelic genes from phaseolin gene
classes II and III. Restriction enzymes used are: R=Eco
RI, B=Bam HI, Bg=Bgl II, and K=Kpn I.*

that we have isolated clones containing five different types
of *Class* II and three different types of Class III phaseolin
genes; they all have different restriction enzyme sites in
their 5' and 3' flanking DNAs. We believe that these clones
contain highly homologous non-allelic variations of
phaseolin genes. We have not yet identified linked
phaseolin genes in a single λ1059 clone, nor have we located
overlapping restriction enzyme fragments which would link
two phaseolin genes. From our initial analysis, it appears
that the individual phaseolin genes may be separated by
distances greater than 20 kb. Whether this is true will
have to await analysis of all the λ1059 phaseolin gene
containing clones, or the use of a Cosmid (Collins, 1979)
type vector system, capable of accepting 30-45 kb DNA
fragments, to finally obtain linked phaseolin genes.

NUCLEOTIDE SEQUENCE OF A PHASEOLIN GENE

The structural outline of the phaseolin genomic gene λ177.4, (Fig. 3), showing intron and exon locations was derived from the nucleotide sequences comparison in Fig. 4. This particular phaseolin genomic gene includes 1990 bp, which consists of 77 bp of 5' untranslated DNA, 1263 bp of protein-coding DNA which is interrupted by five introns, 515 bp of intron DNA, and 135 bp of 3'-untranslated DNA. Thus, the original mRNA transcript of 1990 bp must be processed by five or more RNA splicing reactions to yield a native 1475 bp mRNA molecule, see Figs. 3 and 4.

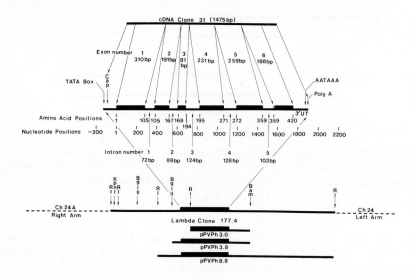

FIGURE 3. Diagram of the structure of the phaseolin gene from clone λ177.4, coding regions (six exons) are shown as raised bars, separated by five introns (IVS).

We have included 100 bp of 5' flanking DNA from λ177.4 in Fig. 4 (bases –1 to –100) which contains several DNA sequences believed important in controlling the transcription of eukaryotic genes, the "TATA" and "CCAAT" box-like sequences (Efstratiadis et al., 1980). Three "TATA" box sequences (overlined in Fig. 4) are located at positions –28, –37 and –39, upstream from the mRNA cap. Two "CCAAT box"-like sequences are located, at positions –67 bp (CCAT)

```
CATATGCGTGTCATCCCATGCCCAAATCTCCATGCATGTTCCAACCACCTTCTCTCTTATATAATACCTATAAATACCTCTAATATCACTCACTTCTTTC
----------+---------+---------+---------+---------+---------+---------+---------+---------+---------+  -1

̧CAP
│
ATCATCCATCCATCCAGAGTACTACTACTCTACTACTATAATACCCCAACCCAACTCATATTCAATACTACTCTACTATGATGAGAGCAAGGGTTCCACT
----------+---------+---------+---------+---------+---------+---------+---------+---------+---------+  100
                                                                            FMetMetArgAlaArgValProLe
CCTGTTGCTGGGAATTCTTTTCCTGGCATCACTTTCTGCCTCATTTGCCACTTCACTCCGGGAGGAGGAAGAGAGCCAAGATAACCCCTTCTACTTCAAC
----------+---------+---------+---------+---------+---------+---------+---------+---------+---------+  200
uLeuLeuLeuGlyIleLeuPheLeuAlaSerLeuSerAlaSerPheAlaThrSerLeuArgGluGluGluGluGluSerGlnAspAsnProPheTyrPheAsn
TCTGACAACTCCTGGAACACTCTATTCAAAAACCAATATGGTCACATTCGTGTCCTCCAGAGGTTCGACCAACAATCCAAACGACTTCAGAATCTTGAAG
----------+---------+---------+---------+---------+---------+---------+---------+---------+---------+  300
SerAspAsnSerTrpAsnThrLeuPheLysAsnGlnTyrGlyHisIleArgValLeuGlnArgPheAspGlnGlnSerLysArgLeuGlnAsnLeuGluA
ACTACCGTCTTGTGGAGTTCAGGTCCAAACCCGAAACCCTCCTTCTTCCTCAGCAGGCTGATGCTGAGTTACTCCTAGTTGTCCGTAGTGGTAAGTAATT
----------+---------+---------+---------+---------+---------+---------+---------+---------+---------+  400
spTyrArgLeuValGluPheArgSerLysProGluThrLeuLeuLeuProGlnGlnAlaAspAlaGluLeuLeuLeuValValArgSerGl◄─
GCTACTGGTATCACTTGTTTCTTCTTGCAGAAATAATGGTAATGAGTTTTTTATAATTTCAGGGAGCGCCATACTCGTCTTGGTGAAACCTGATGATCGC
----------+---------+---------+---------+---------+---------+---------+---------+---------+---------+  500
────(IVS 1, 72 BP)────────────────────────────────────────►LySerAlaIleLeuVaLLeuValLysProAspAspArg
AGAGAGTACTTCTTCCTTACGAGCGATAACCCGATATTCTCTGATCACCAGAAAATCCCTGCAGGAACCATTTTCTATTTGGTTAACCCTGATCCCAAAG
----------+---------+---------+---------+---------+---------+---------+---------+---------+---------+  600
ArgGluTyrPhePheLeuThrSerAspAsnProIlePheSerAspSerHisGlnLysIleProAlaGlyThrIleLeuPheTyrLeuValAsnProAspProLeuSG
AGGATCTCAGAATAATCCAACTCGCCATGCCCGTTAACAACCCTCAGATTCATGTACTGCCTTTTTGTAATACCGAACTAATTTTTTTGTTATTTTAACTTG
----------+---------+---------+---------+---------+---------+---------+---------+---------+---------+  700
LuAspLeuArgIleIleGlnLeuAlaMetProValAsnAsnProGlnIleHisI◄───────────────────────(IVS 2,
CAATTTCTCTCCCAAATGTGATGATAAATGTTTGTCCTGTAGGAATTTTTCCTATCTAGCACAGAAGCCCAACAATCCTACTTGCAAGAGTTCAGCAAGCA
----------+---------+---------+---------+---------+---------+---------+---------+---------+---------+  800
88 BP)────────────────────────────────►lGluPhePheLeuSerSerThrGluAlaGlnAsnGlnSerTyrLeuGlnGluPheSerLysHi
TATTCTAGAGGCCTCCTTCAATGTAAGAAAGAAAACAGCATCTAACTACATATTTGCGTTGCCATTTAGCTAGTACTTTGTCTAAATGTCACACTTGTTG
----------+---------+---------+---------+---------+---------+---------+---------+---------+---------+  900
sIleLeuGluGluAlaSerPheAsnH◄─────────────────────────(IVS 3, 124 BP)───────────
AATTTGTTGAATGATATCATTATATATGTTTGCATGATTTTTATAGAGCAAATTCGAGGAGATCAACAGGGTTCTGTTTGAAGAGGAGGGACAGCAAGAG
----------+---------+---------+---------+---------+---------+---------+---------+---------+---------+  1000
────────────────────────────────►lSerLysPheGluGluIleAsnArgValLeuPheGluGluGluGluGlnGlnGlu
GGAGTGATTGTGAACATTGATTCTGAACAGATTAAGGAACTGAGCAAACATGCAAAATCTAGTTCAAGGAAATCCCTTTCCAAACAAGATAACACAATTG
----------+---------+---------+---------+---------+---------+---------+---------+---------+---------+  1100
GlyValIleValAsnIleAspSerGluGlnIleLysLeuSerLysHisAlaLysSerSerSerArgLysSerLeuSerLysGlnAspAsnThrIleLeuG
GAAACGAATTTGGAAACCTGACTGAGAGGACCGATAACTCCTTGAATGTGTTAATCAGTTCTATAGAGATGGAAGAGGTAAATACAAAGAAAAACCATAT
----------+---------+---------+---------+---------+---------+---------+---------+---------+---------+  1200
LyAsnGluPheGlyAsnLeuThrGluArgThrAspAsnSerLeuAsnValLeuIleSerSerIleGluMetGluGlul
AGACAAACTCAGCAATTGAGTTCTATTATTCACTGTCGTCTTGGTTAGAAAATCTTAGTATTGAGACTATAATTAAATAATGGTTTTTTTTGTTAACAAA
----------+---------+---------+---------+---------+---------+---------+---------+---------+---------+  1300
───────────────────────(IVS 4, 128 BP)─────────────────────
TTTAGGGAGCTCTTTTTGTGCCACACTACTATTCTAAGGCCATTGTTATACTAGTGGTTAATGAAGGAGAAGCACATGTTGAACTTGTTGGCCCAAAAGG
----------+---------+---------+---------+---------+---------+---------+---------+---------+---------+  1400
───►lGlyValAlaLeuPheValPheHisTyrTyrSerLysAlaIleValIleLeuValValAsnGluValGlyuAlaHisValGluGluValGlyProLysGl
AAATAAGGAAACCTTGGAATATGAGAGCTACAGAGCTGAGCTTTCTAAAGACGATGTATTTGTAATCCCAGCAGCATATCCAGTTGCCATCAAGGCTACC
----------+---------+---------+---------+---------+---------+---------+---------+---------+---------+  1500
yAsnLysGluThrLeuGluTyrGluSerTyrArgAlaGluLeuSerLysAspAspValPheValIleProAlaAlaTyrProValAlaIleLysAlaThr
TCCAACGTGAATTTCACTGGTTTCGGTATCAATGCTAATAACAACAATAGGAACCTCCTTGCAGGTATATATATTTATTTATATATGACCATGAATTTGAA
----------+---------+---------+---------+---------+---------+---------+---------+---------+---------+  1600
SerAsnValAsnPheThrGlyPheGlyIleAsnAlaAsnAsnAsnAsnArgAsnLeuLeuAlaGl◄─
TATAGGGGTTGTTGATGGAATTTTTTATTTATAATTGGTAATGCGTGATTGTGATTGTAAATATGAAGGTAAGACGGACAATGTCATAAGCAGCATCGGTA
----------+---------+---------+---------+---------+---------+---------+---------+---------+---------+  1700
────────(IVS 5, 103 BP)────────────────────────►lIleTyrLysThrAspAsnValIleSerSerIleGluGlyA
GAGCTCTGGACGGTAAAGACGTGTTGGGGCTTACGTTCTCTGGGTCTGGTGACGAAGTTATGAAGCTGATCAACAAACAGAGTGGATCGTACTTTGTGGA
----------+---------+---------+---------+---------+---------+---------+---------+---------+---------+  1800
RgAlaLeuAspGlyLysAspValLeuGlyLeuThrPheSerGlyValSerGlyAspGluValMetLysLeuIleAsnLysGlnSerGlyThrTyrPheValAlaAs
TGCACACCATCACCAACAGGAACAGCAAAAGGGAAGAAAGGGTGCATTTGTGTACTGAATAAGTATGAACTAAAATGCATGTAGGTGTAAGAGCTCATGG
----------+---------+---------+---------+---------+---------+---------+---------+---------+---------+  1900
pAlaHisHisHisGlnGlnGlnGlnLysGlyLysLysGlyAlaPheValTyrTER
AGAGCATGGAATATTGTATCCGACCATGTAACAGTATAATAACTGAGCTCCATCTCACTTCTTCTATGAATAAACAAAGGAGTGTTATGAT---PoLy(A)
----------+---------+---------+---------+---------+---------+---------+---------+---------+---------+  2000
```

FIGURE 4. Comparison of nucleotide sequences from phaseolin genomic clone λ177.4 with phaseolin full length mRNA clone cDNA 3l. Only the genomic nucleotide sequence is presented, mRNA start, introns, coding regions and termination region are indicated.

and −74 bp (CCAAAT), see Fig. 4 double overlined nucleotides. Which of these "TATA" and "CCAAT" box-like sequences plays the more important role in controlling phaseolin mRNA transcription is being tested by injecting phaseolin cloned DNAs which have these structures modified or deleted into *Xenopus laevis* oocytes.

The phaseolin structural gene is interrupted by five small introns (IVS 1, 72 bp; IVS 2, 88 bp; IVS 3, 124 bp; IVS 4, 128 bp and IVS 5, 103 bp) creating six protein-coding DNA segments, or exons (see Figs. 3 and 4). Assigning the second methionine amino acid as codon one (nucleotides 81-83) we find that, IVS 1 splits codon 105, IVS 2 is between codons 167-168; IVS 3 is between codons 194-195; IVS 4 is between codons 271-272 and IVS 5 splits codon 359. All five phaseolin introns conform to the universal G-T/A-G splicing rule (Breathnach et al., 1978) and show reasonable homology with the consensus donor (5') and acceptor (3') splicing sequences (Mount, 1982). Comparison of the 3'nucleotide sequences from cDNA 31 and λ177.4 clones reveals a TGA translation terminator codon (position 1855, Fig. 4), indicating that there are 135 bp of 3'-untranslated DNA. The hexanucleotide "AATAAA" which has been suggested to be a signal for poly(A) addition (Proudfoot and Brownlee, 1976), is located 16 bp 5' to the first nucleotide of poly(A).

PHASEOLIN AMINO ACID SEQUENCE ANALYSIS

The phaseolin gene contains coding information for 420 amino acids, not including the F-met initiator codon. The N-terminal region of the phaseolin protein contains 24 hydrophobic and only two hydrophilic amino acids residues (residues 1-26, Fig. 4) which is followed by a highly hydro-philic region (residues 27-35). The presence of the initial hydrophobic region strongly suggest that phaseolin contains a "signal peptide". In the absence of N-terminal amino acid data from native phaseolin, it is difficult to determine the exact length of this signal peptide, it is most likely 21 amino acids long (cleaved after amino acid 21, serine) and certainly cleaved before amino acid 27 (arginine), see Fig. 4. This amino acid sequence also reveals two N-glycoside attachment sites, Asn-X-Ser (or Thr), (Sharon and Lis, 1979), one at nucleotides 1115-1123 (amino acids 251-254) and the other at nucleotide 1510-1518 (amino acids 340-342).

Comparison of the amino acid composition of these 399 amino acids with the amino acid composition found by amino acid analysis of the combined phaseolin protomers (Sun, 1974) shows excellent agreement with a coefficient of correlation of 0.98. If we assume that each oligosaccharide side-chain adds about 2 kd to the characteristic gel migration of a phaseolin protein, the phaseolin protein identified here will have an apparent molecular weight of about 49.2 kd. This calculated molecular weight indicates that our derived phaseolin protein more closely represents a -type phaseolin protein, which has a one-dimensional polyacrylamide apparent molecular weight of 47 kd (McLeester et al., 1973).

TRANSCRIPTION OF A CLONED PHASEOLIN GENE BY *XENOPUS LAEVIS* OOCYTES

Xenopus laevis oocytes have been shown to translate mRNA from plants, namely phaseolin mRNA (Matthews et al., 1981) and zein mRNA (Larkins et al., 1979). The question

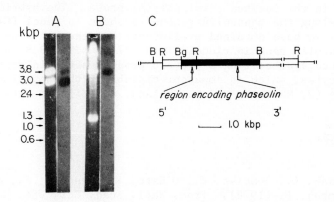

FIGURE 5. Hybridization of ^{32}P-labeled oocyte poly(A) mRNA, isolated from oocytes injected with a cloned phaseolin gene, onto different digest of phaseolin gene containing clones. Panel A, Bam HI and Eco RI digest of pPVPh 3.0 (Fig. 3); Panel B, Bgl II and Bam HI digest of pPVPh 8.8 (Fig. 3); Panel C shows the phaseolin gene region which hybridized to oocyte mRNA, and the location of the phaseolin gene on pPVPh 8.8.

remaining to be answered is, will oocytes injected with an intact plant gene synthesize mRNA from that plant gene and translate it into native protein? We have evidence which partially answers this question, by showing specific transcription, in oocytes, of a cloned phaseolin gene. For this experiment we injected 10 μg of a pBr322 subclone of λ177.4, pPVPh 3.8, (see Fig. 3) into oocyte nuclei as described by Matthews et al. (1981), incubated these oocytes for 24 hours at 19°C, then extracted total poly(A)$_2$ mRNA. This total poly(A) mRNA was end-labeled using [λ-^{32}P] ATP as described by Maizels (1976). Labeled total poly(A) mRNA was then hybridized onto restriction digests of two different subclones of λ177.4 by the method described by Southern (1975). These hybridizations are shown in Fig. 5; Panel (A) is an *Eco* RI- *Bam* HI digest of subclone pPVPh 3.0 and panel (B) is a *Bam* HI-*Bgl* II digest of subclone PVPh 8.8. In both hybridizations the phaseolin gene containing restriction fragment, 3.0 kb in panel (A) and 3.8 kb in panel (B) hybridize strongly to complimentary phaseolin sequences present in the oocytes total poly(A) probe. The hybridizing region along the phaseolin genome is shown in panel (C). Recently we have obtained preliminary evidence showing that the phaseolin gene in clones λPVPh 177.4, pPVPh 3.8, and 8.8 are indeed synthesizing a protein product after injection into oocytes, which is immunopreciptable with phaseolin antibody (J. Matthews and D. Ersland, unpublished data).

REFERENCES

Breathnach, R., Benoist, C., O'Hare, K., Gannon, F., and Chambon, P. (1978). Proc. Natl. Acad. Sci. USA. *75*, 4853.
Collins, J. (1979). *In* "Methods in Enzymology" Vol. 68 (R. Wu, ed.), p. 309. Academic Press, New York.
Efstratiadis, A., Posakony, J. W., Maniatis, T., Lawn, R. M., O'Connell, C., Spritz, R. A., DeRiel, J. K., Forget, B. G., Weissman, S. M., Slightom, J. L., Blechl, A. E., Smithies, O., Baralle, F. E., Shoulders, C. C., and Proudfoot, N. J. (1980). Cell *21*, 653.
Ersland, D. R., Chee, P. Y., Hoffman, L. M., Slightom, J. L. and Hall, T. C. (manuscript in preparation).
Hall, T. C., McLeester, R. C., and Bliss, F. A. (1977). Plant Physiol. *59*, 1122.
Karns, J., Brenner, S., Barnett, L., and Cesareni, G. (1980). Proc. Natl. Acad. Sci. USA *77*, 5172.

Larkins, B. A., Pedersen, K., Handa, A. K., Hurkman, W. J., and Smith, L. D. (1979). Proc. Natl. Acad. Sci. USA *76*, 6448.

Ma, Y., and Bliss, F. A. (1978). Crop Science *17*, 431.

Ma, Y., Bliss, F. A., and Hall, T. C. (1980). Plant Physiol. *66*, 897.

Maizels, N. (1976). Cell *9*, 431.

Matthews, J. A., Brown, J. W. S., and Hall, T. C. (1981). Nature *294*, 175.

McLeester, R. C., Hall, T. C., Sun, S. M., and Bliss, F. A. (1973). Phytochem. *12*, 85.

Mount, S. M. (1982). Nucl. Acids Res. *10*, 459.

Proudfoot, N. J., and Brownlee, G. G. (1976). Nature *263*, 211.

Sharon, N. and Lis, H. (1979). Biochem. Soc. Trans. *7*, 783.

Southern, E. M. (1975). J. Molec. Biol. *98*, 503.

Sun, S. M. (1974). Ph.D. thesis. University of Wisconsin, Madison, WI.

Sun, S. M., Slightom, J. L., and Hall, T. C. (1981). Nature *289*, 37.

35
The Alcohol Dehydrogenase Genes of Maize: A Potential Gene Transfer System in Plants

W. L. Gerlach, H. Lörz, M. M. Sachs, D. Llewellyn, A. J. Pryor,
E. S. Dennis and W. J. Peacock

Division of Plant Industry
CSIRO
Canberra, ACT, Australia

The development of a gene transfer system in plants is important for analysis of gene control and gene action and also for manipulation of genomes of agriculturally important species. Although a number of gene transfer systems exist for animal cells, the only successful approach in plants is based on the Crown Gall inducing organism, *Agrobacterium tumefaciens*. This bacterium contains a plasmid, part of which is incorporated into host chromosomes in the transformation event (1). It is possible to insert exogenous genes into the T-DNA segment of the plasmid and have these cointegrated into host plant chromosomes (2). However this system is limited to certain dicotyledonous species and has additional disadvantages of being associated with abnormal growth patterns, instability of incorporated sequences and low levels of transcription. Some of these difficulties may be overcome but it is not clear that the T-DNA method will be able to be applied to monocotyledons, which include all the important cereal crops.

The alcohol dehydrogenase genes of maize (3) could provide the basis of a selectable gene transfer system in a wide array of plant species, both mono- and dicotyledons. The *Adh* genes, *Adh1* and *Adh2* on chromosomes 1 and 4 respectively, convert acetaldehyde to ethanol, thereby regenerating NAD$^+$, essential for glycolytic ATP production during anaerobiosis (see review, 4). *Adh* gene

MANIPULATION AND EXPRESSION
OF GENES IN EUKARYOTES
ISBN 0 12 513780 x

Copyright © 1983 by Academic Press Australia.
All rights of reproduction in any form reserved.

activity is required for the survival of maize seedlings
when their roots are placed under anaerobic conditions, e.g.
flooding (5). The introduction of Adh^+ genes into maize
cells lacking Adh enzyme activity could ensure the survival
of these cells in an anaerobic environment (Fig. 1).

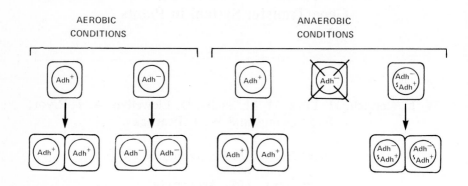

*FIGURE 1. A potential gene transfer selection system
using maize cells of different ADH constitution. Under
aerobic conditions maize cells grow regardless of ADH
phenotype. In anaerobiosis Adh^+ cells will grow and Adh^-
will not. But Adh^- cells could be rescued by addition of a
functional Adh^+ gene.*

$Adh1$ AND $Adh2$ cDNA CLONING

As the first step in the isolation of the Adh genes we
prepared 4,500 cDNA clones from polyA RNA of anaerobic roots
because ADH enzyme activity is induced by these conditions
(6). The clones were plated in replicate and hybridized
with probes from mRNA of anaerobic and aerobic roots. 105
of the clones hybridized anaerobic probe but did not
hybridize aerobic probe.sequences, suggesting they were
derived from mRNAs induced by the anaerobic treatment. Some
of these clones were homologous to mRNAs which coded for a
40,000 d polypeptide in a wheat germ *in vitro* translation
system. The 40,000 d product from one of the sequence
families was identified as $Adh1$ polypeptide because it
comigrated with native ADH1 polypeptide on gel electrophoresis systems, was specifically precipitated by anti-ADH
IgG, and showed the expected, altered electrophoretic
behaviour when an $Adh1$ mutant was used as an mRNA source (7).

Sequencing of regions of ADH1 polypeptide showed amino acid tracts identical to those predicted by the nucleotide sequence.

Another of the families encoding a 40,000 d polypeptide was identified as *Adh2* cDNA. Inserts in this class cross hybridize weakly with the *Adh1* cDNA and do not hybridize to mRNA prepared from a stock of maize containing an *Adh2* null mutation which does not produce any translatable *Adh2* mRNA (8).

INDUCTION OF ADH ACTIVITY

These cDNA probes enabled us to show that the induction of ADH enzyme activity was associated with an increased level of *Adh* mRNA in the root cells of anaerobically treated seedlings. Northern blots established that mRNA for the *Adh1-S* allele was approximately 1650 bases long and mRNA for the *Adh2* gene approximately 1750 bases long (Fig. 2a).

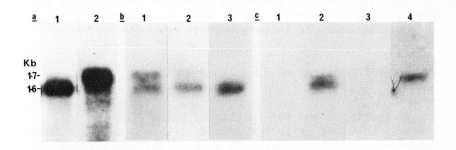

FIGURE 2. Northern blot analysis of RNA from roots of different maize stocks. Hybridization of: a. cDNA probes to RNA from an Adh1-S Adh2-N stock showing Adh1-S (lane 1) and Adh2-N (lane 2) message size classes.   b. Adh1 cDNA probe to RNA from lines containing Adh1-1F, Adh1-78F and Adh1-33F alleles (lanes 1-3 respectively).   c. Adh1 cDNA probe to RNA from aerobic and anaerobically treated seedling roots (lanes 1,2, respectively).   Lanes 3,4 same as 1,2 except that probe was Adh2 cDNA.

Some other alleles of *Adh1* showed two sizes of mRNA, one of approximately 1750 bases in addition to the 1650 base class produced by the *Adh1-S* allele; naturally occurring *Adh1-F* alleles have different proportions of the two mRNA size classes (Fig. 2b). Although the longer mRNA was similar in size to the *Adh2* messenger RNA molecule, stringent hybridizations showed that both the 1750 and 1650 mRNAs were transcripts from the *Adh1* locus. We knew from sequencing that although the coding sequences of the *Adh1* and *Adh2* genes are highly homologous with approximately 85% sequence identity, there is complete divergence between their 3' untranslated regions. A hybridization probe specific for the 3' non-translated region of *Adh1* confirmed that the *Adh1-F* alleles were producing the two mRNA size classes.

Quantitative Northern blots established that anaerobiosis resulted in at least a fifty-fold enhancement of mRNA levels for both of the *Adh* genes in seedlings (Fig. 2c), suggesting that the induction of ADH enzyme activity is under transcriptional control. The maximum level of mRNA in root cells was reached five hours after the onset of anaerobiosis. In the case of the *Adh1* gene this maximal level of mRNA was maintained for more than two days until cell death began in the third day of anaerobiosis. The level of *Adh2* mRNA decreased after only ten   hours of anaerobiosis.

The similarity in the early stages of induction of the two unlinked *Adh* genes suggests that they may have a common control sequence for their anaerobic response.

GENOMIC ORGANIZATION OF *Adh1* AND *Adh2*

The patterns and intensities of bands detected in Southern blots of maize DNA probed with *Adh1* and *Adh2* cDNA sequences were consistent with each of the genes being present in single copy in the genome. We isolated an *Adh2-N* sequence from a partial Sau3A library of maize DNA in λ-Charon28 vector. An *Adh1-S* segment was isolated by Dr J. Bennetzen (pers.comm.) from Bam-digested DNA cloned in λ1059. A comparison of restriction and sequence maps of these genomic segments with our cDNA maps showed intervening sequences in each gene. The gene for *Adh2-N* is more than 3,000 bp and contains at least seven introns. The intron borders follow the GT-AG rule which applies in animal gene systems (8). The *Adh1-S* gene is similar in length and we know of at least five introns. The most 5' intron is

positioned at precisely the same location relative to the
translation start codon in each gene. The sequences within
the introns have diverged, as have both the 5' and 3'
untranslated regions flanking the coding segments.
Sequencing of the 5' region of the *Adh1-S* gene has shown it
to have typical eukaryote transcription signal sequences
(Fig. 3).

5'...GGCCAAACCG CACCCTCCTT

CCCGTCGTTT CCCATCTCTT CCTCCTTTAG AGCTACCACT ATATAAATCA

likely transcription start region
GGGCTCATTT TCTCGCTCCT CACAGGCTCA TCTCGCTTTG GATCGATTCG

5' end of longest cDNA clone
↓
TTTCGTAACT GGTGAAGGAC TGAGGGTCTC GGAGTGGATC GATTTGGGAT

coding region:
TCTGTTCGAA CATTTGCGGA GGGGGGCAAT GGCGACCGCG GGGAAGGTGA..3'

*FIGURE 3. Sequence of 5' region of Adh1-S showing
translation start point and possible CCAAT and TATAA
regulatory regions.*

EXPRESSION AND TRANSFER OF THE CLONED *Adh⁺* GENES

We suggested earlier that the *Adh* genes have potential
as selectable markers for gene transfer in plants. The
rationale was that ADH enzyme activity would be essential
for cells to survive anaerobic conditions, just as it is
for whole plants.
In developing a recipient cell system we established
callus cultures from immature maize embryos. Plants can be
regenerated from small cell aggregates (< 1mm diameter) of
these cultures. We have found that when the cell genotype
includes active alleles of one or both *Adh* genes, the cells
survive anaerobiosis. However cell aggregates from plants
lacking a functional allele at both loci fail to survive
extended anaerobic conditions and cannot regenerate mature
plants (Fig. 4).

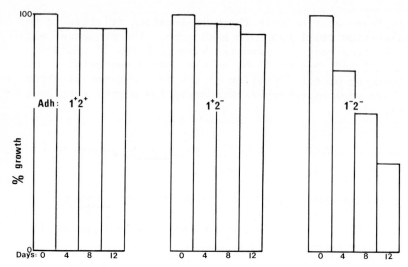

FIGURE 4. Maintenance of (embryo derived) maize cell cultures under anaerobic conditions. Lines of maize containing active Adh⁺ genes survive anaerobic conditions better than the Adh1⁻ Adh2⁻ line which shows inhibited growth. The graph shows callus weight (as a proportion of control aerobic callus) after anaerobic treatments of 0, 4, 8, 12 days.

This defines a regime under which $Adh1^-$ $Adh2^-$ cells may be rescued by introducing cloned Adh^+ genes. We have used microinjection techniques to introduce DNA into the cell aggregates. These were then subjected to the anaerobic selection system and survivors are being cultured and will be used for protein and DNA analysis.

The major difficulty with the system is that it involves cell aggregates. Ideally microinjection would be carried out on single cells where the DNA could be accurately introduced directly into the nucleus. This may be possible in cell cultures of *Nicotiana plumbaginifolia* or *Daucus carota* which regenerate plants readily from single cells and from protoplasts. We are mutagenizing haploid cell lines of these two species and screening with an allyl alcohol selection system for Adh^- mutations (10). Adh^- variants have been detected in *Nicotiana plumbaginifolia* and are being analysed. If confirmed as mutants, these Adh^- cells will be used for microinjection, and also as protoplasts for the introduction of Adh^+ DNA by direct uptake. We anticipate that the maize Adh genes will function in these cell lines since both show an anaerobiosis-inducible ADH system, and because all plant species examined have immunologically cross-reacting ADH polypeptides (4). We may be able to monitor the production of ADH enzyme in single cells using the staining procedure developed for ADH in pollen (11).

We intend analysing the components of the *Adh1* and *Adh2* genes necessary for expression. For example, we should be able to confirm the functional roles of the CCAT and TATA boxes in the *Xenopus* oocyte system (12) and using cultured monkey cells with *Adh* gene component incorporated in SV40 expression vectors (13). We are also constructing Agrobacterium T-DNA segments containing *Adh* genes. Transcription and enzyme activity will be monitored in galls and cultured cells of *Adh⁻* lines of *Nicotiana* and *Daucus*.

One other feature of the maize *Adh* genes which may prove of importance in the development of a gene transfer system is that there are *Adh1* mutants involving the controlling element, Ds, extensively characterized by McClintock (14). Mobility of this Ds element is dependent on the presence of another element, Ac. The Ac/Ds system may be similar to the P factor system which has proven an effective means of incorporating genes into the *Drosophila melanogaster* genome (15). Ds may have sequence characteristics for the insertion of cloned genes into chromosomes of recipient cells. Southern analyses of an *Adh1*-Ds mutant isolated by Osterman and Schwartz (16) suggest a 400 bp insert in the 5' region of the gene. We have now cloned this segment for analysis.

A Ds probe will enable us to determine the number and characteristics of other Ds elements in the maize genome and may also lead to the isolation of Ac.

ACKNOWLEDGMENTS

We are grateful to K. Ferguson, Y. Hort, M. Jeppesen, G. Koci, J. Norman and A. Tassie for their skilled assistance.

REFERENCES

1. Willmitzer, L., De Beickeleer, M., Lemmers, M., Van Montagu, M., and Schell, J., *Nature (Lond.) 287*, 359-361 (1980).
2. Leemans, J., Shaw, C., Deblaere, R., De Greve, H., Hernalsteens, J.P., Maes, M., Van Montagu, M., and Schell, J., *J. Molec. Appl. Genet. 1*, 149-164 (1981).

3. Schwartz, D., *Proc. Natl. Acad. Sci. USA 56*, 1431-1436 (1966).
4. Freeling, M., and Birchler, J.A., in "Genetic Engineering" Vol.III (J.K. Setlow and A. Hollaender, eds.), Plenum, New York (1982).
5. Schwartz, D., *Am. Nat. 103*, 479-481 (1969).
6. Hageman, R.H., and Flesher, D., *Arch. Biochem. Biophys. 87*, 203-209 (1960).
7. Gerlach, W.L., Pryor, A.J., Dennis, E.S., Ferl, R.J., Sachs, M.M., and Peacock, W.J., *Proc. Natl. Acad. Sci. USA 79*, 2981-2985 (1982).
8. Ferl, R.J., Brennan, M.D., and Schwartz, D., *Biochem. Genet. 18*, 681-691 (1980).
9. Breathnach, R., Benoist, C., O'hare, K., Gannon, F., and Chambon, P., *Proc. Natl. Acad. Sci. USA 75*, 4853 (1978).
10. Schwartz, D., and Osterman, J., *Genetics 83*, 63-65 (1976).
11. Freeling, M., *Genetics 83*, 701-717 (1976).
12. Mertz, J.E., and Gurdon, J.B., *Proc. Natl. Acad. Sci. USA 74*, 1502-1506 (1977).
13. Mulligan, R.C., Southern, P.J., Howard, B.H., Yaniv, M., Geller, A.I., and Berg, P., *J. Molec. Appl. Genet.* in press (1982).
14. McClintock, B., *Brookhaven Symp. Biol. 18*, 162-184 (1965).
15. Rubin, G.M., and Spradling, A.C., *Science 318*, 348-353 (1982).
16. Osterman, J.C., and Schwartz, D., *Genetics 99*, 267-273 (1981).

36
Two Partial Copies of Heterogeneous Transposable Element *Ds* Cloned from the *Shrunken* Locus in *Zea mays* L.

Ulrike Courage-Tebbe, Hans-Peter Döring, Martin Geiser,[*] Peter Starlinger, Edith Tillmann, Ed Weck and Wolfgang Werr

```
            Institut für Genetik
            Universität zu Köln
               Köln, FRG
```

Transposable genetic elements were discovered earlier and have been characterized in more genetic and physiological detail than most mobile genetic elements in other organisms (McClintock, 1951, 1956, 1965; Peterson 1953, 1970).

Not only is the activity of genes adjacent to these elements increased or decreased upon their insertion, but this activity can be regulated in trans by related elements. They undergo "changes in state" manifested by an altered time of their excision from their previous location. Even the time at which "changes in state" occur can be pre-programmed. These observations would be sufficient to create an interest in the biochemical study of these elements. Other reasons add to this interest:

1. Transposable genetic elements in maize were uncovered repeatedly after treatment leading to

[*]present address: Friedrich Miescher Institut Basel, Switzerland

MANIPULATION AND EXPRESSION
OF GENES IN EUKARYOTES
ISBN 0 12 513780 x

chromosome breaks in maize lines not showing any previous manifestation of their activity (McClintock, 1951, Peterson, 1953). A general increase in mutability in maize after exposition to ionizing radiation must be interesting because its study will help us to evaluate better the radiation risks.

2. Increased knowledge of the transposition mechanism and its regulation might allow transposon mutagenesis and thus eventually may help breeders in mutating particular genes without exposing all other genes to mutagen treatment (Kleckner et al, 1977).

3. The introduction of isolated DNA into plants is a goal in genetic engineering. Transposable DNA elements might serve as vectors and allow the movement of a cloned gene from injected DNA to the host chromosome.

We have therefore begun to characterize maize transposable elements biochemically. Most schemes for the isolation of specific DNA sequences start from the mRNA encoding the product of the gene to be isolated. No such product is known yet for maize transposable DNA elements. We therefore intended to isolate one of these elements as DNA present in or next to a gene in a mutant caused by the insertion of this element but not in wild type DNA from the same locus. For our studies we chose gene Shrunken encoding endosperm sucrose synthase which is a major enzyme of starch biosynthesis and present in major amounts among endosperm proteins (Chourey and Schwartz, 1971 , Chourey and Nelson, 1976). sh mutants caused by the insertion of the transposable element Ds have been described (McClintock, 1952, 1953). We isolated a cDNA clone complementary to part of Sh-mRNA (Geiser et al, 1980, Woestemeyer et al, 1981), and used this as a probe to identify the size of convenient restriction fragments containing the corresponding genomic DNA (Döring et al., 1981). These results are in agreement with those published by two other laboratories (Burr and Burr, 1981, Chaleff et al., 1981). We then cloned genomic DNA from the shrunken locus from wild type and the two mutants sh-m5933 and

sh-m6233. The following results were obtained with
these clones:
1. Preliminary RNA-DNA hybridization experiments
as well as S1 and exonuclease VII mapping
showed that the gene occupies a region of approx.
4 kb. This is in agreement with the mRNA length
of 2.8 kb and the presence of several small introns.
2. The clones obtained from the mutants share several kb of DNA with the wild type clone. This
region contains most or all of gene sh. Adjacent
to the left of the shared region is a DNA segment
not hybridizing to the wild type clone. This DNA
may either be (part of) Ds, or alternatively, it
may be genomic DNA rearranged under the influence
of Ds. The junction between the "shared" DNA and
the "non-shared" DNA is the insertion point or a
breakpoint in the DNA rearrangement.
3. The junction in the sh-m5933 clone is located
within gene Sh. DNA sequencing around the junction
showed   all reading frames closed. The junction
must therefore be located within an intron. The
junction in the clone of sh-m6233 is located 2.5
kb to the left of the junction in sh-m5933. The
mutant has a shrunken phenotype. This indicates
either the presence of small leader sequences far
to the left of the gene and separated from it by a
large intron or else a negative influence on the
gene over a considerable distance.
4. The DNA segments present in the mutant but not
in the wild type clones share some DNA within a
1.5 kb segment to the left of the junction point.
The remainder of these DNA segments do not hybridize to each other. Even in the region showing
homology by DNA-DNA hybridization, the restriction
maps show considerable differences. If the DNA
segments present in the mutants but not in the
wild type are (part of) Ds, different copies of
Ds are heterogeneous.
5. The above argument is supported by the result
of experiments, in which restriction digests of
genomic DNA were separated by electrophoresis and
hybridized to probes consisting of subcloned fragments of the mutant-specific DNA of the clone obtained from sh-m5933. In these experiments, a limited number of bands (20 to 30) light up. These

DNAs are thus neither unique DNAs, nor do they show the diffuse hybridization pattern typical for repetitive DNA. The patterns obtained with different fragments show similarities and differences, but the approximate number of bands is similar. The result is thus not only obtained with probes spanning the junction in the cloned DNA, but also with internal fragments. If these were part of Ds, and if all copies of Ds were identical, the detection of only one single band would be expected. The presence of many bands indicates that either all of the probes used contain non-Ds DNA, or else that the different copies of Ds are heterogeneous.

6. By DNA-DNA hybridization, restriction mapping and DNA sequencing, the presence of two pairs of inverted repeats in the mutant segments of sh-m5933 DNA was detected. The members of one of the pairs contain a BamHI site. One of these sites is located 184 bp to the left of the junction. One terminus of the inverted repeat is located exactly at the junction. The other terminus has not yet been determined, but it is located at least 500 bp to the left of the BamHI site, bringing the total length of the inverted repeats to > 700 bp. The BamHI site in the other member of the pair is located 2.7 kb to the left of the BamHI site in the first member of the pair.

A second pair of inverted repeats is defined by two SphI sites located 300 bp apart with one HindII site and one PvuII site between. One member of this pair is located between the two members of the first pair, while the other member is located to the left of the two BamHI-containing repeats.

7. The termination of one of the BamHI-containing repeats at the junction suggests a role of this structure in transposition or genetic rearrangement. However, no BamHI site was found near the junction in the clone obtained from sh-m6233. Should a structure similar to the BamHI-containing repeat in sh-m5933 DNA be located at the junction in sh-m6233, it must differ at least to some extent.

Of the SphI-containing repeat, one copy was detected in the clone from sh-m6233. It is located at or near the junction to shrunken DNA.

8. The DNA sequence of the BamHI-containing repeat contains many copies of the hexanucleotide CCGTTT. These hexanucleotides are sometimes directly repeated. In other positions, they are separated by a few other nucleotides. Some imperfect copies of the hexanucleotide differing by one nucleotide are also present. In a certain segment they are all oriented in one direction, while in another segment all of them are oriented in inverted manner.
9. The analysis of the DNA clones presently available does not yet allow us to deduce the structure of the mutations unambiguously. Hybridizing restriction digests of genomic DNA of sh-m5933 DNA to probes of nick-translated fragments of unique DNA of the cloned wild type DNA show that none of the DNA present in the wild type clone is deleted in the mutant. The results are competible with either an insertion of more than 25 kb or with an inversion that has one breakpoint at the junction. In addition, the Southern blotting analysis with sh-m5933 DNA suggests that a DNA segment located at the left end of the shrunken gene or to the left of it is duplicated.

In summary, the DNA rearrangements in two genetically very similar mutants were shown to be heterogeneous and extensive. The DNA adjacent to the junction has some homology to each other, but its extent is limited. The "mutant" DNA in sh-m5933 DNA has two pairs of inverted repeats interspersed with each other, and one of them bordering at the junction. Only one member of the other one of these pairs is clearly present in the other mutant DNA. The BamHI-containing repeat is composed partly of repeats of a small subunit. All of these features distinguish this DNA from most of the transposable elements characterized in Drosophila melanogaster, which are represented by the element copia (Spradling and Rubin, 1981). These elements may be DNA copies of retroviruses (Flavell and Ish-Horowicz, 1981). Potter has pointed out that the fold-back transposon is very different from the retrovirus-like transposons (Potter, 1982). The DNA clone from the Ds-induced mutants resem-

bles FB-DNA in heterogeneity and presence of subunits. The pentanucleotide CGTTT is part of the
CCGTTT subunit of the BamHI-containing repeat of
sh-m5933 and of the dekanucleotide CGTTTGCCCA of
FB-DNA (Potter, 1982). It will be interesting to
see, whether these similarities can be strengthened by additional observations and, if so, whether
it will be an indication of convergent evolution
or common descent.
One copy of FB-DNA is located at one terminus of
the several hundred kb TE-transposon in Drosophila (Goldberg et al, 1982).

We are presently trying to clone the other junction of the shrunken DNA with Ds (or rearranged
DNA)and will be interested to see, whether a
copy of the BamHI-containing repeat is located
at that junction also.

Transposable element Ds is known to undergo "changes in state"and phenotypic reversions without
being removed from its point of insertion (McClintock, 1951, 1965).

The presence of inverted repeats themselves built
up from smaller direct and inverted repeats might
allow partial deletions as well as internal partial inversion. It will be interesting to see,
whether such alterations form the basis of the
in situ alterations of the genetic activity of
Ds. In collaboration with Dr.N.Fedoroff, Baltimore, we are presently characterizing the DNA structure of several phenotypic revertants of sh-m5933.

Final proof that the DNA cloned from the mutants
is (part of) Ds will be obtained only after the
complete segment between the two breakpoints in
the DNA rearrangement is cloned and characterized
as a continuous segment and the demonstration
of the complete loss of the mutant specific DNA
in a revertant. An alternative would be the detection of (part of) the cloned DNA that is present in the mutant but not in the wild type clone
of the shrunken gene in DNA clones of a Ds-induced
mutation at another locus. As a first step in this

direction, we isolated a putative Ds-induced
mutation at the Adh1 locus encoding alkohol dehydrogenase. This mutant was isolated in collaboration with Drs. Salamini and Motto, Bergamo, Italy
and is presently being characterized in collaboration with Dr.M.Freeling, Berkeley, USA. Present
genetic tests indicate that the mutant is unstable
only in the presence of another controlling element Ac. This property is diagnostic for Ds-deri-
ved mutants (McClintock, 1951). We will try to
clone this mutant, using an Adh1 genomic clone
kindly provided by Dr.J.Bennetzen, San Carlos, USA.

Further work will have to include both the biochemical characterization of DNA and the isolation
of new mutants.

ACKNOWLEDGEMENT
This work was supported by the Landesamt für For-
schung des Landes Nordrhein-Westfalen and the
Kommission der Europäischen Gemeinschaften through
contract Nr.BIO-396-81 D (B).

We thank Drs. N.Fedoroff, Baltimore and M.Freeling,
Berkeley for the communication of unpublished results.

REFERENCES

Burr, B., Burr, F (1981) Genetics 98, 143-156
Chaleff, D., Mauvais, J., McCormick, S., Shure, M.,
    Wessler, W., Fedoroff, N. (1981) Carnegie Inst.
    Wash.Yearbook 80, 158-174
Chourey, P., Nelson, O. (1976) Biochem.Genetics
    14, 1041-1055
Chourey, P., Schwartz, D. (1971) Mutat.Res. 12,
    151-157
Döring, H.P., Geiser, M., Starlinger P., (1981)
    Mol.Gen.Genet. 184, 377-380
Flavell, A.J., Ish-Horowicz, D.(1981)Nature 292,
    591-595
Geiser, M., Döring, H.P., Wöstemeyer, J., Behrens,
    U., Tillmann, E., Starlinger, P. (1980) Nucl.
    Acids Res. 8, 6175-6188
Goldberg, M.L., Paro, R., Gehring, W.J. (1982)
    EMBO Journal 1, 93-98

Kleckner, N., Roth, J., Botstein, D. (1977)
    J.Mol.Biol. 116, 125-159
McClintock, B. (1951) Cold Spring Harbor Symp.
    Quant.Biol. 16, 13-47
McClintock, B. (1952) Carnegie Inst.Wash.Yearbook
    51, 212-219
McClintock, B. (1953) Carnegie Inst.Wash.Yearbook
    52, 227-237
McClintock, B. (1956) Cold Spring Harbor Symp.
    Quant.Biol. 21, 197-216
McClintock, B. (1965) Brookhaven Symp.Biol. 18,
    162-184
Peterson, P.A. (1953) Genetics 38, 682
Peterson, P.A. (1970) Genetics 41, 33-56
Potter, S.S. (1982) Nature 297, 201-204
Spradling, A.C., Rubin, G.M. (1981) Ann.Rev.
    Genet. 15, 219-264
Wöstemeyer, J., Behrens, U., Merckelbach, A.,
    Müller, M., Starlinger, P. (1981) Eur.J.
    Biochem. 114, 39-44

37
Chlorophyll a/b Binding Proteins and the Small Subunit of Ribulose Bisphosphate Carboxylase are Encoded by Multiple Genes in Petunia

Pamela Dunsmuir and John Bedbrook*

CSIRO Division of Plant Industry
Canberra City, Australia

INTRODUCTION

Two major chloroplast proteins - the small subunit (SSU) of ribulose (-1,5-) bisphosphate carboxylase, and the chlorophyll a/b binding protein (Cab) of the thylakoid light harvesting complex, are encoded by nuclear genes (Kawashima and Wildman, 1972; Von Wettstein, 1981). It is well established for both SSU and Cab polypeptides, in pea, spinach, and chlamydomonas, that the mRNAs are translated on cytoplasmic polyribosomes to yield precursor peptides (containing 45-50 additional N-terminal amino acids) which undergo post translational transport into intact chloroplasts, followed by, or coupled with proteolytic cleavage to produce mature peptides in the chloroplast (Cashmore, 1976: Dobberstein $et\ al.$, 1977; Chua and Schmidt, 1979; Highfield and Ellis, 1978, Schmidt $et\ al.$, 1979). The mature SSU peptide is combined with the chloroplast encoded large subunit peptide to form the holoenzyme ribulose (-1,5-) biphosphate carboxylase. The mature Cab peptides are non-convalently bound with chlorophylls a and b, which are synthesized in the chloroplast, and incorporated into the thylakoid membrane (Boardman $et\ al.$, 1978). We show here that both SSU and Cab proteins are specified by multiple genes. Many of the individual genes differ in nucleotide sequence. Further, several of these different genes for both SSU and Cab polypeptides are transcribed in petunia leaves.

*Advanced Genetic Sciences Inc., P.O.Box 3266, Berkeley, California, U.S.A.

MANIPULATION AND EXPRESSION
OF GENES IN EUKARYOTES
ISBN 0 12 513780 x

RESULTS AND DISCUSSION

In order to understand the structure and expression of the *Cab* and *SSU* genes we have focused our attention on the Mitchell Petunia strain.   This plant has a low C value (ca.   1 pg) is a doubled haploid and thus genetically homozygous (Mitchell, 1979), and is amenable to propagation from protoplast culture. We isolated five *Cab* and five *SSU* cDNA clones from the set constructed using Petunia leaf poly A$^+$RNA (Fig.1).   The alignment of these cDNAs with respect to each other and to the end of the mRNA has been determined from DNA sequence analyses and comparison with the published sequence for the *SSU* mRNA in pea (Bedbrook *et al.*, 1980) and *Cab* mRNA sequence in pea (Coruzzi *et al.*, 1982).

Although several of the *Cab* cDNA clones have a number of conserved restriction endonuclease cleavage sites, no single site is present in all five examples.   Thus each cDNA is unique on the basis of physical mapping.   For the *SSU* cDNA clones, *pSSU* 117 and *pSSU* 71 differ by virtue of EcoRI and HpaII sites, and *pSSU* 51 may be distinguished from these by a HinfI site.   Although *pSSU* 103 and *pSSU* 41 are not distinguished by unique restriction enzyme cleavage sites, they do differ at the nucleotide sequence level from the other *SSU* cDNAs (Dunsmuir and Bedbrook, submitted).

On the basis of these characterized cDNA clones we expect that at least five different *Cab* peptide genes and four different *SSU* peptide genes are transcribed in green leaf tissue.

The autoradiographs of Figure 2 illustrate that a petunia *Cab* cDNA probe hybridizes to fourteen EcoRI fragments, labelled A-M and ranging in size from 17 kb to 1 kb, and a petunia *SSU* cDNA probe hybridizes to twelve EcoRI fragments, labelled A-L and ranging in size from 12 kb to 2 kb in the Petunia Mitchell DNA.   To isolate and characterize these hybridizing genomic sequences in petunia DNA we cloned nuclear DNA partially digested with EcoRI into the lambdoid phage vector Charon 28 (Rimm *et al.*,1980).   Petunia cDNAs for *Cab* and *SSU* were used as probes to identify the recombinant phage containing the corresponding genomic coding regions (Benton and Davis, 1977).   To localize the regions of these genomic fragments which correspond to the *Cab* and *SSU* genes we initially hybridized a cDNA clone (of known length) corresponding to the 3' end of the coding region, to recombinant phage DNAs digested with appropriate enzymes and transferred to nitrocellulose (Southern, 1975).   A comparison between these

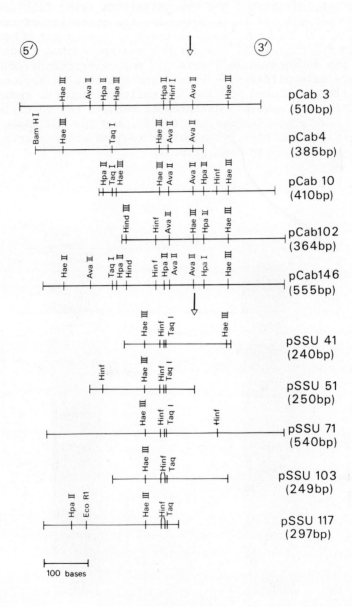

Fig. 1. Restriction endonuclease cleavage site maps for Petunia cDNA clones. The vertical arrow indicates the translation termination points.

3' end hybridizations, and hybridizations using full length
cDNA probes allowed us to determine the direction of transcription and also make some estimate of the extent of homology
between the *Cab* or *SSU* transcripts and the cloned fragment -
these are designated in Figure 3. The restriction endonuclease
cleavage site differences within the hybridizing regions on
the different cloned *Cab* and *SSU* nuclear fragments indicated
that there are sequence differences between the multiple
genes. *Cab* phage fragments 9 and 22, and the *SSU* phage
fragment 1, each have two separate regions which hybridize
to the corresponding cDNA probes.

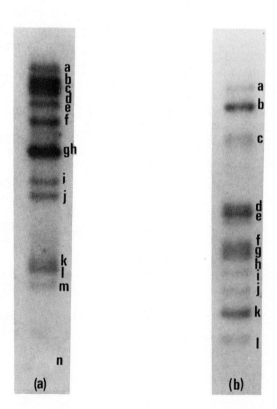

Fig. 2. *Petunia DNA digested with EcoRI and hybridized
with (a)* pCab *146 cDNA (b)* pSSU *117 cDNA.*

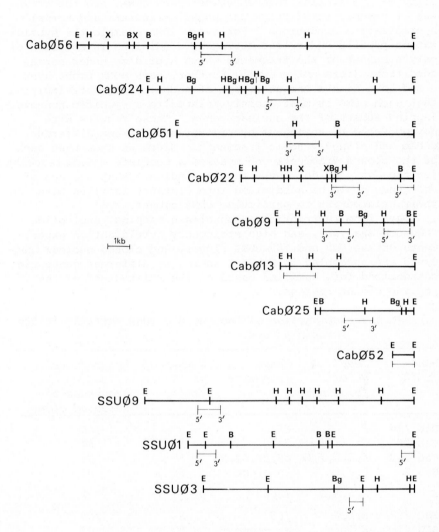

Fig.3. Restriction endonuclease maps of Petunia nuclear fragments cloned in Charon 28 phage. Sites are abbreviated, E-EcoRI, B-BamHI, H-HindIII, Bg-BglII and X-XhoI. Regions corresponding to Cab *and* SSU *coding sequences are indicated under the maps.*

In an attempt to correlate the different *Cab* and *SSU* cDNA clones with the different nuclear genes we compared the stability of the hybrids formed between each cDNA, and the known set of genomic EcoRI fragments under stringent and normal hybridization conditions. For every cDNA example the hybridization profile under more stringent conditions is composed of only a subset of the fragments which hybridize under normal conditions (data not shown). Further, each cDNA probe hybridized best with a unique subset of genomic fragments implying that each cDNA is most closely related to a specific non-overlapping subset of the nuclear genes. These results were corroborated by analogous hybridizations between different cDNAs and cloned genomic fragments. Since we know that each of the cloned fragments encompasses a complete coding sequence for the *Cab* peptide, these results indicate that the set of *Cab* genes could be subdivided into distinct families based on their relatedness to particular cDNA clones.

Data from restriction endonuclease mapping, nucleotide sequence analyses, and high stringency hybridization experiments on the *Cab* and *SSU* cDNA clones, and cloned nuclear fragments are assembled in Table 1 where the different genes are categorized into families based on the relatedness of the peptide coding regions.

TABLE 1. *Designation of* Cab *and* SSU *gene families in the Petunia genome.*

Gene family	EcoRI genomic fragments	Phage fragments	Number of genes per family	Relatedness[a] of coding sequences to selected clone
Cab *146*	F	22	2	*100*
Cab *10*	G	9	2	*91*
Cab *102*	A,B,C,I,K	56,24,51,25	5	*89*
Cab *3*	D,J,L	No clone	3	*91*
X	E,H,M,N	13,52	?	?
SSU *117*	B,E,K	1,3	3	*100*
SSU *41*	E	No clone	?	*93*
SSU *51*	B,C,D,F,H I,L	No clone	?	?
SSU *71*	D	9	?	?

[a]*The sequences compared are those encoding the 60 N-terminal amino acids. All* Cab *clones were compared to* pCab *146, and* SSU *clones were compared to* pSSU *117.*

The family of *Cab* 146, corresponds to the genes of nuclear fragment F, cloned in *Cab* Ø22, which we know encompasses 2 distinct genes (Fig.3). The *Cab* 102 family contains genes on the nuclear fragments A (*Cab* Ø56), B (*Cab* Ø24), C (*Cab* Ø51), I (*Cab* Ø25) and K (no genomic clone isolated). Each of these cloned fragments encompasses a single coding region and although we have not yet isolated the genomic fragment K, we assume from its relative hybridization intensity that it also contains a single gene. Thus the *Cab* 102 family comprises 5 separate genes. We do not know which nuclear fragment relates specifically to cDNA clone *102* nor whether the genes are identical in sequence. The *Cab* 10 family corresponds to the genes on nuclear fragment G (*Cab* Ø9). The two complete genes on *Cab* Ø9 hybridize equally well to the cDNAs *Cab* 10, and *Cab* 4 under stringent conditions. The fourth family, *Cab* 3 is composed of the genes on fragments D, J and L, however these genomic clones have not been isolated. The hybridization intensity of these fragments indicates that they each contain a single region.

The nuclear fragments H (*Cab* Ø13), and N (*Cab* Ø52) cannot be ascribed to any of the known *Cab* families on the basis of high stringency hybridization experiments with the characterized cDNA probes, however from experiments with total leaf cDNA as a probe to *Cab* phage DNA we know that the genes of fragments H and N are transcribed. *Cab* Ø13 appears to contain a complete coding region and *Cab* Ø52 contains part of a gene, thus we postulate that there is at least one additional *Cab* gene family which is composed of a minimum of two genes. We have no information relating to the expression of the genes of fragments E and M which we have not isolated.

We have sequenced each of the *Cab* cDNA clones (Dunsmuir and Bedbrook, submitted) and these data are summarised in Table 1. The cDNA clones *Cab* 4 and *Cab* 3 are more closely related to the *Cab* 146 clone with 16 nucleotide changes in the common region, 15 of which are silent third codon position changes. For the *Cab* 102 cDNA, 14 of the 20 nucleotide differences are silent.

The *SSU* cDNA clones also classify the nuclear genes into distinct families, however we have cloned only six of the *SSU* genomic *EcoRI* fragments, thus it is not possible to specify the number of genes of each family except for the *SSU* 117 family. This *SSU* gene family contains 3 genes, two of which are contained in the *SSU* Ø1 (genomic fragments B and K) and the third in fragment E (*SSU* Ø3). The analysis of the *SSU* genes is complicated by the presence of EcoRI sites internal to a number of the nuclear gene sequences. Although we do not yet have as extensive data relating to the *SSU* genomic fragments as we do for the *Cab* genes, the evidence is strongly

suggestive that like the *Cab* genes, the nuclear genes which
code for the *SSU* peptide may be classified into discrete gene
families which are represented in the leaf mRNA population
and, if translated, would specify distinct peptides. We conclude then, that both the chlorophyll a/b binding protein of
the light harvesting complex and the small subunit of ribulose
bisphosphate carboxylase are encoded by multiple expressed
genes. Each set of genes is divisible into families based on
their sequence relatedness to independent cDNA clones and at
least one member of each family is transcribed in Petunia
leaf tissue.

REFERENCES

Boardman N.K., Anderson J.M., and Goodchild D.J. (1978).
    Current Topics in Bioenergetics *8*, 35-109.
Bedbrook J.R., Smith S.M., and Ellis R.J.   (1980). *Nature*
    287, 692-697.
Benton W.D., and Davis R.W. (1977). *Science 196*, 180-182.
Cashmore A.R. (1976). *J. Biol. Chem. 251*, 2848-2853.
Chua N.-H., and Schmidt G.W. (1979). *Proc. Natl. Acad. Sci.
    USA 75*, 6110-6114.
Coruzzi G., Broglie A., Cashmore A., and Chua N.-H. (1982)
    J. Biol. Chem. (in press).
Dobberstein B.G., Blobel G., and Chua N.-H. (1977) *Proc. Natl.
    Acad. Sci. USA 74*, 1082-1085.
Highfield P.E., and Ellis R.J. (1978). *Nature 271*, 420-424.
Kawashima N., and Wildman S.G. (1972). *Biochim. Biophys.
    Acta 262*, 42-49.
Mitchell A.Z. (1979). "Anther Culture in Petunia". Thesis
    B.A. (Hons.), Harvard University.
Proudfoot N.J., and Brownlee G.G. (1976). *Nature 263*, 211-214.
Rimm D.L., Horness D., Kucera J., and Blattner F.R. (1980)
    Gene 12, 301-309
Schmidt G.W., Devilliers-Thiery A., Desruisseaux H., Blobel G.,
    and Chua N.-H. (1979). *J. Cell Biol. 83*, 615-622.
Southern E.M. (1975). *J. Mol. Biol. 98*, 503-517.
Von Wettstein D. (1981). *In* "International Cell Biology
    1980-81". (H.G. Schweiger, ed.), pp. 250-272, Springer,
    Berlin.

38

Gene Expression in Barley during Infection by the Powdery Mildew Fungus (*Erysiphe graminis* f. sp. *hordei*)

J. M. Manners, A. K. Chakravorty and K. J. Scott

Department of Biochemistry, University of Queensland
St. Lucia, Australia

The powdery mildew fungi are obligate parasites whose host specificity is sometimes so strict that compatibility is determined by single genes in the host and pathogen. It is now being recognised that changes in host metabolism occur following infection by these fungi which, in turn, lead to the induction of a compatible or incompatible reaction depending on the plant and fungal genotypes. The aim of our research is to elucidate the molecular basis of such changes in barley during powdery mildew infection. The alterations in gene expression of the susceptible host induced by the pathogen which were reported at this Congress are outlined below. The changes observed can almost certainly be attributed to the host as all surface fungal structures were removed prior to sampling. Results from infected material were related to non-inoculated controls.

1. PROTEIN SYNTHESIS *IN VIVO*

Analysis by gel electrophoresis and fluorography of leaf polypeptides labelled in a 1 h pulse of ^{35}S-methionine indicated no changes 24 h after inoculation. In contrast, inoculation of the non-green tissues of the coleoptile resulted in increased isotope incorporation into several polypeptides (approx. mol. wts., 16, 19, 24, 36 and 94 Kd) within 18 h of inoculation. At 72 h after inoculation a marked reduction in the incorporation of isotope into several size classes of polypeptides was observed in diseased leaves. This included polypeptides corresponding to the rapidly labelled 32 Kd chloroplast protein, the light harvesting proteins and the small subunit of ribulose 1-5 bisphosphate carboxylase/oxygenase (RuBpC) indicating effects on both chloroplast and cytoplasmic protein synthesis. This was further confirmed by feeding in the presence of L-threo chloramphenicol (CAM). When the 1 h pulse of ^{35}S-methionine was followed by a chase in unlabelled amino acids for 5 h, the decreased incorporation into many polypeptides was still evident in diseased leaves. However, increased incorporation into other polypeptides was observed in leaves between

237

MANIPULATION AND EXPRESSION
OF GENES IN EUKARYOTES
ISBN 0 12 513780 x

72-168 h after inoculation. These results indicate that alterations in both the rates of protein synthesis and turnover occur in diseased leaves.

2. POLYSOMAL mRNA POPULATIONS

By using the inhibitors CAM, lincomycin and cycloheximide we have shown that the translational activity of either chloroplast or cytoplasmic polysomes can be discerned in a mixed polysome population with either *E. coli* (CAM, lincomycin sensitive) or wheat germ (cycloheximide sensitive) soluble factors. Polysomes isolated from leaves at 24, 72 and 120 h after inoculation showed a marked decrease (50%) in translational activity per A_{260} unit with the *E. coli* system whilst little change was observed with the wheat germ system. The activity of thylakoid bound polysomes was also assessed, this subpopulation of chloroplast polysomes was unaffected by infection at 24 h but a decrease of 31% was detected at 72 h after inoculation. The decrease in translational activity of total polysomes with the *E. coli* system occurred only in the susceptible host and not in a near isogenic resistant variety.

3. POLY A$^+$ RNA

Poly A$^+$ RNA was isolated by the method of Chirgwin *et al.* (1979) and translated in a rabbit reticulocyte cell free system. Increased synthesis of a polypeptide of mol. wt. 31 Kd and reduced synthesis of another of 58 Kd were observed at 24 h following inoculation. At 72 and 120 h, total translational activity of poly A$^+$ RNA was reduced (20-30%) in infected leaves, including a decreased incorporation into the major product (mol. wt. 20 Kd) tentatively identified as the precursor to the small subunit of RuBpC. This was associated with an increase in the labelling of the 5' end of the poly A$^+$ RNA by T4 polynucleotide kinase with $[\gamma-^{32}P]$-ATP indicating that mildew infection decreased the capping of leaf poly A$^+$ RNA.

As the establishment of a functional haustorium and the inception of secondary hyphal development requires only 24 h for these fungi, it seems likely that those changes in host gene expression observed within this period may be determinant events in the susceptibility of barley to powdery mildew and merit further study.

REFERENCES

Chirgwin J.M., Przybyla A.E., MacDonald R.J. and Rutter W.J.
(1979) *Biochemistry 18*, 5294.

39
Variation in the Ribosomal RNA Genes of *Lilium henryi*[1]

Laurie von Kalm and D. R. Smyth

Department of Genetics
Monash University
Clayton, Australia

We have examined the organization of rRNA genes in *Lilium henryi* by Southern blot analysis of restricted genomic DNA using a cloned wheat 18S-25S rDNA unit as probe (Gerlach and Bedbrook, 1979).

The full repeating unit in lily is approximately 11 kbp, with two size families differing in length by about 300 bp (Fig. 1). As in other plants, such as wheat (Gerlach and Bedbrook, 1979), rice (Oona and Sugiura, 1980) and lemons (Fodor and Beridze, 1980), the variation seems to be localized to the NTS (in this case between the *Hpa*I and *Kpn*I sites). The functional significance of such variation remains obscure. Its origin may involve unequal crossing over between intra-spacer repeats as has been proposed in *Drosophila* (Coen *et al.*, 1982).

In plants, methylation of CG and CXG sequences is extensive. This is also true in *Lilium* rDNA, as shown by comparative digests using the isoschizomers *Bst*NI and *Eco*RII. Only the former enzyme will cut methylated CCXGG sequences. While there are many *Bst*NI sites in genomic rDNA, *Eco*RII cuts at only one (Fig. 1), and only in a few repeats. This site is in the same region as the only unmethylated *Hpa*II and *Pst*I sites. This small, relatively undermethylated region is in the NTS, and is analogous to a similar region found in *Xenopus laevis* rDNA (Bird and Southern, 1978).

[1]*This work was supported by the Australian Research Grants Scheme and a Monash University Special Research Grant.*

MANIPULATION AND EXPRESSION
OF GENES IN EUKARYOTES
ISBN 0 12 513780 x

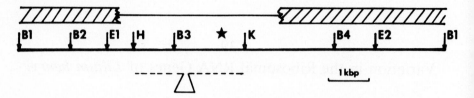

FIGURE 1. *Restriction map of rDNA repeat units in*
L. henryi. *The shaded bar represents regions homologous to
the cloned wheat rDNA probe. Regions lacking homology are
presumably NTS.* B - BamHI, E - EcoRI, H - HpaI, K - KpnI,
★ - *region of undermethylation containing single sites for*
EcoRII, HpaII *and* PstI.

There are four sites within lily rDNA which can be cut by
BamHI (Fig. 1). Two of these are always restricted (Bl, B4)
but the other two (B2, B3) are not always cut. The two size
families show similar patterns of *BamHI* digestion. The *BamHI*
site variation may arise through incomplete methylation (Bird
and Southern, 1978; Gerlach and Bedbrook, 1979), or DNA
sequence variation between repeat units.

A *KpnI* site is also variably present (Fig. 1). It seems
that repeats in which the *KpnI* site is not cut are limited to
those in which the *Bam* site B3 is restricted. This particular
repeat type is not interspersed with those containing *KpnI*
sensitive sites.

In situ hybridization has shown that there are only two
sites for the rRNA genes in the *L. henryi* genome. If current
cloning experiments reveal that variable *BamHI* and *KpnI*
digestion results from sequence divergence, then more than
one rDNA family must be present at one or both of these
chromosomal locations.

REFERENCES

Bird, A.P., and Southern, E.M. (1978). *J. Mol. Biol. 118,* 27.
Coen, E., Strachen, T., and Dover, G. (1982). *J. Mol. Biol.
    158,* 17.
Fodor, I., and Beridze, T. (1980). *Biochem. Int. 1,* 493.
Gerlach, W.L., and Bedbrook, J.R. (1979). *Nucl. Acids Res. 7,*
    1869.
Oona, K., and Sugiura, M. (1980). *Chromosoma 76,* 85.

40
Experiments on Transformation of Plant Cells Using Synthetic DNA Vectors

Miriam Fischer [1]

Department of Agriculture
New South Wales

H. Lörz
Philip J. Larkin
William R. Scowcroft
John Langridge

Division of Plant Industry
CSIRO, Canberra

Recombinant DNA technology might facilitate the transfer of specific genes from one species to another and so enhance the scope for traditional plant breeding. This article outlines our use of synthetic DNA vectors to introduce exogenous DNA into plant cells, followed by selection of lines for antibiotic drug resistance genes present in the vectors.

I. THE VECTORS

The principles of vector construction were described by Langridge (1981). The main vector used, C91, contains an origin of replication from the $ColE_1$ plasmid, prokaryotic genes coding for trimethoprim (Tn 7) and kanamycin (Tn 903) resistance, a replicator from tobacco chloroplast DNA and the yeast his-3 gene. The last two sequences were inserted into the non-coding region of Tn 7 (Fig. 1). The chloroplast replicator was isolated by testing for replication of the vector in yeast.

[1]

Present address: Division of Plant Industry, CSIRO, Canberra

MANIPULATION AND EXPRESSION
OF GENES IN EUKARYOTES
ISBN 0 12 513780 x

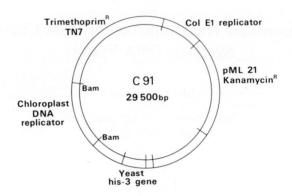

FIGURE 1.   Diagram of vector C91.

II.   RECIPIENT CELLS AND SELECTION SCHEME

As recipient plant cells we have used cell lines for
which protoplast isolation and culture is efficient and
reproducible.   These lines include, W 38 and NIA 115 of
Nicotiana tabacum, Hm-1 of *Hyoscyamus muticus* and WC-2 of
Daucus carota.   The plating efficiency of cultured protoplasts
from these lines is routinely 40-60%.

Sensitivity levels for trimethoprim (T), kanamycin (K),
G 418 (G) and methotrexate (MTX) have been determined from
day 0 to day 5 after culture initiation.   Optimal concentrations were defined as those which allowed some initial
development, i.e. cell wall regeneration and first cell
division in some protoplasts, but which inhibited sustained
development.   The levels selected were: trimethoprim 200
µg/ml, kanamycin 400 µg/ml, G 418 20 µg/ml and methotrexate
1.5 µg/ml added at day 1 or 2 of culture.

III.   DNA UPTAKE

Several methods which have proved efficient in other
eukaryotic transformation systems, were tested.   5-10 µg of

vector DNA per 10^6 protoplasts was used in most of the experiments. The procedures include protoplast fusion inducing conditions with polyethylene glycol (PEG) or Ca^{++} and high pH, incubation with poly-L-ornithine (PLO) and Zn^{++} and the widely applied Ca-coprecipitation method. Other less extensively used regimes include bacterial spheroplasts-protoplasts fusion for "in vivo" vector transfer into plant cells, transitory incubation of protoplasts with cell wall inhibitors, and inclusion of carrier DNA with vector DNA. An average of 5-8 x 10^6 protoplasts were used for each experiment and in all cases one half were treated with vector DNA and the other half used as control (no treatment or unspecific DNA).

IV. SELECTED LINES

The vector, C 91, has been used in seven experiments involving about 25 x 10^6, treated protoplasts. The experiments in which vector C 91 was used are summarized in Table I. Preliminary selection includes all the colonies which were subcultured from the original plates. Long-term cultures are selected lines showing consistently good growth after several subcultures in media containing antibiotics.

TABLE I.   **TRANSFORMATION EXPERIMENTS**

EXPERIMENT (5-8x10⁶pp)	CELLS	VECTOR	METHOD	SELECTION (μg / ml)	PRELIMINARY SELECTION + vDNA (no vDNA)	LONG TERM CULTURES + vDNA (no vDNA)
1	NIA	C91	PEG,PLO	T, 100-300	6 (9)	6 (3)
2	NIA	C91	PLO	T, 100-300	5 (16)	3 (2)
3	NIA	C91	PEG, Ca^{++}	T, 75-300	6 (0)	5 (0)
4	NIA	C91	Ca-prec	T, 200	44 (3)	9 (1)
5	NIA	C91	Ca^{++}, Ca-prec	G, 10-25	18 (21)	2 (1)
6	NIA	W38	Ca-prec	G, 20-30	27 (0)	11 (0)
7	NIA	W38	Ca^{++}, Ca-prec	G, 12-25	7 (4)	7 (3)

In all of the experiments so far conducted 16% of selections following vector treatment compared to only 6.2% of the spontaneous resistant lines retained drug resistance after prolonged subculture.

A comparison of different selection schemes, different vectors or different recipient cell lines did not show any

significant difference with respect to the number of putative
transformants or the number of spontaneously adapted lines.
The selected lines displayed variable growth rates under
standard selection and also in their ability to grow on
increased antibiotic concentrations. Extensive variability
was also found when lines were tested for either cross-
resistance or stability of the resistance after a period
without any drug selection. These data were only partially
encouraging, as only about half of the selected lines (+
vector DNA) showed cross resistance. Resistance of the lines
selected on trimethoprim as well as on G418 is rather stable.

V. FURTHER ANALYSIS

Analysis of the putative transformed lines at the DNA
level is very preliminary and restricted to only a few lines
from the first experiments. Southern blot analysis indicated
that less than 30-50 pg of vector DNA could be detected in a
standard assay of DNA isolated from plant cell lines. This
would limit the detection level in our hybridization studies
to about one copy of the vector per plant cell. Lower copy
numbers could not be analysed with accuracy. Southern blotting data have indicated the presence of the vector in DNA
isolated from transformed plant cells, but these results are
still uncertain because of difficulties in reproducibility.
Confirmation of positive transformation is also being sought
by "back-transforming" *E. coli* or yeast with isolated plant
DNA from controls, spontaneous resistant lines and putative
transformants.

Further analysis at the DNA level is needed to substant-
iate the results on transformation with these recombinant DNA
vectors. Improvement is also expected by the use of new
vectors containing prokaryotic protein-coding sequences fused
to plant gene promotors and nuclear and chloroplast DNA
sequences which allow vector replication in yeast. More
recent transformation experiments are also utilising protoplast culture systems from which plants can be regenerated.

REFERENCES

1.   Langridge, J. (1981). In: Wheat science - today and
     tomorrow. (L.T. Evans and W.J. Peacock, eds.) Cambridge
     University Press, pp. 91-95.

Part V
Organelles

Introduction

Eukaryotic cells are characterized by the presence of membranous cytoplasmic organelles which, in the case of mitochondria and chloroplasts, have well defined organelle DNA genomes (mtDNA and ctDNA, respectively). This part is concerned with the organelle genomes which specify a relatively small number of proteins that are translated on organelle ribosomes. In many cases these proteins join with nuclear-coded and cytosolically synthesized components to form enzyme complexes that are necessary for the metabolic, bioenergetic and biosynthetic functions of the organelles. The organelle genomes also code for ribosomal and transfer RNA components of the protein biosynthetic apparatus in the organelle.

The prokaryotic characteristics of organelle protein synthesis have been well documented for more than 15 years. Chapter 41 demonstrates that the sequence organization of chloroplast genes in spinach (ctDNA genome 150 kb) and bacteria are remarkably similar, except for large intervening sequences in certain chloroplast tRNA genes. The evolution of mammalian mtDNA genomes (16 kb) is discussed in Chapter 51; here we see an organelle genome with very few sequences not directly coding for a gene product.

By contrast, the mitochondrial genome of *Saccharomyces cerevisiae* (75 kb), has extended A,T-rich sequences between coding regions and within some genes. The molecular genetic techniques available to study this simple eukaryote have facilitated detailed analyses of specific mitochondrial genes. Chapters 43 and 44 describe features of the DNA sequence of three mitochondrial genes in wild type and in mutant or variant strains. DNA replication functions have been assigned to particular regions between genes; these *ori* sequences are documented in Chapter 45. This chapter, and others in this part, discuss some roles of the short G,C-rich clusters found embedded in the A,T-rich sequences of *S. cerevisiae* mtDNA.

Respiration deficient petite mutants play a very important role in all the above-mentioned studies on *S. cerevisiae* mtDNA. The *rho⁻* class of petites has had extensive deletions of wild type mtDNA sequences. The residual segment of the mtDNA genome in each *rho⁻* petite is a biologically cloned segment of the yeast mitochondrial genome. Such petite mtDNA segments are not only useful in studies on *S. cerevisiae* itself, but have also been applied successfully in mapping (by DNA–DNA hybridization) the mtDNA genomes of other organisms, as shown in Chapter 49 for *Leishmania* maxicircle DNA (the mtDNA genome that is found in the kinetoplast of this trypanosome). Other approaches to studying mitochondrial gene expression in different eukaryotes are described in Chapters 48 and 50.

Analyses of mtDNA expression and replication in *S. cerevisiae* could be extended by the development of a transformation system for introducing defined mtDNA segments back into *rho*^{0} (mtDNA-less) hosts (Chapter 47). This approach is still in its infancy compared with the successful cloning of genes in yeast using the several types of nuclear-replicating vectors now available.

Clearly, the problem of organelle biogenesis can be understood fully only when the interaction of the products of the two contributing genetic systems (organelle and nuclear) are evaluated. Until now the emphasis has been on the organelle genes and their products, but recently attention has turned to the nuclear genes. Other parts of this book describe the isolation and analysis of nuclear genes that code for particular chloroplast proteins (Chapter 37) or mitochondrial proteins (Chapters 25 and 26).

41

Features Revealed by the Sequencing of Chloroplast Genes Highlight the Prokaryote-like Nature of These Organelles

Paul R. Whitfeld, Gerard Zurawski[1] and Warwick Bottomley

Division of Plant Industry
CSIRO
Canberra, Australia

I. INTRODUCTION

Evidence supporting the concept that chloroplasts are pro-karyote in origin has been extensively documented (See (1) for references). One of the more compelling arguments has been that chloroplast ribosomes are 70S in size and their protein synthetic activities are sensitive to the same specific inhibitors as are *E. coli* ribosomes, and in both these properties the organelle ribosomes are clearly distinguishable from plant cytoplasmic ribosomes (1). More recently this line of reasoning has been strengthened by the observation that the nucleotide sequences of the genes for the 16S and 23S rRNAs of maize and tobacco chloroplast ribosomes are 67%-74% homologous with the corresponding sequences of *E. coli* ribosomes (2-5). In this paper we ask whether the resemblance between chloroplast and bacterial systems is also apparent at the level of the structure and organization of protein-coding and tRNA genes and their flanking sequences. We conclude that, for most of the features examined, bacterial and chloroplast genes are remarkably similar.

II. RESULTS AND DISCUSSION

Two regions of the 150 kb spinach *(Spinacia oleracea)* chloroplast DNA have been studied in detail. The location of

[1]Present address : DNAX Research Institute, Palo Alto, California.

MANIPULATION AND EXPRESSION
OF GENES IN EUKARYOTES
ISBN 0 12 513780 x

these on the spinach chloroplast DNA map is shown in Figure 1.
The first region encompasses the gene for the large subunit of
ribulose bisphosphate (RuBP) carboxylase *(rbcL)*, the genes for
the β and ε subunits of chloroplast ATPase *(atpBE)* and the genes
for tRNA(Met) and tRNA(Val) (6,7). RuBP carboxylase is the most
abundant protein in leaf tissue and is responsible for the
fixation of CO_2. It is a soluble enzyme located in the stroma
phase of the chloroplasts and is composed of eight identical
large subunits which are coded by a chloroplast gene and eight
small subunits which are coded by (a) nuclear gene(s)(see (8)
for references). The ATPase component (CF_1) of chloroplast ATP
synthase is composed of five subunits, three of which (α,β and
ε) are encoded by chloroplast DNA and two (γ and δ) are nuclear
DNA encoded (see (8) for references). The organization of this
region is detailed in Figure 2A.

The second region studied includes the gene for a thylakoid
membrane protein *(psbA)* (9) and tRNA(His). The membrane protein,
usually referred to as the $32000-M_r$ protein, is the most
rapidly labeled membrane protein in chloroplasts but it does not
accumulate to any extent because of its high turnover rate (see
(8) for references). It is part of Photosystem II and is
involved in electron flow. The size of the transcribed region
and the relationship of *psbA* to the tRNA(His) gene is shown in
Figure 2B.

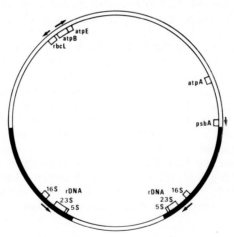

FIGURE 1. *Map of spinach chloroplast DNA showing the location
of the genes for the ribosomal RNAs, the large subunit of RuBP
carboxylase* (rbcL), *the α, β and ε subunits of ATPase* (atpA,B,E)
and the 32,000-M_n thylakoid membrane protein (psbA). *The solid
section of the circle indicates the region of the inverted
repeat sequence. Arrows show the direction of transcription.*

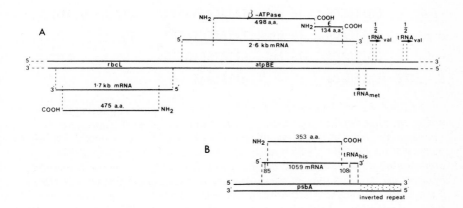

FIGURE 2. *Diagrams of the organization, transcription and translation of genes in the vicinity of* rbcL *and* atpBE(A) *and* psbA(B) *of spinach chloroplast DNA.*

A. 5' Leader Regions and Promoter Sites

The genes *rbcL* and *atpBE* are transcribed divergently from opposite strands of the DNA, their 5' ends being approximately 150 bp apart (6,7). Transcription initiation sites for both these genes and for *psbA* have been determined by S1 nuclease mapping. Scanning of the DNA sequences immediately upstream from each of these sites (Figure 3) reveals the existence of regions analogous to the two regions which are most extensively conserved between promoters in *E. coli*. These are the Pribnow box centred about -7 and the recognition site centred about -35 with respect to transcription initiation (10). The close similarity of the chloroplast and *E. coli* DNA sequences upstream from the transcription start point suggests that the recognition signal for chloroplast RNA polymerase will prove to be very similar to that for the bacterial enzyme. It also explains why *E. coli* RNA polymerase transcribes chloroplast DNA sequences efficiently (11) and why the chloroplast *rbcL* gene is expressed when introduced into *E. coli* cells (12).

B. Transcription Termination Regions

A feature of prokaryote genes which is known to be important for efficient termination of transcription is the presence of sequences having dyad symmetry just prior to the transcription termination point (10). Regions immediately preceding the

```
rbcL      GGTTGCGC...16 bp...TATACAATAATGA...177 bp...ATG...
                                            ▼
atpB      TCTTGACA...19 bp...TATATCCTAGAT....454 bp...ATG...
                                          ▼
psbA      GGTTGACA...16 bp...GTTATACTGTTGA....85 bp...ATG...
                                          ▼
E. coli   tgTTGaca            gtTAtaaT
```

FIGURE 3. *Sequences of putative promoter regions for three
spinach chloroplast genes. Sequences upstream from the
transcription initiation sites (marked with an arrow) which are
comparable to the '-35' and 'Pribnow box' regions of the
consensus sequence (10) of* E. coli *promoters are underlined.
Higher case letters in the consensus sequence indicate that a
base appears more frequently in that position of promoters than
bases indicated by lower case letters (10). The number of base
pairs between the transcription initiation site and the
translation start (methionine) codon is also indicated.*

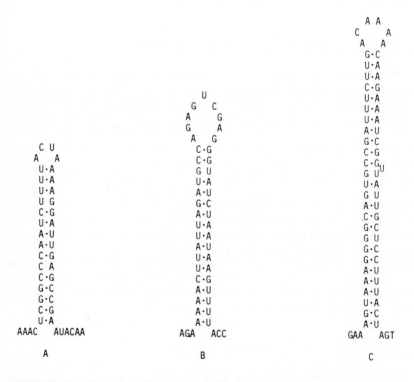

FIGURE 4. *Proposed stem and loop structures at the 3' end of
the mRNA transcripts from three spinach chloroplast genes.
Sequences are for A,* rbcL; *B,* atpBE; *C,* psbA.

```
          -20                    -10
rbcL    ...GAAGTTGTAGGGAGGGACTTATG...
atpB    ...GACATACTTTACTATATATTATG...
psbA    ...AAATAAACCAAGATTTTACCATG...
              HO-UUUCCUCCAC...
```

*FIGURE 5. Sequences immediately upstream from the
translation initiation (methionine) codon of three spinach
chloroplast genes. The 'Shine-Dalgarno' sequence (13) in* rbcL
*is underlined. The terminal ten nucleotides of maize
chloroplast 16S rRNA (2) are shown in 3' to 5' orientation
below the gene sequences.*

3' end of the mRNA for *rbcL, atpBE* and *psbA* in spinach chloroplast DNA all exhibit dyad symmetry. RNA transcribed from these
regions can therefore form stem and loop structures (Figure 4).
It is likely that, as is the case for *E. coli* RNA polymerase,
such structures will prove to be important recognition signals
for termination of transcription by chloroplast RNA polymerase.

C. Translation Initiation

The ribosome binding sites in bacterial mRNAs usually contain
sequences that are complementary to those at the 3' end of 16S
rRNA. These 'Shine-Dalgarno' sequences (13) are located a short
distance upstream from the initiating methionine codon. Except
for the terminal nucleotide, the last 15 nucleotides of maize
and tobacco chloroplast 16S rRNAs are identical to those of *E.
coli* 16S rRNA (2,4). Thus one might expect that 'Shine-Dalgarno'
type sequences would feature in translation initiation regions
of chloroplast genes. It can be seen in Figure 5 that this is
indeed the case for *rbcL* where the sequence GGAGG occurs 6 bp
before the first ATG codon. However, no such sequences occur
prior to the translation start points of *atpBE* or *psbA* (Figure
5) and their significance is as yet unresolved.
Translation of total chloroplast RNA in an *in vitro E. coli*
system yields the large subunit of RuBP carboxylase as a major
product (14) presumably because the messenger for the protein is
an abundant RNA species and contains a strong 'Shine-Dalgarno'
sequence. Relatively minor amounts of the 32,000-M_r protein
are formed in an *E. coli* system even though its mRNA is an
abundant species. In chloroplasts, however, both proteins are
synthesized rapidly suggesting that there are factors other
than the presence or absence of a 'Shine-Dalgarno' sequence that
determine the efficiency of translation *in vivo.*

```
β SUBUNIT...  LYS   LEU   LYS   LYS   STOP
          ... A A A T T A A A G A A A T G A C C T T A A A T C T T ...
                                    MET   THR   LEU   ASN   LEU  ...  ε SUBUNIT
```

*FIGURE 6.   Nucleotide and amino acid sequence showing the
overlapping translation stop/start codons at the junction of the
β and ε subunits of spinach chloroplast ATPase.*

D. Dicistronic mRNA and Overlapping Genes

As indicated in Figure 2A the genes coding for the β and ε
subunit of spinach chloroplast ATPase are cotranscribed into a
single large ($\sim$ 2.6 kb) RNA species (7). This was established by
showing that probes specific for sequences coding for the ε
subunit hybridized to the same RNA species as did probes specific
for the β subunit gene (7). Hybridization to smaller RNA species
was not detected. Sequence analysis of the two coding regions
showed that they are read in different reading frames and that
the genes overlap. The A of the ATG initiation codon of the ε
subunit gene is also the third nucleotide of the terminal lysine
codon of the β subunit gene (Figure 6). Overlapping reading
frames have been reported in bacterial, viral and mitochondrial
genomes. Overlapping translation stop and start codons between
genes in a bacterial operon are thought to play a role in
coupling the translation of two proteins destined to be assembled
in equimolar amounts into the same enzyme complex (15). It is
doubtful that this will turn out to be the case for the β and ε
subunit genes of chloroplast ATPase because our sequencing data
suggests that a reassessment of the stoichiometry of the subunit
composition is called for and that the revised ratio between the
β and ε subunits is likely to be 3:1. We also have sequenced
the equivalent region of pea chloroplast DNA and found that the
atpB and *atpE* genes are separated by 20 bp. Thus translation of
the ε subunit protein probably does not involve read-through by
the ribosomes from the β subunit region but relies on the initiation by ribosomes just upstream from the ε subunit AUG codon.
The presence of a GGAG 'Shine–Dalgarno' sequence 15 nucleotides
prior to this initiation codon makes this proposition plausible
(7).

Although we have not shown explicitly that both the β and ε
subunit proteins are translated from the same 2.6 kb RNA species
it appears probable that they are in view of the overlapping
codons and the fact that a small RNA species specific to *atpE*
sequences has not been detected in chloroplasts. We have
evidence of large RNA transcripts from other regions of spinach
chloroplast DNA and are of the opinion that polycistronic mRNAs
will prove to be a feature of the chloroplast system.

E. *Comparison of the β and ε Subunits of Chloroplast and* E. coli *ATPase*

A direct comparison of the amino acid sequences of the β subunit protein from spinach chloroplast and *E. coli* ATPase shows the sequences to be 67% homologous (7). Some regions have over 91% homology, whereas others, such as the amino-terminal 150 residues, have only 36%. The ε subunit protein of spinach chloroplast ATPase is only 26% homologous with the *E. coli* ε subunit (7). This contrasting degree of divergence probably reflects a basic difference in the roles the two subunits play in the ATPase complex. The β subunit is involved in the catalytic reaction and has sites for binding nucleotides and Mg^{++}. Constraints on its amino acid sequence may therefore be fairly severe. The ε subunit on the other hand has only a structural/regulatory role and constraints on its sequence may be less demanding.

F. *Structure and Organization of Chloroplast tRNA Genes*

The location of three tRNA genes is shown in Figures 2A and 2B. The proximity of the tRNA(Met) gene to the end of the *atpBE* gene and of the tRNA(His) gene to the end of the *psbA* gene is reminiscent of the arrangement of tRNA genes relative to protein-coding genes in animal mitochondrial genomes. However there is no tRNA gene at the 3' end of *rbcL* and so this is not a general rule for the arrangement of tRNA genes on the chloroplast genome.

Chloroplast tRNAs exhibit greater homology with the analogous tRNAs from *E. coli* than with those from the plant cytoplasm (16). However, the chloroplast tRNA genes do not encode the terminal - CCA of the tRNA and in this respect they are different from the bacterial tRNA genes. In one other respect chloroplast tRNA genes differ from their bacterial counterparts. In Figure 2A it can be seen that the tRNA(Val) gene is split by an intron of approximately 550 bp. Other chloroplast tRNA genes have been shown to contain introns (17) which can range in size up to 949 bp. Why some chloroplast tRNA genes contain intervening sequences and others do not is at present a mystery.

III. CONCLUSION

We have examined the extent of the similarity between chloroplasts and bacteria at the level of the structure and organization of a number of individual genes. Features of chloroplast genes which can be seen to resemble the equivalent

features of *E. coli* genes include (i) sequences upstream from
the transcription initiation point are similar to the consensus
sequence of *E. coli* promoters; (ii) sequences close to the
end of 3' untranslated regions can base pair to form typical
prokaryote-like transcription terminator stem-loop structures;
(iii) cotranscription of adjacent genes into a dicistronic
mRNA; (iv) the occurrence of overlapping translation stop/start
codons; (v) extensive homology between chloroplast and *E. coli*
β subunits of ATPase; (vi) homology between the sequences of
chloroplast and *E. coli* tRNA genes. In two respects chloroplast
genes do not resemble their *E. coli* counterparts. Chloroplast
tRNA genes do not encode the terminal –CCA of the tRNA and a
number of chloroplast tRNA genes contain large intervening
sequences.

REFERENCES

1. Gray, M.W. and Doolittle, W.F., *Microbiol. Rev.* 46, 1 (1982).
2. Schwarz, Z. and Kössel, H., *Nature* 283, 739 (1980).
3. Edwards, K. and Kössel, H., *Nucleic Acids Res.* 9, 2853 (1981).
4. Tohdoh, N. and Sugiura, M., *Gene* 17, 213 (1982).
5. Takaiwa, F. and Sugiura, M., *Eur. J. Biochem.* 124, 13 (1982).
6. Zurawski, G., Perrot, B., Bottomley, W. and Whitfeld, P.R., *Nucleic Acids Res.* 9, 3251 (1981).
7. Zurawski, G., Bottomley, W. and Whitfeld, P.R., *Proc. Natl. Acad. Sci. USA* In press (1982).
8. Bottomley, W. and Bohnert, H.J., *in* "Encyclopedia of Plant Physiology : Nucleic Acids and Proteins in Plants" (D. Boulter, B. Parthier, eds) Vol. 14B. Springer-Verlag, Berlin (In press).
9. Zurawski, G., Bohnert, H.J., Whitfeld, P.R. and Bottomley, W. *Proc. Natl. Acad. Sci. USA* In press.
10. Rosenberg, M. and Court, D., *Annu. Rev. Genet.* 13, 319 (1979).
11. Bottomley, W. and Whitfeld, P.R., *Eur. J. Biochem.* 93, 31 (1979).
12. Gatenby, A., Castleton, J.A. and Saul, M.W., *Nature* 291, 117 (1981).
13. Shine, J. and Dalgarno, L., *Nature* 254, 34 (1975).
14. Hartley, M.R., Wheeler, A. and Ellis, R.J., *J. Mol. Biol.* 91, 67 (1975).
15. Oppenheim, D.S. and Yanofsky, C., *Genetics* 95, 785 (1980).
16. Weil, J.H. and Parthier, B., *in* "Encyclopedia of Plant Physiol. Nucleic Acids and Proteins in Plants I" (D. Boulter, B. Parthier, eds.) Vol. 14A, p.65. Springer-Verlag, Berlin (1982).
17. Koch, W., Edwards, K. and Kössel, H., *Cell* 25, 203 (1981).

42

Chloroplast and Nuclear DNA Levels in Dividing and Expanding Leaf Cells of Spinach

N. Steele Scott and J. V. Possingham

CSIRO Division of Horticultural Research
Adelaide, South Australia

Second order renaturation kinetics have been used to measure the proportion of chloroplast DNA (ctDNA) in total DNA extracts from spinach leaves. During the growth of the primary leaves of seedlings from 1 to 7 mm there is a 50-fold increase in cell number and plastid number per cell is in the range 10 to 17. On average, the cells of these leaves are approximately 3C and ctDNA is 7-10% of total DNA. Plastome copy numbers per plastid are approximately 100 throughout and on a per cell basis approximately 1500.

There is a pronounced gradient of cellular development in 20 mm long leaves (Table I), with dividing cells mainly confined to a basal zone and with cell size increasing towards the distal end of the leaf. The cells of these leaves average 3-4C and ctDNA levels vary from 14% at the base to 21% at the tip while chloroplast numbers per cell change from 14 to 51. Plastomes per cell increase from an average of 1800 in the dividing cells of the basal zone to over 5000 in the expanding cells of the leaf tip, the same as found in the fully expanded cells of leaves 100 mm long which contain around 170 chloroplasts (Scott and Possingham, 1980).

On a per chloroplast basis, plastid copy numbers vary from about 100 in the chloroplasts of dividing cells to 200 in the chloroplasts of cells in which division has recently ceased. The chloroplasts of fully expanded cells have about 30 plastome copies (Table 1).

The tissues of expanding leaves are not completely homogeneous with respect to growth by cell expansion, cell division, DNA synthesis, etc., but we suggest that our results, as summarized in Table I, show three distinguishable stages of chloroplast and ctDNA synthesis. In dividing cells there are few (10-15) chloroplasts per cell and ctDNA synthesis keeps pace with nuclear DNA synthesis, maintaining a supply of chloroplasts which

MANIPULATION AND EXPRESSION
OF GENES IN EUKARYOTES
ISBN 0 12 513780 x

TABLE I.   Expansion of spinach leaves. In 2 mm leaves and the basal
(B) area of 20 mm leaves growth is by cell division. In the top half
of 20 mm leaves (T) cell expansion becomes the main determinant of
growth.

Leaf length (mm)	cell area (μm^2)	chloroplasts per cell	DNA per cell (pg)	plastome copies per	
				cell	chloroplast
2	150	10	2.7	1500	150
20(B)	280	14	2.0	1750	125
20(T)	1150	23	3.9	5500	190
100(T)	7200	171	3.4	4700	27

contain relatively high levels of ctDNA for daughter
cells. Chloroplasts at this stage are small with an area of
less than 6-8 μm^2. During the change from cell division to
cell expansion seen along the length of 20 mm leaves, there
is a three-fold increase in the amount of ctDNA per cell.
The level of ctDNA increases relative to nuclear DNA, and
the average plastome copy number rises to 195.   During
subsequent cell expansion, no ctDNA synthesis occurs and
ctDNA levels remain at about 5,000 plastome copies per
cell.   In the major phase of chloroplast formation as 3
rounds of chloroplast division take place, chloroplasts
expand to their full size with an area of about 30 μm^2
although both cell division and DNA synthesis have stopped.
    The large difference between the 5000 chloroplast
plastome copies relative to the 4-8 nuclear genome copies
per cell leads to a similar disparity in the number of mRNA
genes for cooperatively synthesized proteins such as RuBP
carboxylase.   However this large difference results in a
similar number of rRNA genes (about 10,000) being available
for the synthesis of ribosomes in the chloroplast and
nuclear systems at the time of rapid synthesis of proteins
for dividing and developing chloroplasts.   A similar
amplification mechanism for rRNA genes has been found in the
Xenopus oocyte (Brown and Dawid, 1968).

REFERENCES

Brown, D.D. & Dawid, I. (1968).   Science 160, 272.
Scott, N.S. & Possingham, J.V. (1980).   J. Exp. Bot., 31,
    1081.

43

Features of Nucleotide Sequences in the Region of the *oli2* and *aap1* Genes in the Yeast Mitochondrial Genome

Charles E. Novitski, Ian G. Macreadie, Ronald J. Maxwell,
H. B. Lukins, Anthony W. Linnane and Phillip Nagley

Department of Biochemistry, Monash University
Victoria, Australia

The mitochondrial *oli2* gene in *Saccharomyces cerevisiae* codes for subunit 6 of the mitochondrial ATPase (mtATPase) (Roberts *et al.*, 1979; Linnane *et al.*, 1980). Subunit 6 (M_r about 20,000) is a hydrophobic F_0-sector component involved in the coupling of respiration and oxidative phosphorylation (Murphy *et al.*, 1980b). These conclusions were made possible by analysis of *mit⁻* mutants that contain altered mtATPase and whose mutations map within the *oli2* gene. New *mit⁻* mutants carrying mutations affecting the mtATPase have recently been found (Macreadie *et al.*, 1982); the mutations were mapped between the *oxi3* and *oli2* genes where no functional sequences had previously been identified. We have now determined the complete nucleotide sequence of a 2.8 kb segment of the mitochondrial DNA (mtDNA) from a wild-type yeast strain (Fig. 1) extending from the 3' end of the *oxi3* gene through and beyond the *oli2* gene, and including the entire 1.8 kb mtDNA segment lying between the *oxi3* and *oli2* genes. DNA sequencing studies using several classes of mutants (drug resistant, petite, *mit⁻*) have provided information concerning the role of particular sequences in this region which we consider in this paper: (i) the nature of oligomycin resistance mutations in the coding region of the *oli2* gene; (ii) a new gene, denoted *aap1* (ATPase associated protein), which is located between the *oxi3* and *oli2* genes; (iii) G,C-rich clusters found in this region, and the possible roles of such sequences in DNA replication and RNA processing.

MANIPULATION AND EXPRESSION
OF GENES IN EUKARYOTES
ISBN 0 12 513780 x

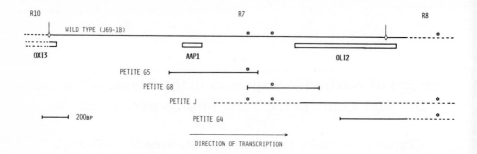

Fig. 1. The region of the mitochondrial genome between the oxi3 *and* oli2 *genes. The wild-type (J69-1B) mtDNA nucleotide sequence was determined by analysis of EcoRI (◇) fragments R7 (2.57kb) and R8 (1.75 kb) cloned into pBR325, and propagated in* E. coli. *The mtDNA genomes of petites G5, G8, J and G4 are aligned with the segments of wild-type sequence from which they originate. Solid lines represent DNA segments for which sequence was obtained. The locations of the* oxi3 *gene, the* oli2 *gene and the* aapl *gene are indicated. Asterisks mark the positions of G,C-rich clusters.*

OLIGOMYCIN RESISTANCE MUTATIONS IN THE *oli2* GENE

   The wild-type *oli2* gene sequence we have determined is from mtDNA of strain J69-1B, the parent of many mutant strains used in this laboratory. The EcoRI fragments R7 and R8 of J69-1B mtDNA (Fig. 1) were cloned in *E. coli* and were used for sequence analysis. The sequence of the 780 nucleotide coding region of the *oli2* gene in strain J69-1B is identical to that of the *oli2* gene in strain D273-10B/A1 (Macino and Tzagoloff, 1980) except for a single base difference at nucleotide 678 (a C/T polymorphism) which as a C in J69-1B constitutes the fourth base in the MboI site that lies very close to the EcoRI site within the *oli2* gene (Linnane *et al.*, 1980). This nucleotide sequence polymorphism does not change the predicted amino acid sequence. The A,T-rich sequences flanking the *oli2* gene exhibit considerably more divergence (about 1 change per 25 nucleotides) than the coding region. Substantial differences were observed between J69-1B and petite DS14 (derived from D273-10B) in and around the two G,C-rich clusters upstream of the *oli2* gene whose location is indicated in Fig.1 (manuscript in preparation).
   The *oli2* gene was initially recognized by study of oligomycin resistant mutants (see Dujon, 1981). We have determined the *oli2* gene sequence in a particular oligomycin resistant strain 70M [*rho$^+$ oli2-23r*] (previously *oli23-r*) derived from

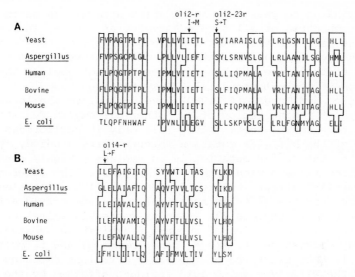

Fig. 2. Amino acid sequence comparison of portions of ATPase subunit 6 from diverse organisms. A. Yeast mtATPase subunit 6 amino acid residues 155-197 are compared with residues 152-194 in Aspergillus *mtATPase subunit 6 (Grisi et al., 1982), with residues 128-170 in the mammalian mtATPase subunit 6 from human (Anderson et al., 1981), bovine (Anderson et al., 1982) and mouse (Bibb et al., 1981), and with residues 179-221 in the* E. coli *ATPase subunit 6 (uncB gene product) (Gay and Walker, 1981). B. Similar comparison to that in A, except that the following residues of mtATPase subunit 6 are compared: yeast (231-254),* Aspergillus *(228-251), human (201-224), bovine (201-224), mouse (201-224), and* E. coli *subunit 6 residues (243-266). Amino acids at each position are boxed in if they are present in at least three of the four major groups (yeast* Aspergillus, *mammalian,* E. coli*). The arrows above the sequences indicate amino acid positions at which particular mutations in yeast mtDNA lead to amino acid substitutions producing an oligomycin resistant phenotype.*

J69-1B (Murphy *et al.*, 1980a). The 70M sequence was obtained by analysis of mtDNA from the petites G8, J and G4 (Fig. 1), all of which were derived spontaneously from 70M. The single base change which is responsible for oligomycin resistance is at nucleotide 523 (T→A), which converts a serine to a threonine at amino acid 175 of mtATPase subunit 6. This serine is present in a region of the ATPase subunit 6 whose sequence is highly conserved among *S. cerevisiae*, *Aspergillus nidulans* and mammalian mitochondria, and *E. coli* (Fig. 2). The residue

corresponding to yeast subunit 6 amino acid 175 is serine in
the mitochondria of all five eukaryotes as well as in *E. coli.*
Nevertheless, the fact that 70M can grow well on non-fermentable substrates indicates that the conservative change of
serine to threonine at that position does not abolish the
function of the mitochondrial ATP synthetase.

Macino and Tzagoloff (1980) have determined the location
in the *oli2* coding region of two other oligomycin resistance
mutations. One of these (*oli2-r*) is only 10 nucleotides upstream of the *oli2-23r* mutation which we have studied; the
effect of the *oli2-r* mutation is to replace a highly conserved
isoleucine residue with methionine (Fig. 2). The other mutation (*oli4-r*) affects nucleotide 696 (i.e. 173 nucleotides
downstream of the *oli2-23r* mutation). The amino acid affected
in this case is residue 232 (leucine → phenylalanine) which is
found in a second region of subunit 6 exhibiting substantial
homology between yeast, *Aspergillus,* mammals, and *E. coli*
(Fig. 2). For all three of the *oli-r* mutations described
above, the wild-type amino acid residues affected by the mutations are identical in those species that exhibit oligomycin
sensitivity (yeast, *Aspergillus* and mammals) and they occur
within the two regions of the subunit 6 protein which have
been the most highly conserved during evolution. The particular components of the mtATPase with which oligomycin interacts
are not well established, although it has been reported that
oligomycin may bind only to the proteolipid subunit 9 (Enns
and Criddle, 1977). One or both regions of amino acid sequence homology, in which the *oli2-r, oli2-23r* and *oli4-r*
mutations are located, may be involved directly in oligomycin
binding. Alternatively, these conserved regions may be involved in the association between subunits 6 and 9 in the F_o-
sector in the inner mitochondrial membrane.

A NEW GENE *(aap1)* UPSTREAM OF THE *oli2* GENE

The mtATPase complex contains two components in the F_o-
sector coded by mtDNA, namely subunits 9 and 6, specified by
the *oli1* and *oli2* genes, respectively (reviewed by Dujon, 1981).
The *mit⁻* mutants reported by Macreadie *et al.* (1982) affecting
the mtATPase define a third genetic locus lying between the
oxi3 and *oli2* genes. The data we present here focus on one
such *mit⁻* mutant, M26-10. The preliminary localization of the
mutation in M26-10 was made possible by use of petite G5
(Fig. 1), which when crossed with M26-10, produced respiratory
competent diploid progeny. The site of the mutation in M26-10
thus lies within the mtDNA sequences contained in G5. The

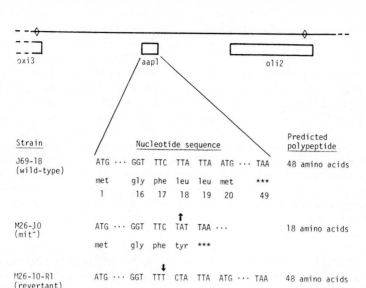

Fig. 3. Nucleotide sequences within the aapl *gene coding region. The wild-type (J69-1B) sequence is shown over a region that includes the sites of deletion (↑) leading to the* mit⁻ *mutant M26-10 (aapl-1) and the insertion (↓) leading to the revertant M26-10-R1 (aapl-1-R1). Amino acid residues in the predicted* aapl *gene product are numbered. Predicted polypeptide lengths are given. Key to the restriction map on the top line are given in Fig. 1.*

length and location of the position of the wild-type mtDNA genome contained in G5 was initially determined by restriction enzyme digestion of mtDNA and Southern blotting analysis (Macreadie *et al.*, 1982). The complete nucleotide sequence of G5 mtDNA has now been determined, which when compared to the J69-1B sequence (Fig. 1), showed petite G5 to contain a single continuous 681 bp segment of the wild-type mitochondrial genome. Within this region, there exist two overlapping reading frames. One is capable of coding for a polypeptide 48 amino acids long with a codon usage typical of yeast mitochondrial genes by virtue of the rare occurrence G or C as the third base of codons; the other could code for a polypeptide 53 amino acids long, but as its codon usage is uncharacteristic of yeast mitochondrial genes, it is not considered further here.

Study of mtDNA derived from mutant M26-10 has demonstrated that the mutation in this strain, designated *aapl-1*, lies within the 48 amino acid reading frame (Fig. 3). This was determined by use of a petite, 620M, that was isolated from

mit⁻ M26-10 and was shown genetically to carry the *aap1-1*
mutation.   The 620M mtDNA was sequenced through more than 95%
of the region of the mitochondrial genome retained by petite
G5, including the entire 48 amino acid reading frame and
more than 300 nucleotides on each side of the reading frame.
Only one difference was noted between 620M and wild-type
sequence (Fig. 3).   That change is the deletion of a T in the
18th codon in the 48 amino acid reading frame; the deletion
would result in an 18 amino acid product due to premature termination.   A revertant of M26-10, named M26-10-R1, was selected and the mtDNA sequence determined in the vicinity of the
aap1-1 mutation by directly sequencing restriction fragments
of mtDNA purified from strain M26-10-R1.   The only difference
observed between M26-10-R1 mtDNA and M26-10 mtDNA is the insertion of a T in the 17th codon (Fig. 3);   the insertion re-
opens the reading frame and restores the predicted 48 amino
acid product except for the replacement of leucine by threonine
at residue 18.   We conclude that the mutation in mutant M26-10
lies in a new mitochondrial gene coding for a protein. The
predicted product of the *aap1* gene is a highly hydrophobic
polypeptide of molecular weight 5800 with the following amino
acid sequence:

MPQLVPFYFM NQLMYGFLLM ITLLILFSQF FLPMILRLYV SRLFISKL

Biochemical analysis of the *mit⁻* mutant M26-10 has
indicated that the *aap1-1* mutation affects the mtATPase complex
but the available data have not yet allowed the unambiguous
identification of the *aap1* gene product among the observed
mitochondrial proteins.   The mitochondrial translation products
associated with immunoprecipitates of the mtATPase complex of
wild-type strains are subunit 6, subunit 9 and usually a 10
kilodalton (kd) protein.   In mutant M26-10 the 10 kd protein
band is consistently missing, both from sodium dodecyl sulphate
lysates of mitochondria and from immunoprecipitates of Triton
X-100 mitochondrial extracts made with anti-holo-mtATPase
(Macreadie *et al.*, 1982); subunit 6 was not found in these
immunoprecipitates.   More recent experiments have shown that
subunit 6 is present, though at reduced levels, among the
total mitochondrial translation products in strain M26-10.
Thus the *aap1-1* mutation not only affects the synthesis of the
10 kd protein, but also affects subunit 6, at the same time
perturbing the assembly and function of the mtATPase complex.
What is the gene product of the *aap1* gene?   It may
code directly for the protein with an apparent size of
10 kd (based on its gel mobility).   This protein might be
involved in the assembly of the ATPase complex, either as a
subunit or as a transiently associated component.   It is not

possible to predict with confidence the gel mobility of a
hydrophobic protein having the sequence shown above for the
theoretical gene product (5.8 kd) of the *aap1* gene, especially
if some post-translational modification occurs. These con-
siderations are thus not definitive in determining whether the
aap1 gene does code directly for the 10 kd protein. On the
other hand, the *aap1* gene may code for an unrecognized regulatory protein whose role is to control the synthesis and function of both the 10 kd ATPase-associated protein and subunit 6
of the ATPase. The *aap1* region is clearly not an exon of the
oli2 gene since subunit 6 is synthesized in mutant M26-10,
which carries a frameshift mutation in the *aap1* gene.

The organization of sequences upstream of the gene coding
for ATPase subunit 6 in yeast mitochondria is strikingly
similar to that in mitochondria from the fungus *Aspergillus
nidulans* and mammalian cells, as well as in the bacterium
Escherichia coli. A closely related organism to *S.cerevisiae*
in this group is *Aspergillus*, in whose mtDNA an unassigned
reading frame (URF) that could code for a hydrophobic protein
48 amino acids long has recently been reported (Grisi *et al.*,
1982). This URF lies about 100 bases upstream of the subunit 6
coding region and its predicted product is 50% homologous to
the predicted *aap1* gene product. In the frugally organised
mitochondria of mammalian species (Anderson *at al.*, 1981;
Anderson *et al.*, 1982; Bibb *et al.*, 1981) there is a reading
frame (URF A6L) upstream of, and overlapping with, the coding
region for subunit 6. In the *E. coli unc* operon (Gay and
Walker, 1981; Kanazawa *et al.*, 1981) there is an URF located
eight nucleotides upstream of *uncB* gene, coding for subunit 6.
There is a small amount of homology between the predicted
polypeptide products of the *aap1* reading frame and the
mammalian URF A6L. In addition to an extensive hydrophobicity
in each of the predicted products, the most notable feature in
the eukaryotes is the common sequence MPQL at the N-terminus
of each polypeptide.

Further analogies among these mitochondrial and bacterial
systems occur in the transcriptional organisation of the
ATPase subunit 6 genes. In yeast mitochondria the most abundant *oli2* transcript (Cobon *et al.*, 1982), presumably the *oli2*
mRNA, includes the sequence of the *aap1* gene. Likewise, in
both mammalian mitochondria and in *E. coli* the subunit 6 gene
transcript includes in its 5' leader the sequences of the upstream URF. The location of a functional reading frame on
the mRNA upstream of the ATPase subunit 6 coding region in
these organisms could reflect the need for a hydrophobic polypeptide whose synthesis at a particular time and place is
specifically required for the assembly of subunit 6 into the
ATPase complex.

G,C-RICH CLUSTERS UPSTREAM OF THE *oli2* GENE

Apart from the *aap1* gene coding region, the vast majority of sequences between the *oxi3* and *oli2* genes consists almost exclusively of A and T residues, except for two G,C-rich clusters at the positions indicated by asterisks in Fig. 1. In the course of our studies, we have determined the complete nucleotide sequence of two petite mtDNA genomes (petite G5, 681 bp; petite G8, 566 bp). Petite G5 contains one G,C-rich cluster (Fig. 4) corresponding to the left hand asterisk in Fig. 1, while petite G8 contains the right hand G,C-rich cluster upstream of the *oli2* gene (Fig. 4, G8-I) as well as portion of the left hand G,C-rich cluster (Fig. 4, G8-II). These three G,C-rich clusters can be drawn in the form of stem and loop structures, each of which has at least one of the following characteristics (Fig. 4): (i) 5'-GGNCC-3' is present in the stem (G5, G8-I), where N is unpaired or part of an A-T pair, and (ii) the base of the stem is bounded by 5'-ATAG-3' and 5'-GGAG-3' (G5, G8-II).

It has been suggested by Bernardi and colleagues that a particular set of sequences in yeast mtDNA containing G,C-rich clusters act as origins of replication; these are denoted *ori* sequences (de Zamaroczy *et al.*,1981; see also Bernardi *et al.*, 1982). Within *ori* sequences, there are regions denoted A and B which are well conserved among the seven *ori* regions so far analysed and which would be available to base pair with each other in a stem and loop structure (de Zamaroczy *et al.*, 1981). The A region, as illustrated for *ori6* in Fig. 4, contains the sequence 5'-GGTCC-3' at the base of the stem. In view of the observation that many petites do not contain such *ori* sequences within their mtDNA, Goursot *et al.* (1982) have recently pointed out a particular sequence, *ori^s*, which is bounded by 5'-ATAG-3' and 5'-GGAG-3' and which they propose to be a surrogate origin of DNA replication (Fig. 4). A further class of sequences, called *ori^s*-related, are defined as those bounded by 5'-ATAG-3' and 5'-GGAG-3' but otherwise dissimilar to *ori^s*; two of the clusters in our petites (G5, G8-II) are *ori^s*-related sequences by these criteria.

One should bear in mind, however, that *ori^s* and *ori^s*-related sequences may be only one subset of sequences which can substitute for replication origins in yeast petite mutants. There are other criteria by which the three G,C-rich clusters included in petites G5 and G8 can be compared with *ori* and *ori^s* sequences. There are two common features in *ori6*, *ori^s* and petite G5: the sequence 5'-GGNCC-3' in the stem (N may be unpaired), and the sequence 5'-GGA-3' at the base of the stem on

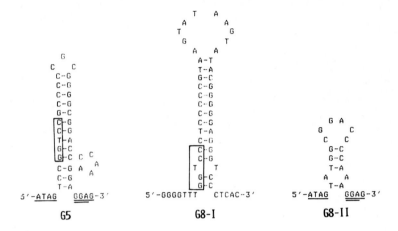

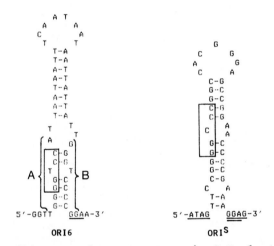

Fig. 4. Possible secondary structure in G,C-clusters. The G,C-rich clusters in mtDNA of petite G5 and petite G8 (G8-I) are located about 300 bp and about 100 bp upstream of the oli2 *coding region, respectively. The second cluster in G8 (G8-II) is derived from sequences at the site of deletion of this petite segment from wild-type mtDNA: the 17 nucleotides at the 3' end of the sequence shown here (G8-II) are identical to a sequence in the G5 cluster (cf. Fig. 1). A stem and loop structure from the* ori6 *sequence (de Zamaroczy et al.,1981) is shown with intramolecular hybridization of sequences of the A and B regions of* ori6. The ori^S *sequence is from Goursot et al. (1982). The common sequence 5'-GGNCC-3' in the stem is boxed in and flanking sequences 5'-ATAG-3', 5'-GGAG-3' and 5'-GGA-3 below the stem are underlined.*

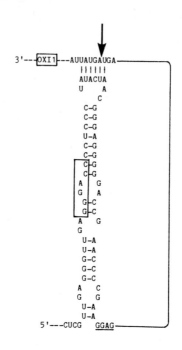

Fig. 5.   *Possible role of G,C-rich cluster in processing of* oxi1 *gene transcript.   Sequences are taken from Coruzzi et al. (1981).   The regions for which sequences are shown for the inferred RNA transcript are nucleotides -291 to -246 and -57 to -48; the nucleotide immediately 3' to the initiation codon AUG of the* oxi1 *coding region (boxed) is denoted -1.   The arrow indicates the cleavage site that generates the 5' end of the mature* oxi1 *transcript (presumed mRNA) (Coruzzi et al., 1981). The solid line indicates continuous sequence in between nucleotides -246 to -57.   Flanking sequences 5'-GGAG-3' below the stem and 5'-GG(GA)CC-3' in the stem are indicated (cf. Fig. 4).*

the 3' side.   In considering the two clusters in petite G8, each has just one of these features (Fig. 4).   It is of further interest that cluster G8-I has the sequence 5'-CCCTCCCCC-3' which appears in the C region of each of the seven canonical *ori* sequences (de Zamaroczy *et al.*, 1981; Bernardi *et al.*, 1982).   In our view both clusters I and II in petite G8 are candidates for the mtDNA replication origin(s) in this petite.

The possibility should also be considered that the stem and loop structures in many of the G,C-rich clusters in wild-type strains may represent sites for protein-nucleic acid interactions in wild-type mtDNA other than initiation of mtDNA replication. For example, they could be involved in processing RNA transcripts, as pointed out by Tzagoloff *et al.* (1981) who suggested that a double strand specific nuclease (such as RNase III) may cleave precursor RNAs within the base-paired stem of such structures. It is interesting to note that upstream of the *oxi1* gene (Coruzzi *et al.*, 1981) there occurs a G,C-rich cluster (Fig. 5) similar to those shown in Fig. 4. The loop above the stem contains 9 nucleotides, of which 6 are exactly complementary to a sequence some 200 nucleotides downstream on the same polynucleotide strand. Within this downstream sequence is the RNA processing site at which the cleavage in the pre-mRNA occurs to give rise to the mature *oxi1* mRNA. One can envisage that in the pre-mRNA, hybridization between the loop of the G,C-cluster and the downstream complementary sequence may provide the substrate recognized by the RNA processing enzyme. Tests of the applicability of this scheme to the processing of *oli2* transcripts (see Cobon *et al.*, 1982) awaits a precise determination of the 5' ends of the various *oli2* transcripts observed.

REFERENCES

Anderson S., Bankier A.T., Barrell B.G., de Bruijn M.H.L., Coulson A.R., Drouin J., Eperon I.C., Nierlich D.P., Roe B.A., Sanger F., Schreier P.H., Smith A.J.H., Staden R., and Young I.G. (1981). *Nature 290,* 457-465.

Anderson S., de Bruijn M.H.L., Coulson A.R., Eperon I.C., Sanger F., and Young I.G. (1982). *J. Mol. Biol. 156,* 683-717.

Bernardi G., Baldacci G., Colin Y., Faugeron-Fonty G., Goursot R., Goursot R., Huyard A., Le Van Kim C., Mangin M., Marotta R., and de Zamaroczy M. (1982). *In* "Manipulation and Expression of Genes in Eukaryotes". (P. Nagley, A.W. Linnane, W.J. Peacock, and J.A. Pateman, eds.), Academic Press, Australia (this volume).

Bibb M.J., van Etten R.A., Wright C.T., Walberg M.W., and Clayton D.A. (1981). *Cell 26,* 167-180.

Cobon G.S., Beilharz M.W., Linnane A.W., and Nagley P. (1982). *Curr. Genet. 5,* 97-107.

Coruzzi G., Bonitz S.G., Thalenfeld B.E., and Tzagoloff A. (1981). *J. Biol. Chem. 256,* 12780-12787.

Dujon B. (1981). *In* "The Molecular Biology of the Yeast Saccharomyces: Life Cycle and Inheritance". (J.N. Strathern, E.W. Jones, and J.R. Broach, eds.), pp. 505-635, Cold Spring Harbor Laboratory, Cold Spring Harbor, New York.

Enns R.K., and Criddle R.S. (1977). *Arch. Biochem. Biophys.* *182*, 587-600.

Gay N.J., and Walker J.E. (1981). *Nucleic Acids Res.* *9*, 3919-3926.

Goursot R., Mangin M., and Bernardi G. (1982). *EMBO J.* *1*, 705-711.

Grisi E., Brown T.A., Waring R.B., Scazzocchio C., and Davies R.W. (1982). *Nucleic Acids Res.* *10*, 3531-3539.

Kanazawa H., Mabuchi K., Kayano T., Noumi T., Sekiya T., and Futai M. (1981). *Biochem. Biophys. Res. Commun.* *103*, 613-620.

Linnane A.W., Astin A.M., Beilharz M.W., Bingham C.G., Choo W.M., Cobon G.S., Marzuki S., Nagley P., and Roberts H. (1980). *In* "The Organization and Expression of the Mitochondrial Genome". (A.M. Kroon and C. Saccone, eds.), pp. 253-263, Elsevier/North-Holland Biomedical Press, Amsterdam.

Macino G., and Tzagoloff A. (1980). *Cell*, *20*, 507-517.

Macreadie I.G., Choo W.M., Novitski C.E., Marzuki S., Nagley P., Linnane A.W., and Lukins H.B. (1982). *Biochem. Int.* *5*, 129-136.

Murphy M., Choo K.B., Macreadie I.G., Marzuki S., Lukins H.B., Nagley P., and Linnane A.W. (1980a). *Arch. Biochem. Biophys.* *203*, 260-270.

Murphy M., Roberts H., Choo W.M., Macreadie I.G., Marzuki S., Lukins H.B., and Linnane A.W. (1980b). *Biochem. Biophys. Acta 592*, 431-444.

Roberts H., Choo W.M., Murphy M., Marzuki S., Lukins H.B., and Linnane A.W. (1979). *FEBS Lett. 108*, 501-504.

Tzagoloff A., Nobrega M., Akai A., and Macino G. (1980). *Curr. Genet. 2*, 149-157.

de Zamaroczy M., Marotta R., Faugeron-Fonty G., Goursot R., Mangin M., Baldacci G., and Bernardi G. (1981). *Nature 292*, 75-78.

44
The Unusual Organization of the Yeast Mitochondrial
var1 Gene

Ronald A. Butow
W. Michael Ainley
H. Peter Zassenhaus

Division of Molecular Biology
Department of Biochemistry
The University of Texas
Health Science Center at Dallas
Dallas, Texas

Michael E. Hudspeth
Lawrence I. Grossman

Department of Cellular and Molecular Biology
Division of Biological Sciences
The University of Michigan
Ann Arbor, Michigan

Most of the well characterized products encoded by mitochondrial genomes of eukaryotes are the same. These gene
products include polypeptides which function in the electron
transport and oxidative phosphorylation machinery and RNA
species (tRNAs and rRNAs) which are part of the mitochondrial
protein synthesis apparatus (Borst and Grivell 1978; Anderson
et al., 1982). In spite of the conservation of these gene
products, there is a striking lack of conservation among
species in the size of mitochondrial genomes, and in the
arrangement of mitochondrial genes and their internal
organization. This variability is most evident for genes
encoding the apoprotein of cytochrome *b*, subunit I of cytochrome oxidase and the large mitochondrial rRNA: in lower

MANIPULATION AND EXPRESSION
OF GENES IN EUKARYOTES
ISBN 0 12 513780 x

eukaryotes, these genes sometimes contain intervening
sequences (Borst and Grivell, 1978), whereas such sequences
are absent from their counterparts on mtDNA of higher
eukaryotes (Anderson *et al.*, 1982). Even in a single organism such as *Saccharomyces cerevisiae*, some of these intervening sequences are optional; i.e. they are present in
some wild-type alleles, but not others (Sanders *et al.*,
1977; Grivell, 1982).

One of the conspicious mitochondrial translation products in lower eukaryotes is a polypeptide found associated
with the mitochondrial small ribosomal subunit (Terpstra,
et al., 1979; Terpstra and Butow, 1979). In yeast, this
ribosome-associated protein is called var1 (variant 1)
polypeptide because it shows a strain-dependent size polymorphism (Douglas and Butow, 1976). However, unlike most
examples of protein polymorphism which are the result of
point mutations, the var1 protein polymorphism is due to
quite large differences in molecular weight between the
various forms (Table I).

Table 1.   Forms of Var1 Protein

Var1 Allele[a]	Strain	Genotype
40.0	ID 41 - 6/161	$a^- b^-$
41.8	-	$a^+ b^-$
42.0	D 273 - 10B	$a^- b_p^+$
42.2	-	$a^- b^+$
43.8	D6	$a^+ b_p^+$
44.0	5DSS	$a^+ b^+$

[a] *Designated by the apparent molecular weight (kd) of the
var1 protein as estimated by SDS-PAGE.*

Table I presents the different molecular weight forms
of var1 protein we have identified which are grouped into
six genotypic classes according to their genetic behavior.
Using the var1 size difference as a genetic marker, we
have identified a region of the yeast mitochondrial genome
called the *var1* determinant region, located between *oli1*

and *ery*, which uniquely specifies the molecular weight of
the var1 protein (Perlman *et al.*, 1977). These *var1*
alleles have been analyzed further by both genetic and
physical mapping procedures (Strausberg and Butow, 1977;
Vincent *et al.*, 1980; Strausberg and Butow, 1981) and the
results of these studies have led to the following
conclusions:

1. The *var1* alleles map physically to a DNA segment,
about 1.8 kb in length, within fragment 10 of a Hinc II
digest of wild-type mtDNA.
2. All *var1* alleles can be grouped according to the
genotypes shown in Table I, assigned on the basis of the
presence or absence of the recombining elements *a, b,* and
b_p. Combinations of these elements define any and all *var1*
alleles we have observed to date.
3. The *a, b,* and b_p elements recombine with characteristic frequencies by a process comparable to asymmetric
gene conversion, whereby *minus* alleles are converted pre-
ferentially to *plus* alleles. The *a* element is genetically
indivisible, whereas the *b* element can be subdivided
further into smaller, independently recombining units
designated b_p.
4. Restriction analysis suggest that these elements
correspond to DNA insertions within *var1*.

IDENTIFICATION OF THE *VAR1* GENE AND THE MOLECULAR BASIS FOR
VAR1 PROTEIN POLYMORPHISM

We have gained considerable understanding of the
molecular basis of var1 protein polymorphism from DNA
sequencing of *var1* alleles and the analysis of var1 transcripts and the purified var1 protein. The DNA sequence of
the *var1* region of an a^- b^- allele reveals the presence of
an open reading frame sufficient to encode a protein of
396 amino acids (Hudspeth *et al.*, 1982). The location
of this open reading frame is shown in Fig. 1. The amino
acid composition of a protein encoded by this DNA sequence
agrees very closely (> 90% homology) with the amino acid
composition of the purified var1 protein (Hudspeth *et al.*,
1982). From these results, and the extensive genetic and
physical mapping data we have obtained, we conclude that
the *var1* region contains the complete structural gene for
the var1 protein.

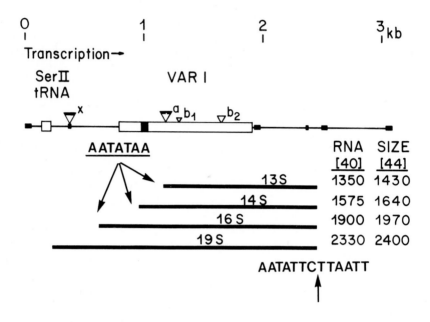

Figure 1: Transcripts and physical map of the var1 region. The location of the var1 and the seryl tRNA genes within fragment 10 of a Hinc II digest of 1D41-6/161 [var1 40.0] mtDNA is shown by the open bars. The solid bars denote the positions of GC-rich clusters. Also shown is the location of DNA inserts found in strains with larger var1 alleles (b₁ + b₂ comprise the b element). The map position of the four, major var1 transcripts is drawn below, with their length (in nucleotides) indicated to the right for RNAs from [var1 40.0] and [var1 44.0] alleles. By S1 mapping and DNA sequencing, the sequence of the non-coding strand flanking the ends of the RNAs has been determined. Note that all the var1 transcripts are 3' co-terminal.

The amino acid composition data (Hudspeth *et al.*, 1982) and S1 mapping of stable var1 transcripts also suggest that the *var1* gene is not extensively interrupted by intervening sequences. Remarkably, the a^- b^- allele is about 90% AT, and is one of the most AT-rich genes known. Moreover, nearly 40% of the total GC residues in the a^- b^- allele are concentrated in a 46 bp palindromic GC-rich cluster in register 186 bp down stream from the initiator codon. This GC-rich

cluster, which we term the *common GC cluster*, is present
in all *var1* alleles, is transcribed, and found within a
stable 16S *var1* transcript (Fig. 1), the most abundant of
four *var1* transcripts found in wild-type cells. We consider
the 16S RNA to be a good candidate for the var1 protein mRNA
(Fig. 1) (Farrelly *et al.*, 1982). Taken together, these
findings prompt some reevaluation of the previously held
views which depict the yeast mitochondrial genome as organized into genes of modest GC content flanked by AT-rich
"spacer" DNA, punctuated by GC-rich clusters (Bernardi,1976;
Bernardi, 1982). Although our findings on the high AT content and presence of long GC-rich clusters within the coding
sequence of the *var1* gene by no means rule out the validity
of this model, they do suggest the possibility that the
informational content of yeast mtDNA may be greater than
has been previously recognized.

THE a, b, AND b_p ELEMENTS

DNA sequence comparisons between the a^- b^- allele and
other *var1* alleles (Tzagoloff *et al.*, 1980; Hudspeth *et al.*,
in preparation) have provided some insight into the molecular nature of the a, b, and b_p elements. In a^+ strains
the *common GC cluster* is duplicated 146 bp downstream in
register but in an *inverted* orientation. The consequence of
this duplication is to generate an additional 16 amino acids
in the protein, 9 of which represent a unique amino acid
sequence. Like the *common GC cluster*, the duplicated GC
cluster is also transcribed into an abundant, stable RNA in
wild-type cells (Fig. 1).
A possible mechanism for insertion of the GC cluster
in the conversion of a^- to a^+ alleles is suggested from a
comparison of the DNA sequences flanking the site cluster
(in a^+ alleles) and the sequence into which the site cluster
inserts (the 20 bp "recipient" sequence in a^- alleles,
Fig. 2). In a^+ alleles, both GC clusters are flanked by the
same pure AT sequences, eleven bp on each side, which are
unique in the *var1* gene (Fig. 2). [These sequences are not
unique on yeast mtDNA, however, and are found near *oli1*
flanking a GC cluster identical to the *common GC cluster*
(Tzagoloff *et al.*, 1980)]. In a^+ x a^- crosses, the a^+ GC
cluster may insert into the 20 bp "recipient" sequence by a
transpostion-like process in which a staggered two bp
cleavage and duplication occurs within the recipient
sequence, thus generating the 22 bp flanking sequence

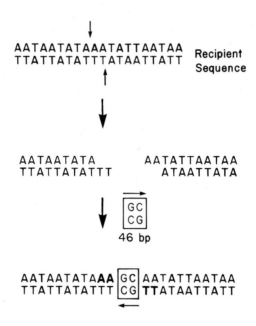

Figure 2: *Hypothetical transposition of a GC cluster.*
Flanking sequence duplication of nucleotides is indicated in
bold face.

Such flanking sequence duplications are known to occur as a
consequence of the transposition of prokaryotic (Grindley an
Sherrat, 1978) and eukaryotic (Roeder *et al.*, 1980) transposable elements. Since the location of GC clusters within
specific sequences of the yeast mtDNA appears to be variable
and strain-dependent (Dujon, 1980; Tzagoloff *et al.*, 1980;
Sor and Fukuhara, 1982; Hudspeth *et al.*, in preparation),
"jumping" GC clusters may be an inherent property of the
yeast mitochondrial genome.

Sequence analysis of b^-, b^+ and b_p^+ *var1* alleles shows
that within the coding region there are runs of AAT

(asparagine) codons whose length is allele-specific. The
var1 protein is unusually high in aspartic + asparagine
content ($\sim$ 35 mole %) (Hudspeth *et al.*, 1982) and in b^+
alleles, the var1 protein is extended further by the addition of some 10-15 asparagine residues. We tentatively
conclude that in the conversion of b^- to b^+, all of these
"extra" AAT's are inserted into the b^-allele. In some
cases, however, the conversion is not complete -- a
situation which would then give rise to the b_p^+ classes of
var1 alleles.

DISCUSSION

The unusual genetic properties, structure, and polymorphism of the *var1* gene and its product, the var1 protein,
reflect in part the striking variability and dynamic properties of fungal mitochondrial genomes in general, and
yeast in particular. We may naturally ask what is the
underlying biological significance of this dynamic variability? In the case of *var1* (and for that matter, other polymorphic mitochondrial genes, e.g. those with optional
introns), the phenotypic consequences of these polymorphisms
are not readily apparent. Possibly, these polymorphisms
may simply reflect a balance between the marked recombina-
tional activity of the yeast mitochondrial genome and some
degree of flexibility in both gene regulation and the
structure of gene products.
Although var1 is intimately associated with the small
ribosomal subunit (Terpstra *et al.*, 1979), we do not yet
have a clear idea as to its function. Preliminary data
suggest that it may play a role in ribosome assembly
(Terpstra and Butow, 1979). If this is indeed the case, we
could envisage that the structural requirement for such a
function might not be particularly stringent; thus localized size polymorphisms could be tolerated. Relevant to
this point is that in *Neurospora,* the small mitochondrial
ribosomal protein, S-5, like var1, is synthesized in mitochondria and appears also to have a role in ribosome
assembly. (LaPolla and Lambowitz, 1981a). Yet a comparison
of the amino acid composition of var1 (Hudsepth *et al.*, 1982)
and S-5 (LaPolla and Lambowitz, 1981b) indicates that there
is little homology between these proteins. Possibly, var1
and S-5 are divergent products of an ancestral mitochondrial
ribosomal protein and are no longer recognizable as evolu-
rionarily related proteins. Another alternative we presently

276   *Ronald A. Butow et al.*

favor is that var1 and S-5 are not products of divergent evolution, but are different ribosomal proteins. Thus we may speculate that in the evolution of the semi-autonomous mitochondrial genetic system within the eukaryotic cell, a control point in the assembly of the small mitochondrial ribosomal subunit could have been fulfilled by any one of a number of ribosomal proteins, whether translated in the mitochondrion or not. Indeed, species differences in the location of genes for mitochondrial proteins is not without precedent. For example, the gene for a proteolipid subunit of the mitochondrial ATPase is located in the nuclear genome in *Neurospora* (Jackl and Sebald, 1975) and on the mitochondrial genome in yeast (Tzagoloff *et al.*, 1980). This allows for the possibility that in mammalian cells either all mitochondrial ribosomal proteins are encoded in the nuclear genome, or alternatively, that one of the numerous unassigned mitochondrial translation products (Attardi *et al.*, 1980) could be a ribosome-associated protein but different structurally from var1 or S-5 such that no homology is evident either at the protein or DNA level.

ACKNOWLEDGEMENTS

We thank Marie Rotondi for her expert help with this manuscript. This work was supported by National Institutes of Health Grants GM26546 and GM22525 and by a grant (I-642) from The Robert A. Welch Foundation.

REFERENCES

Anderson S., Bankier A.T., Barrel B.G., DeBruijn M.H.L., Coulson A.R., Drovin J., Eperon I.C., Nierlich D.P., Roe B.A., Sanger F., Schreier P.H., Smith A.J.H., Staden R. and Young I.G. (1982) *In* "Mitochondrial Genes" (P. Slonimski, P. Borst, and G. Attardi, eds.), pp. 5-43. Cold Spring Harbor Laboratory, Cold Spring Harbour, New York.
Attardi G., Cantatore P., Ching E., Crews S., Gelford R., Merkel C., Montoya J., and Ojala D. (1980) *In* "The Organization and Expression of the Mitochondrial Genome" (A.M. Kroon and C. Saccone, eds.), p. 103, Elsevier/North-Holland Biomedical Press, Amsterdam.
Bernardi G. (1976) *In* "Genetics and Biogenesis of Chloroplasts and Mitochondria" (T. Bucher, W. Neupert, W. Sebald, S. Werner, Eds.) pp. 503-509, North Holland Publishing Co., Amsterdam.

Bernardi G. (1982) *In* "Mitochondrial Genes" (P. Slonimski,
    P. Borst, and G. Attardi, eds.), pp. 269-278. Cold Spring
    Harbor Laboratory, Cold Spring Harbor, New York.
Borst P. and Grivell L.A. (1978) *Cell 15,* 70
Douglas M.G. and Butow R.A. (1976) *Proc. Natl. Acad. Sci.
    USA 73,* 1982.
Dujon B. (1980) *Cell 20,* 185·
Farrelly F., Zassenhaus H.P., and Butow R.A. (1982) *J. Biol.
    Chem. 257,* 6581.
Grindley N.D.F., and Sherrat D. (1978) *Cold Spring Harbor
    Symp. Quant. Biol. 43,* 1257.
Grivell L.A. (1982) *In* "Genetic Maps. A Compilation of
    Linkage and Restriction Maps of Genetically Studied
    Organisms". (S.J. O'Brien, ed.). Vol. 2, Lab. of Viral
    Carcinogenesis, National Cancer Institute, Nat. Inst. of
    Health, Frederick, MD (in press).
Hudspeth M.E.S., Ainley W.M., Shumard D.S., Butow R.A., and
    Grossman L.I. (1982) *Cell* (in press)·
Hudspeth M.E.S., Vincent R.D., Treisman L.O., Shumard D.S.,
    Perlman P.S., and Grossman L.I. (1982) in preparation.
Jackl G., and Sebald W. (1975) *Eur. J. Biochem. 54,* 97.
LaPolla R.J., and Lambowitz A.M. (1981a) *J. Mol. Biol. 116,*
    189.
LaPolla R.J., and Lambowitz A.M. (1981b) *J. Biol. Chem. 256,*
    7064.
Perlman P.S., Douglas M.G., Strausberg R.L., and Butow R.A.
    (1977) *J. Mol. Biol. 115,* 675·
Roeder G.S., Farabaugh P.J., Chaleff D.T., and Fink G.R. (1980)
    Science 209, 1375.
Sanders J.P.M., Heyting C., Verbeet M.Ph., Meijlink F.C.P.W.,
    and Borst P. (1977) *Mol. Gen. Genet. 157,* 239·
Sor F., and Fukuhara H. (1982) *Nucl. Acids Res. 10,* 1625
Strausberg R.L., and Butow R.A. (1977) *Proc. Natl. Acad. Sci.
    USA 74,* 2715·
Strausberg R.L., and Butow R.A. (1981) *Proc. Natl. Acad. Sci.
    USA 78,* 494.
Terpstra P., Zanders E., and Butow R.A. (1979) *J. Biol. Chem.
    254,* 12653.
Terpstra P., and Butow R.A. (1979) *J. Biol. Chem. 254,* 12662
Tzagoloff A., Norbrega M., Akai A., and Macino G. (1980)
    Current Genetics 2, 149·
Vincent R.D., Perlman P.S., Strausberg R.L., and Butow R.A.,
    (1980) *Current Genetics 2,* 27.

45

The Origins of Replication of the Mitochondrial Genome of Yeast

Giorgio Bernardi, Giuseppe Baldacci, Yves Colin, Godeleine Faugeron-Fonty, Regina Goursot, René Goursot, Alain Huyard, Caroline Le Van Kim, Marguerite Mangin, Renzo Marotta and Miklos de Zamaroczy

Laboratoire de Genetique Moleculaire
Institut de Recherche en Biologie Moleculaire
Paris, France

The mitochondrial genomes of the vast majority of spontaneous petites are exclusively derived from the tandem amplification of a DNA segment excised from any region of the parental wild-type genome (Faugeron-Fonty et al., 1979; Figs. 1,2). Therefore, either the wild-type genome contains several origins of replication and at least one of them is present on the excised segment (Prunell and Bernardi, 1977), or sequences other than the origins of replication of the wild-type genome are used as surrogate origins of replication. In fact, both situations have been found to occur, although with very different frequencies.

THE CANONICAL AND THE SURROGATE ORIGINS OF REPLICATION

Considering that the first explanation was the more likely one, when we first sequenced (Gaillard and Bernardi, 1979; Gaillard et al., 1980) the repeat units of two petite genomes excised from the same region of the wild-type genome, we looked for an origin of replication in the segment shared by them and we found a region characterized by two short GC clusters, A and B, flanking a palindromic AT sequence, p, and a short AT segment, s; and one long GC cluster, C, separated from B by a long AT segment, l (see Fig.3). The potential secondary structure of the A-B region, the primary structure of cluster C and the general arrangement of the whole region

MANIPULATION AND EXPRESSION
OF GENES IN EUKARYOTES
ISBN 0 12 513780 x

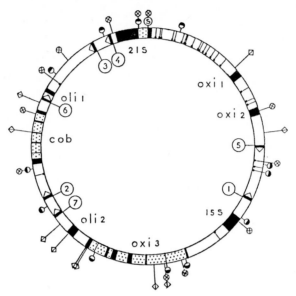

Fig. 1.   Physical and genetical map of the mitochondrial genome unit of wild-type yeast (strain A). Some restriction sites are indicated: Hinc II (☻), Hha I (☻), EcoRI (◈) Sal I (☺). Circled numbers indicate the location of ori sequences 1-7 (arrowheads point in the direction cluster C to cluster A; see Fig. 3). Black and dotted areas correspond to exons and introns of mitochondrial genes, respectively. Thin radial lines indicated tRNA genes. White areas correspond to long AT spacers embedding short GC clusters (Modified from de Zamaroczy et al., 1981).

are remarkably similar to those found in other mitochondrial origins of replication (Crews *et al.*, 1979; Gillum and Clayton, 1979; Fig.4).

An *ori* sequence like the one just described was found in almost all (see below) the mitochondrial genomes of spontaneous petite mutants. Restriction mapping and hybridization of petite genomes on restriction fragments of wild-type genomes (Bernardi *et al.*, 1980; de Zamaroczy *et al.*, 1981) provided evidence for the existence of seven such *ori* sequences in the mitochondrial genome of wild-type cells. The primary structure of the *ori* sequences shows that they are extremely homologous to each other, particularly in their GC clusters; some of them, *ori* 4, 6 and 7, contain additional GC clusters, β and γ identical in sequence and position (Fig. 3). All these *ori* sequences have been precisely localised

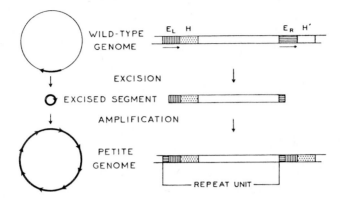

*Fig. 2. A. Scheme depicting the excision-amplification
process leading to the formation of the genome of a spontaneous petite mutant. A segment of a unit of a wild-type
mitochondrial genome is excised and tandemly amplified into
a defective genome unit. This then replicates and segregates
into the buds to form the genome of a petite mutant; the
petite genome can undergo further excisions leading to the
formation of secondary petite genomes.*
       *B. Scheme showing the left and right (E_L, E_R)
excision sequences as found on the parental wild-type genome
region from which the repeat unit of the petite genome was
excised. H, H' indicate sequences flanking the excision
sequences and sharing a significant yet imperfect homology
(from de Zamaroczy et al., 1982).*

and oriented on the physical map of the wild-type genome
(Fig.1).
      It should be noted: a) that some *ori* sequences display
one orientation on the wild-type genome, other ones the
opposite one; b) that *ori* 2 and 7, and *ori* 3 and 4 are close
to each other and tandemly oriented; c) that *ori* 4 is absent
in a wild-type strain (G. Faugeron-Fonty, personal comm.);
d) that *ori* sequences containing the γ cluster and not
accompanied by other *ori* sequences have either not been found
(*ori* 6, *ori* 7) or have been found only once (*ori* 4) in
extensive screenings of spontaneous petite genomes.
      Ori° petites, lacking a canonical *ori* sequence, have also
been found, although very rarely. An investigation of the
mitochondrial genomes of eight such *ori*° petites (Goursot *et
al.*, 1982) has revealed that their repeat units contain,
instead of canonical *ori* sequences, one or more *ori*[S]
sequences. These 44-nucleotide long surrogate origins of
replication are a subset of GC clusters characterized by a

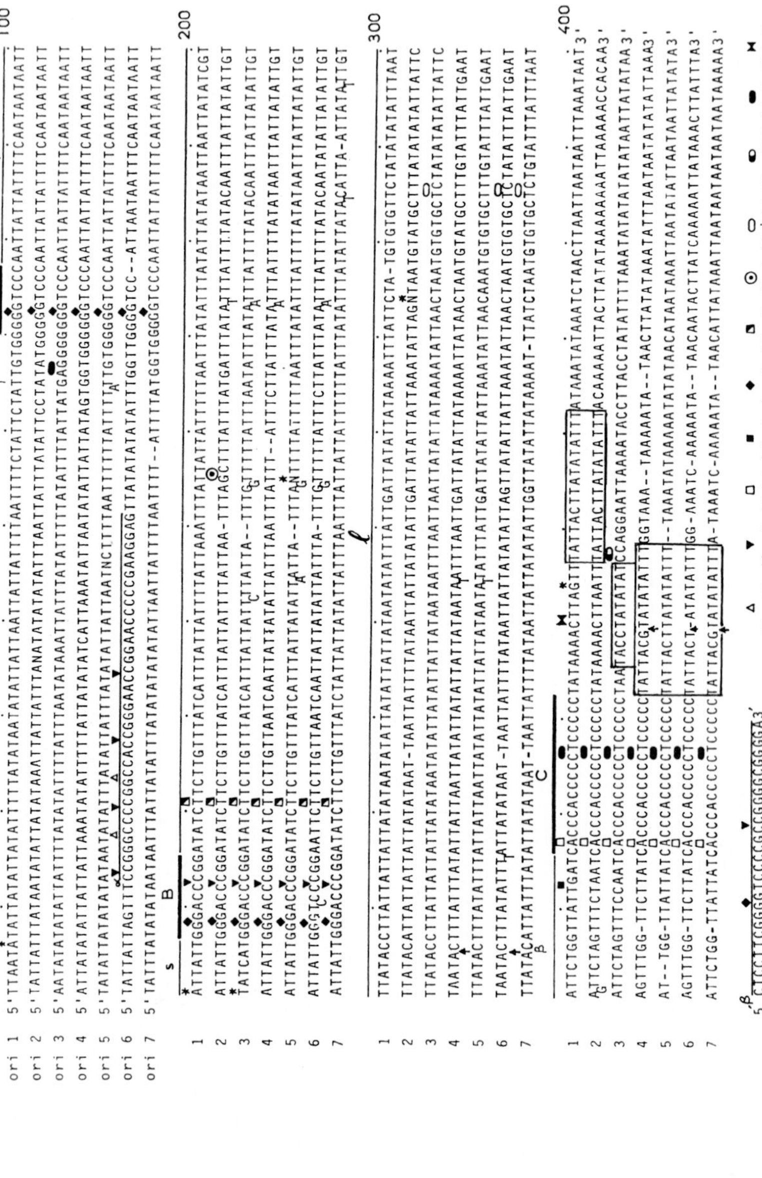
HaeIII
HpaII
HphI
MboI
AvaII
MboII
AluI
HgiAI
EcoRII
MnlI
DdeI

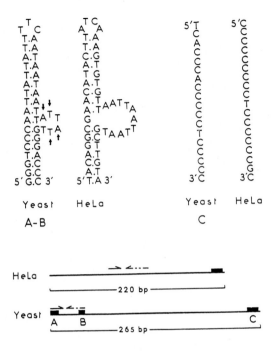

Fig. 4. *Comparison of ori sequences of mitochondrial genomes from yeast (de Zamaroczy et al., 1981) and Hela cells (Crews et al., 1979). Homology of potential secondary structure is found for the inverted repeats in cluster A - cluster B region; arrows indicate the base changes found in this region in different petite genomes. Homology of primary structure is found for cluster C. The bottom compares the two ori sequences; the arrows indicate the inverted repeats of the A-B region, the broken line corresponding to the looped-out sequence; bp, base pairs. (From de Zamaroczy et al., 1981).*

Fig. 3. *Primary structure of the ori sequences and their flanking sequences. Thick lines indicate GC clusters A, B, and C, thin lines AT regions p, s and l. The positions and the sequences of extra GC clusters β and γ are given (as well as that of cluster α, which is located outside the ori sequence). Restriction sites are indicated by the symbols shown. The sequences homologous to initiation transcription sites at the right of cluster C, are indicated by boxes. (From de Zamaroczy at al., 1981; Baldacci and Bernardi, 1982; and unpublished results of M. de Zamaroczy).*

potential secondary fold with two sequences ATAG and GGAG inserted in AT spacers; these sequences are followed by two AT base pairs, a GC stem (broken in the middle and in most cases also near the base, by non-paired nucleotides) and a terminal loop (Fig.5). This structure is reminiscent of that of GC clusters A and B from canonical *ori* sequences (Fig.4). Like the latter, *ori*^{S} sequences are present in both orientations, are located in intergenic regions and can be used as excision sequences when tandemly oriented. *Ori*^{S} sequences are homologous with many other subsets of GC clusters (one of these subsets, the *ori*^{S}-like sequences, is shown in Fig.5) some or all of which might perhaps also act as surrogate origins of replication, possibly still less efficiently than *ori*^{S} sequences.

THE REPLICATION OF PETITE GENOMES AND THE PHENOMENON OF SUPPRESSIVITY

A functional evidence that *ori* sequences were indeed involved in the replication of the mitochondrial genome came from observations on the outcome of crosses of spontaneous petites, characterized in their mitochondrial genome and their suppressivity, with wild-type cells (de Zamaroczy *et al.*, 1979, 1981; Goursot *et al.*, 1980). Crosses of spontaneous, highly suppressive petites having mitochondrial genomes formed by very short repeat units (400-900 base pairs) with wild-type cells produced diploids only harboring the unaltered mitochondrial genomes of the parental petite, which was called supersuppressive (de Zamaroczy *et al.*, 1979; Goursot *et al.*, 1980). If petites having different degrees of suppressivity were used in the crosses, again the genomes of diploid petites issued from the crosses had restriction maps identical with those of parental haploid petites, the very few exceptions corresponding to new excision processes affecting either parental genome.

A closer look at the results revealed two clear correlations between the *ori* sequence of the petite used in the cross and its degree of suppressivity. The first one was that, all other things being comparable (namely, the intact state of the *ori* sequence and the total amount of mitochondrial DNA per cell), the lower the overall density of *ori* sequences on the genome units, the lower the suppressivity. The second one (see Table 1) was that partial deletions of the *ori* sequences, their rearrangement and their absence affect the suppressivity of the petites carrying them in their mitochondrial genomes: a) *Ori*⁻ petites, which show a

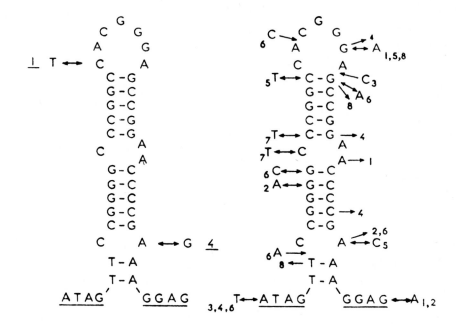

Fig. 5. *Potential secondary structure of (a) the* ori[S]
sequences (b) the ori[S]*-like sequences. All sequences are*
drawn in the same orientation ATAG → GGAG. Double-headed
arrows indicate base exchanges, arrows pointing towards, or
away from, the structure indicate insertions and deletions,
respectively. Numbers indicate the ori[S]*, or* ori[S]*-like,*
sequences presenting these changes. (From Goursot et al.,
1982).

partial deletion of the *ori* sequence show a decreased
suppressivity relative to *ori*[+] petites carrying intact *ori*
sequences, the loss of cluster C with its flanking sequence
having a much more dramatic effect that the loss of cluster
A; b) *ori*[r] petites, which show an inverted orientation of two
ori sequences within the same repeat units (the latter having,
in turn, an alternate inverted and tandem orientation) have a
very low degree of suppressivity; c) *ori*[o] petites, which lack
the *ori* sequence altogether but contain *ori*[S] sequences
instead, have low to minimal degrees of suppressivity.

These results provide a molecular basis for a replicative
advantage being the explanation for suppressivity. This
hypothesis can be traced back to Ephrussi *et al.* (1966) but
received a particular attention (Rank 1970 a,b) after the
work of Mills *et al.* (1967) on the *in vitro* replication of
Qβ RNA. It is quite possible that a replicative advantage

TABLE 1. Replication and Transcription of Petite Genomes

Petite class	Particular petite	Suppressivity[a] (%)	Transcription[d]
Ori[+]	ori 1	> 95	+
	ori 2	> 95	+
	ori 3	85	±
	ori 4	b c	−
	ori 5	90	+
	ori 6	b	n.d.
	ori 7	b	n.d.
Ori[−]	ori 1 A[−]	80	+
	ori 1 C[−]	n.d.	−
	ori 3 C[−]	< 5	−
Ori°	a-15/4/1/10/3	∿ 1	−
	a-3/1/B4	c	−
Ori[r]		< 5	n.d.

[a]For ori[+] petites suppressivity values found for petite genomes having repeat units ∿ 900 (ori 1, 2) or 1800 (ori 3, 5) base pairs long.

[b]Ori 4 was only found once, ori 6 and 7 were never found alone in the extensive screenings of spontaneous petite genomes.

[c]Diploid.

[d]Not determined.

also is the explanation for the phenomenon of amplification of the original monomeric genomes of petites.

THE *ORI* SEQUENCES AS TRANSCRIPTION INITIATION SITES

Two recent results (Baldacci and Bernardi, 1982) on the transcription initiation sites in petite genomes should be mentioned here because of their relevance for some of the issues discussed above. The first one is that transcription initiation efficiency parallels replication efficiency. Petite genomes containing some canonical *ori* sequences (*ori* 1, 2, 5 and, to a lesser extent, *ori* 3) are transcribed very actively; other ones, containing *ori* 4, or deleted in their

C clusters (but not those deleted in A clusters), or lacking
canonical *ori* sequences (*ori*° petites), are not. These
results (Table 1) bear on two different points: a) different
ori sequences are not functionally equivalent to each other:
those containing a γ cluster (*ori* 4, 6 and 7; two of these
are in tandem with *ori* 3 and *ori* 2, respectively) probably
are not very efficient in DNA replication, as suggested by
the fact that they are very rarely or never found in extensive screenings of spontaneous petites and may even be absent
(*ori* 4) in some wild-type genomes; likewise, a petite genome
containing *ori* 4 is not transcribed; b) since transcriptionally active *ori* sequences (see above) are present in
both orientations on the wild-type genome, it is very likely
that both strands are transcribed, although the non-sense
strand might be transcribed in a poorer way (Baldacci and
Bernardi, 1982); this conclusion, supporting previous independent evidence (Li and Tzagoloff, 1979; Coruzzi *et al.*, 1981;
Linnane *et al.*, 1980; Beilharz *et al.*, 1982) puts the transcription of the mitochondrial genome of yeast in line with
that of the animal mitochondrial genome; similarly replication might proceed unidirectionally from some *ori* sequences
(possibly *ori* 2 and 5 for one strand, and *ori* 1 and 3 for the
other); if this is the case, the replication of the mitochondrial genome of wild-type yeast cells would be analogous
with the replication of unicircular dimers of the mammalian
mitochondrial genome (Clayton, 1982).

The second result is that transcription initiates next to
the oligo-pyrimidine stretch of cluster C, at a sequence
(Fig.3) showing a very great homology (Fig.6) with the transcription initiation sequences of rRNA genes (Osinga and Tabak,
1982), and proceeds in a cluster C to cluster A direction. As
already mentioned, the insertion of cluster γ in the middle
of this sequence (as in *ori* 4) or the loss of cluster C and
of this sequence (as in *ori* C⁻ petite genomes) is accompanied
by a loss of transcriptional activity. This suggests that
ori 2 and *ori* 5 might be among the initiation sites used for
transcribing the sense strand, *ori* 1 and *ori* 3 among those
used for the transcription of the other strand. A small
number of other sequences largely homologous to the transcription initiation sites of tRNA genes have also been found
and might play a role in the multipromotor transcription
(Levens *et al.*, 1981) of the mitochondrial genome of wild-
type cells, postulated years ago (Prunell and Bernardi,1977).

In conclusion, the molecular genetical investigations
outlined in this paper have provided answers to the questions
raised many years ago by the work of Ephrussi on suppressivity.
In fact, they have done more than this, since they have
opened a way to a fine analysis of replication, recombination

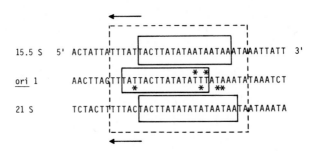

Fig. 6. Comparison of the transcription initiation sequences of ori 1 and of the 15.5 and 21s rRNA genes. Solid-line boxes indicate the transcription initiation sequences (as read on the coding strand to insure consistency with Fig. 3); the broken-line box indicates the region of homology among the three sequences; asterisks indicate base differences, the arrows the starts of tRNA transcripts. (From Baldacci and Bernardi, 1982.)

and expression in the mitochondrial genome of yeast, and have shed some light on the general problem of genome evolution (see Bernardi, 1982 for a recent discussion on the latter point).

REFERENCES

Baldacci G., and Bernardi G. (1982). *EMBO J. 1*, 987-994.
Beilharz M.W., Cobon G.S., and Nagley P. (1982) *Nucleic Acid Res. 10*, 1051-1070.
Bernardi G., Baldacci G., Bernardi G., Faugeron-Fonty G., Gaillard C., Goursot R., Huyard A., Mangin M., Marotta R., and de Zamaroczy M. (1980). *In* "The Organization and Expression of the Mitochondrial Genome" (Kroon A.M., and Saccone C., eds.), pp. 21-31, Elsevier/North Holland, Amsterdam.
Bernardi G. (1982). *In* "Mitochondrial Genes" (Slonimski P.P., Borst P., and Attardi G., eds.), pp. 269-278, Cold Spring Harbor Laboratory, New York.
Clayton D.A. (1982). *Cell, 28,* 693-705.
Coruzzi G., Bonitz S.G., Thalenfeld B.E., and Tzagoloff A. (1981). *J. Biol. Chem. 256,* 12782-12787.
Crews S., Ojala D., Posakony J., Nishiguchi J., and Attardi G. (1979). *Nature, 277,* 192-198.

Ephrussi B., Jacob H., and Grandchamp S. (1966). *Genetics,* *54,* 1-29.

Faugeron-Fonty G., Culard F., Baldacci G., Goursot R., Prunell A., and Bernardi G. (1979). *J. Mol. Biol. 134,* 493-537.

Gaillard C., and Bernardi G. (1979). *Mol. Gen. Genet. 174,* 335-337.

Gaillard C., Strauss F., and Bernardi G. (1980). *Nature, 283,* 218-220.

Gillum A.M., and Clayton D.A. (1979). *J. Mol. Biol. 135,* 353-368.

Goursot R., de Zamaroczy M., Baldacci G., and Bernardi G. (1980). *Current Genet. 1,* 173-176.

Goursot R., Mangin M., and Bernardi G. (1982). *EMBO J. 1,* 705-711.

Levens D., Ticho B., Ackerman E., and Rabinowitz M. (1981). *J. Biol. Chem. 256,* 5226-5232.

Li M., and Tzagoloff A. (1979). *Cell, 18,* 47-53.

Linnane A.W., Astin A.M., Beilharz M.W., Bingham C.G., Choo W.M., Cobon G.S., Marzuki S., Nagley P., and Roberts H. (1980). *In* "The Organization and Expression of the Mitochondrial Genome." (Kroon A.M., and Saccone C., eds.), pp. 253-263, Elsevier/North Holland, Amsterdam.

Mills D.R., Peterson R.L., and Spiegelman S. (1967). *Proc. Natl. Acad. Sci. USA, 58,* 217-224.

Osinga K.A., and Tabak H.F. (1982). *Nucleic Acid Res. 10,* 3617-3626.

Prunell A., and Bernardi G. (1977). *J. Mol. Biol. 110,* 53-74.

Rank G.H. (1970a). *Can. J. Genet. Cytol. 12,* 129-136.

Rank G.H. (1970b). *Can. J. Genet. Cytol. 12,* 340-346.

de Zamaroczy M., Baldacci G., and Bernardi G. (1979). *FEBS Letters, 108,* 429-432.

de Zamaroczy M., Marotta R., Faugeron-Fonty G., Goursot R., Mangin M., Baldacci G., and Bernardi G. (1981). *Nature, 292,* 75-78.

de Zamaroczy M., Faugeron-Fonty G., and Bernardi G. (1982). Submitted for publication.

46

Occurrence of a Super-abundant Petite Genome in Different *mit⁻* Mutants of Yeast

Christopher G. Bingham and Phillip Nagley

Department of Biochemistry, Monash University
Victoria, Australia

The *mit⁻* strain Mb12 of *Saccharomyces cerevisiae* carries a mutation in the *oli2* gene (Roberts *et al.*, 1979). We have recently observed that strain Mb12 produces exceptionally high frequencies of a particular type of petite mutant (Bingham and Nagley, submitted). About 70% of cells in Mb12 cultures are petite mutants; a mtDNA segment 880 bp in length with a single MboI site was found to be present in 80% of these spontaneous petites. These petites all showed high suppressiveness due to the *oril* replication origin present in this mtDNA segment (cf. de Zamoroczy *et al.*, 1981). Following digestion of mtDNA from cultures of Mb12 with MboI and subsequent analysis on agarose gels, an intense 880 bp band was seen to be superimposed on the pattern of MboI fragments characteristic of mtDNA of the parent wild-type strain J69-1B. The 880 bp band did not correspond to a wild-type fragment.

We show here that production of this super-abundant petite genome is not due to a change peculiar to the *mit⁻* mutant Mb12. Gel analysis of MboI digested mtDNA was used to screen other *mit⁻* mutants derived from J69-1B for the presence of this super-abundant petite mtDNA genome. The results (Table I) show that a superstoichiometric 880 bp MboI fragment is found in many *mit⁻* mutants affected in three different loci controlling the mito-chondrial ATPase *(oli1, oli2, aap1;* see Novitski *et al.*, this volume). Not all *oli2 mit⁻* mutants, however, show this pattern. Some contain the 880 bp fragment at substoichiometric levels whilst others do not show it at all (Table I). In the case of the parent *rho⁺* strain J69-1B, we have shown that although the 880 bp band does not appear on the stained gel, it can neverthe-less be detected by probing Southern blots of the MboI digest with ³²P-labelled 880 bp band DNA (Bingham and Nagley, submitted). The low representation of this 880 bp band in total J69-1B mtDNA is in part due to the low petite frequency (∿ 5%) in cultures of J69-1B. Cultures of mutant Ma30 contain 70% petites, but Ma30

MANIPULATION AND EXPRESSION
OF GENES IN EUKARYOTES
ISBN 0 12 513780 x

TABLE I.   *Appearance of 880 bp fragment in MboI digests of mtDNA from* mit⁻ *mutants, analysed on 2% agarose gels stained with ethidium bromide.*

Intensity of 880 bp fragment in stained gel	Strain	Mutant Locus	Petite Frequency (%)
Superstoichiometric	M26-11	oli2	89
	Mb12	oli2	70
	M18-5	oli2	46
	M31	aap1	60
	M26-10	aap1	56
	M68-5	aap1	38
	5102	oli1	54
	51223	oli1	49
	2242	oli1	43
Substoichiometric	C58	oli2	34
	M13-20	oli2	43
	M11-28	oli2	8
Not detectable	Ma30	oli2	70
	M9-36	oli2	18
	2008	oxi2	26
	M21	oxi3	13
	J69-1B	–	5

mtDNA fails to show the 880 bp band; therefore this petite genome must be present, at most, in very low amounts in this *oli2 mit⁻* strain.

The two *mit⁻* mutants we have tested carrying mutations affecting cytochrome oxidase *(oxi2, oxi3)* do not show this 880 bp band, but more mutants need to be screened to enable us to decide whether the production of this super-abundant petite mtDNA genome is related specifically to *mit⁻* mutants with altered mtATPase function.

REFERENCES

Roberts H., Choo W.M., Murphy M., Marzuki S., Lukins H.B., and Linnane A.W. (1979). *FEBS Lett.* 108, 501-504.

de Zamoroczy M., Marotta R., Faugeron-Fonty G., Goursot R., Mangin M., Baldacci G., and Bernardi G. (1981). *Nature* 292, 75-78.

47
Transformation of Yeast with Mitochondrial DNA

Phillip Nagley, Paul R. Vaughan, Sam W. Woo
and Anthony W. Linnane

Department of Biochemistry, Monash University,
Victoria, Australia

The ability to transform yeast with mitochondrial DNA
(mtDNA) can provide new approaches for genetic and molecular
analyses of the structure and expression of the mitochondrial
genome. In our laboratory we have been studying various
strategies for the uptake of mtDNA by yeast cells, and this
paper will review our recent studies in this area. In the
first instance we used genetic tests to detect changes in the
mitochondrial genome of the recipient cells; this has led to
the subsequent molecular demonstration of the presence of
mtDNA in transformed cells.

SHOULD MITOCHONDRIAL DNA BE COVALENTLY LINKED TO A VECTOR?

We initially reasoned (Atchison et $al.$, 1980; Nagley et
$al.$, 1980) that to obtain effective transformation, the mtDNA
should be ligated to a yeast DNA molecule that replicated
autonomously in yeast cells. The yeast 2μ circular plasmid
was chosen, as its properties of autonomous replication had
been exploited for the successful cloning of yeast nuclear
genes (Beggs, 1978; Struhl et $al.$, 1979; Broach et $al.$, 1979).
A ligation mixture containing the following components was
prepared: (i) a linearized segment of mtDNA, namely an 8.6
kilobase (kb) DNA fragment from petite 70M-J (obtained by
PstI digestion) in which the $oli2$ gene (Novitski et $al.$,1982)
and an associated oligomycin resistance determinant was
located; (ii) 2μ plasmid DNA, linearized at its unique PstI
site. This mixture, treated with DNA ligase, was used to transform spheroplasts of yeast strain X4005-11A [rho^+ oli^S],

293

TABLE I. *Transformation of grande strain X4005-11A a leu2 met5 [rho⁺ oliˢ] to oligomycin resistance.*

DNA used in transformation	oli^R cells/10^6 rho⁺ cells	
	Experiment 1	Experiment 2
None	5.9	2.8
HD1-B linear + 2μ	25.3	10.9
HD1-B self-ligated + 2μ	29.5	27.0
HD1-B linear	-	3.2
HD1-B self-ligated	-	4.9
2μ	7.4	4.2
pSW1	7.9	-
pSW22	-	5.1
YEp13	4.9	-

to produce significant numbers of oligomycin resistant transformants. Cytoplasmic inheritance of oligomycin resistance was shown for most of the transformants (Atchison *et al.*,1980).

As the 2μ plasmid probably replicates in the nucleus (see Broach, 1981) we have now re-examined our assumption that the active DNA element in the transformation was recombinant molecules consisting of mtDNA ligated to 2μ DNA. For this purpose we prepared a mtDNA segment carrying the *oli2* gene: a 2.2 kb HaeIII fragment derived from petite DS14 (Macino and Tzagoloff, 1980). This fragment was joined to BamHI linkers for purposes of manipulation in various plasmids, such as pBR322. The free fragment with BamHI ends is called HD1-B. Table I shows that when HD1-B was mixed with 2μ circles (without ligation) increased levels of oligomycin resistant cells in strain X4005-11A were found, some 3-4 fold above the background level of spontaneous mutants. Little difference in the transformation frequency was observed if HD1-B was self-ligated to generate circular mtDNA molecules, before mixing with 2μ DNA. Neither mtDNA alone (linear or self-ligated), nor 2μ DNA alone, gave a significant increase in the frequency of oligomycin resistant cells (Table I,Experiment 2).

We also examined the behaviour in this system of mtDNA covalently linked to 2μ DNA. A recombinant plasmid pSW1 was prepared by introducing mtDNA fragment HD1-B into the BamHI site of YEp13, which is a yeast vector carrying pBR322 sequences, portions of 2μ plasmid, and the yeast nuclear *LEU2* gene (Broach *et al.*, 1979). The BamHI site of YEp13 lies within the pBR322 segment. In studies with strain X4005-11A, plasmid pSW1 failed to yield oligomycin resistant transformants above the background (Table I, Experiment 1). Indeed, individual Leu⁺ transformants of X4005-11A carrying pSW1 showed no increase in the frequency of oligomycin resistant cells compared to Leu⁺ cells carrying YEp13 or X4005-11A not transformed at all (data not shown). We prepared a further recombinant plasmid (pSW22) in which the mtDNA was ligated within 2μ sequences themselves. This plasmid was prepared by ligating the 2.2 kb HaeIII fragment (the *oli2* segment of DS14 without BamHI linkers) into the HpaI site of the 2μ portion of pSW8, which consists of the complete 2μ plasmid inserted at its PstI site into pBR322. Plasmid pSW22 did not transform X4005-11A to oligomycin resistance (Table I, Experiment 2).

Plasmids such as pSW1 and pSW22 contain 2μ sequences which constrain them to replicate with the resident 2μ population of the recipient cells, probably in the nucleoplasm (Broach,1981). This same constraint on the mtDNA sequences included in these plasmids prevents the genetic effects occurring which are seen when cotransformation mixtures of mtDNA and 2μ DNA are used. The reason why one must use mixtures of mtDNA and 2μ DNA to produce genetic changes in the recipient cells are not clear. The cotransformation strategy is expanded upon in subsequent sections of this article.

A STRATEGY FOR COTRANSFORMATION USING HOSTS THAT LACK mtDNA

The use of host strains for cotransformation that initially lack mtDNA [*rho*°] provides a greater scope for genetic analysis of transformants. The transformants may also be examined for the newly acquired mtDNA. We have used *rho*° strains which also carry *leu2* mutations in the nuclear genome. Such cells are cotransformed with a mixture of mtDNA and the plasmid YEp13 which is used to indicate those cells which are competent for transformation. The Leu⁺ transformants which arise are tested for concomitant uptake of mtDNA by genetic and biochemical procedures.

MARKER RESCUE OF MIT⁻ TESTER STRAINS BY COTRANSFORMANTS

The rationale behind the experiments described in this section is that informational mtDNA taken up by *rho*° recipient cells can be used to rescue, in crosses, the small lesion in *mit*⁻ tester strains so as to generate respiratory competent diploid progeny. This is denoted a marker rescue event. In our first experiments (Vaughan *et al.*, 1980), mtDNA from petite DS14 carrying the *oli2* gene was mixed with YEp13, and Leu⁺ transformants of a *leu2* [*rho*°] strain were found to

TABLE II. Rescue of mit⁻ tester strains by Leu⁺ transformantsᵃ

DNA used in transformationᵃ	Tester mit⁻ strain		
	M18-5 (oli2)	2422 (oli1)	M21 (oxi3)
No DNA	4	0	1
YEp13	4	0	0
YEp13 + DS14	17	1	1
YEp13 + 23-3	4	6	0
pSW1	4	0	0
pSW3	3	0	1
YEp13 + DS14 (HaeIII)	14	1	0
YEp13 + 23-3 (HaeIII)	3	6	0

ᵃSpheroplasts of DC5-E2 a leu2 his3 [rho°] were incubated with the DNA indicated. Leu⁺ transformants were selected and mated to mit⁻ tester strains. Data show the number of respiratory competent diploid patches arising from 122 tested Leu⁺ haploids in each case, or from 122 single clones of untreated DC5-E2.

rescue *oli2 mit⁻* testers. An example of the type of data obtained in such an experiment is shown in Table II. Rescue of tester strain M18-5 occurred after crosses to 17 Leu⁺ transformants made with DS14 mtDNA and YEp13. The background frequency of *rho⁺* occurrence (4 events scored) in control crosses to DC5-E2 (not transformed) or in crosses to DC5-E2 transformed to Leu⁺ with YEp13 is considered to be due to the spontaneous reversion of the tester, M18-5. No rescue of *mit⁻* testers with mutations in the *oli1* gene (2422) or *oxi3* gene (M21) took place with Leu⁺ transformants made with DS14 mtDNA and YEp13. By contrast Leu⁺ transformants made with a cotransforming mixture of mtDNA from petite 23-3 (carrying the *oli1* gene) with YEp13 did lead to rescue of the *oli1 mit⁻* tester strain (2422) but failed to rescue the *oli2* or *oxi3* testers (Table II). This result suggests that there is a regional specificity of rescue by cotransformants.

It would be predicted from the data presented in Table I concerning the recombinant plasmid pSW1 (YEp13 carrying the *oli2* gene) that mtDNA covalently inserted into YEp13 would be incapable of eliciting the genetic changes observed in cotransformants. To test this prediction the plasmid pSW1 and a cognate plasmid pSW3 (carrying a covalent insert of the *oli1* gene in YEp13) were used to generate Leu⁺ transformants of DC5-E2. Transformants carrying these plasmids did not lead to marker rescue events above the background for any *mit⁻*

tester with which they were crossed (Table II). Control cotransformants made with HaeIII digests of DS14 mtDNA or of
23-3 mtDNA (in each case containing the DNA fragment that had
been cloned into YEp13) produced about the same frequency of
marker rescue as when the parent undigested mtDNA had been
used in cotransformation with YEp13 (Table II).

Further features of the marker rescue pattern manifested
by cotransformants have been defined by Vaughan *et al.* (1980).
The ability to rescue markers is an unstable character: on
subsequent retesting some cotransformants no longer rescue the
mit⁻ tester strains they previously had done on the initial
screening. The pattern of marker rescue is also variable;
that is, some cotransformants rescue different *mit*⁻ testers
from those observed to be rescued previously. This instability
and variability was particularly evident when subclones of
cotransformants were tested. Indeed, of some parent transformant clones which were found not to rescue any of the *mit*⁻
testers used at the first screening, the subclones were found
to rescue several of the *mit*⁻ testers (Vaughan *et al.*, 1980).
These observations indicate that the properties of the cotransformants are different from the properties expected if they
contained mtDNA established in mitochondria like the mtDNA of
rho⁻ petites. In the case of *rho*⁻ petites, the crosses with
mit⁻ testers show rescue for those *mit*⁻ strains whose
mutations lie within the segment of the wild type mtDNA
genome retained by the petite. Where the *mit*⁻ mutation lies
outside the mtDNA genome of the *rho*⁻ petite, rescue does not
take place. The difference between *rho*⁻ petites and the cotransformants analysed here is seen most strikingly where
mtDNA fragments carrying incomplete portions of the *oli2* gene
are used in cotransformation with YEp13. The data in Table III
indicate the rescue of *mit*⁻ tester strains carrying mutations
outside the cotransforming mtDNA segment. This was achieved
by use of two adjacent restriction fragments from a petite G4
which itself retains an incomplete portion of the *oli2* gene
(Linnane *et al.*, 1980; Novitski *et al.*, 1982).

The mtDNA from petite G4 is cut into two fragments with
MboI. One fragment, 380 bp in length, lies wholly within the
oli2 gene; the other fragment is 1200 bp in length. These
restriction fragments were purified after agarose gel electrophoresis of MboI digests of petite G4 mtDNA and each used
in cotransformation (with YEp13) of host strain DC5-E2. It is
seen in Table III that both fragments of G4 rescue the same
set of *mit*⁻ testers. It is highly unlikely that any one
mutation lies in sequences common to both fragments of G4.
The testers rescued by these cotransformants include those
rescued by petite G4 itself (Group A, Table III), as well as

TABLE III.   Use of adjacent restriction fragments from petite G4 to demonstrate that marker rescue can occur outside transforming mtDNA segment[a]

mit⁻ tester strain	Mutant group[b] (mit⁻)	DNA used in transformation		
		YEp13 + 380bp MboI fragment	YEp13 + 1200 bp MboI fragment	YEp13 alone
M27-13	A	12	7	0
M27-14	A	11	10	0
M5-32	A	12	7	0
M30-01	A	10	19	0
M11-28	A	2	3	0
M14	A	6	5	0
M26-11	B	29	16	0
M10-7	B	14	11	0
M26-10	C	8	7	0

[a]122 Leu⁺ transformants of DC5-E2 were tested in each case.
[b]Mutant groups (mit⁻) are as follows: A, mutants rescued after crosses to petite G4; B, mutants in the oli2 gene not rescued by petite G4; C, mutants in the aap1 gene (also not rescued by petite G4).

those not rescued by petite G4, carrying mutations both within the *oli2* gene (Group B) and outside the *oli2* gene in the *aap1* gene that lies some 500 bp upstream of *oli2* (Novitski *et al.*, 1982).   This rescue of markers outside the region of homology with the transforming mtDNA is not however a general phenomenon common to all *mit⁻* testers.   The results in Table II showed a regional specificity of marker rescue in that *oli2* mtDNA segments did not rescue *oli1 mit⁻* testers and *vice versa*.   We have further tested cotransformants of DC5-E2 made with DS14 (*oli2*) and YEp13 for their ability to rescue many *mit⁻* testers, each mutated in one of seven distinct genetic loci.   Apart from *oli2* testers and M26-10 (*aap1*), no rescue was observed for 40 *mit⁻* testers in five other loci (data not shown).

What we suggest happens in the case of the *oli2* gene and the *aap1* gene is that there is an increased mutation frequency in the mtDNA genome of the *mit⁻* tester strains after crosses to cotransformants; such mutations can lead to the restoration of respiratory competence.   This is consistent with the marker rescue patterns observed above, in that the rescue is variable and not all *mit⁻* testers in a group are rescued at one screening (see Vaughan *et al.*, 1980).   We consider that the regional specificity of marker rescue is dependent upon

homologous interactions between the transforming mtDNA and the
mit⁻ tester mtDNA. A possible explanation can be proposed for
the rescue of markers in a limited region defined by this
homologous segment and sequences in its immediate vicinity.
Establishment of recombinant structures may be initiated
through the invasion of the *mit⁻* mtDNA by 3' ends of linear
transforming mtDNA segments (or from the 3' end of a strand
found at a nick in this transforming mtDNA) followed by repair
synthesis involving strand displacement as described by
Orr-Weaver *et al*. (1981) for interaction between linear fragments of nuclear genes and the host chromosomal sequences. The
introduction of mutations leading to correction of the *mit⁻*
mtDNA function may occur by error-prone repair polymerization
activity on the *mit⁻* mtDNA template strands, using transforming mtDNA as priming strand.

THE USE OF SUPPRESSIVENESS AS A TOOL FOR THE ANALYSIS OF
COTRANSFORMANTS

   The preceding studies describe the genetic consequences of
the cotransformation of yeast with mtDNA and an indicator
vector. We have not yet been able to define the location
within the haploid transformant of the introduced mtDNA
sequences. In order to enhance the ability of the introduced
mtDNA to replicate in the recipient *rho°* strain we have carried
out cotransformation using mtDNA from petites carrying a
mitochondrial replication origin. Such petites are highly
suppressive: they produce almost quantitatively petite diploid
cells after being mated to a *rho⁺* strain, as a result of the
very efficient replication of the petite mtDNA (see
Bernardi, 1982).
   In our experiments we used mtDNA from petite K45, a highly
suppressive petite carrying the *ori3* mtDNA replication origin
(de Zamaroczy *et al.*, 1981) in a 1.7 kb petite mtDNA genome
(P.R. Vaughan, unpublished). Strain LL20-El, a *rho°* petite
with zero suppressiveness (cf. Nagley and Linnane, 1970), was
cotransformed with a mixture of K45 mtDNA and YEpl3. The Leu⁺
transformants which arose were tested for suppressiveness. The
results (Fig.1) show that many Leu⁺ transformants have
acquired positive suppressiveness values, ranging up to 12% in
the particular group shown in Fig.1, panel A. The control Leu⁺
transformants made just with YEpl3 showed suppressiveness
values (Fig.1, Panel B) not significantly different from the
untransformed LL20-El (defined as having zero percent
suppressiveness; see legend to Fig.1).

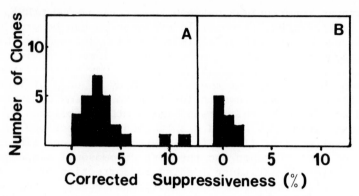

Fig. 1.   *Suppressiveness distribution of Leu⁺ cotransfor-mants made with K45 mtDNA and plasmid YEp13.   Strain LL20-E1 α leu2 his3 [rho°] was transformed with a mixture of K45 mtDNA and YEp13 (Panel A), or YEp13 alone (Panel B).   Leu⁺ transfor-mants were selected and mated to L2200 a ade1 lys2 trp1 [rho⁺] to determine the suppressiveness, as described by Nagley* et al. *(1973).   The suppressiveness values shown here were corrected for the observed percentage (1.9%) of petite zygote clones (diploids) in control crosses between untreated LL20-E1 cells and strain L2200.*

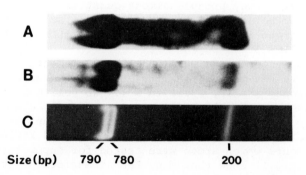

Fig. 2.   *Demonstration of mtDNA in transformant LYK45-12-1. Track A: Cellular DNA from clone LYK45-12-1 was digested with HpaII, run on a 2% agarose gel and probed with ³²P-labelled (nick translated) K45 mtDNA.   Track B: Cellular DNA from petite K45 treated as in Lane A.   Track C: Purified mtDNA from petite K45 digested with HpaII, run on an agarose gel and stained with ethidium bromide.*

We have analysed several series of cotransformants of
LL20-El made with K45 mtDNA and YEpl3. The maximal suppressiveness we observed was 34% in a clone denoted LYK45-12. Subclones of this cotransformant had suppressiveness values around
18-20% and these were tested for the presence of mtDNA. One
such subclone LYK45-12-1 was found to contain significant quantities of K45 mtDNA, detected by probing digests of cellular
DNA with labelled K45 mtDNA (Fig.2). It was estimated by DNA
titration tests, that LYK45-12-1 contained about 500 copies
of K45 mtDNA per cell, i.e. about one-fifth of that in the
K45 parent petite. The suppressiveness of LYK45-12-1 was 10%
at the time of DNA analysis, and this value declined further
over a period of weeks to undetectable suppressiveness levels.
The mtDNA was lost concomitantly with this decrease in
suppressiveness. Due to the inherent instability of transformants such as LYK45-12-1 we have not yet been able to
establish the subcellular location of the acquired mtDNA
of these transformants.

Current experiments are aimed at increasing the stability
of cotransformants carrying mtDNA sequences, and improving the
genetic selection systems to identify more rapidly the cotransformants carrying mtDNA.

REFERENCES

Atchison B.A., Devenish R.J., Linnane A.W., and Nagley P.
    (1980). *Biochem. Biophys. Res. Commun. 96*, 580-586.
Bernardi G., Baldacci G., Colin Y., Faugeron-Fonty G.,
    Goursot R., Goursot R., Huyard A., Le Van Kim C.,
    Mangin M., Marotta R., and de Zamaroczy M. (1982). *In*
    "Manipulation and Expression of Genes in Eukaryotes".
    (P. Nagley, A.W. Linnane, W.J. Peacock, and J.A. Pateman,
    eds.), Academic Press, Australia (this volume).
Beggs J.D. (1978). *Nature 275*, 104-109.
Broach J.R. (1981). *In* "Molecular Biology of the Yeast
    Saccharomyces: Life Cycle and Inheritance" (J.N. Strathern,
    E.W. Jones, and J.R. Broach, eds.), pp. 445-470,
    Cold Spring Harbor Laboratory, Cold Spring Harbor,
    New York.
Broach J.R., Strathern J.N., and Hicks J.B. (1979). *Gene 8*,
    121-133.
Linnane A.W., Astin A.M., Beilharz M.W., Bingham C.G.,
    Choo W.M., Cobon G.S., Marzuki S., Nagley P., and
    Roberts H. (1980). *In* "Organization and Expression of
    the Mitochondrial Genome." (A.M. Kroon and C.Saccone, eds.),
    pp. 253-263, Elsevier/North-Holland, Amsterdam.

302   *Phillip Nagley et al.*

Macino G., and Tzagoloff A. (1980). *Cell 20*, 507-517.
Nagley P., Atchison B.A., Devenish R.J., Vaughan P.R., and
    Linnane A.W. (1980). *In* "Organization and Expression of
    the Mitochondrial Genome". (A.M. Kroon and C. Saccone,
    eds.), pp. 75-78, Elsevier/North-Holland, Amsterdam.
Nagley P., Gingold E.B., Lukins H.B., and Linnane A.W. (1973).
    J. Mol. Biol. 78, 335-350.
Nagley P., and Linnane A.W. (1970). *Biochem. Biophys. Res.
    Commun. 39*, 989-996.
Novitski C.E., Macreadie I.G., Maxwell R.J., Lukins H.B.,
    Linnane A.W., and Nagley P. (1982). *In* "Manipulation
    and Expression of Genes in Eukaryotes". (P. Nagley,
    A.W. Linnane, W.J. Peacock, and J.A. Pateman, eds.),
    Academic Press, Australia (this volume).
Orr-Weaver T.L., Szostak J.W., and Rothstein R.J. (1981).
    Proc. Natl. Acad. Sci. USA 78, 6354-6358.
Struhl K., Stinchcomb D.T., Scherer S., and Davis R.W. (1979).
    Proc. Natl. Acad. Sci. USA 76, 1035-1039.
Vaughan P.R., Woo S.W., Novitski C.E., Linnane A.W. and
    Nagley P. (1980). *Biochem. Int. 1*, 610-619.
de Zamaroczy M., Marotta R., Faugeron-Fonty G., Goursot R.,
    Mangin M., Baldacci G., and Bernardi G. (1981). *Nature 292*,
    75-78.

48

Multiple "Start-Transcription" Sites in *Torulopsis glabrata* Mitochondrial DNA

K. S. Sriprakash and G. D. Clark-Walker

Genetics Department
Research School of Biological Sciences
Australian National University
Canberra, Australia

The sizes of mitochondrial DNA (mtDNA) from different petite positive yeasts vary from 18.9 Kbp to 107.6 Kbp (Clark-Walker *et al.*, 1981). The diversity in length of mtDNA is largely due to presence or absence of intragenic sequences (Clark-Walker and Sriprakash, 1981). A comparison of the mitochondrial genome organization in four different yeasts *Torulopsis glabrata* (18.9 Kbp), *Saccharomyces exiguus* (23.7 bp), *Kloeckera africana* (27.1 Kbp) and *Saccharomyces cerevisiae* (75 Kbp) revealed, in each case that although the relative positions of mitochondrial genes in different yeasts are not the same, the genes are transcribed from one strand. The only exception to this observation is a threonyl tRNA gene in *S. cerevisiae* mtDNA (Li and Tzagoloff, 1979). This constraint in the orientation of mitochondrial genes in yeasts could be explained by either (a) there is a unique promotor from which a polycistronic RNA is transcribed or (b) there are multiple start-transcription sites which are recognized as the DNA moves through a membrane bound RNA polymerase. Our studies communicated in this paper are aimed at resolving these possibilities by anlysing transcription products in *T. glabrata* mitochondria. This yeast was chosen because of (1) compactness of the mitochondrial genome, (2) the ease with which stable mtRNA could be isolated, (3) the ability to obtain mitochondrial petite mutants, and lastly, repressibility of mitochondrial expression by glucose.

MANIPULATION AND EXPRESSION
OF GENES IN EUKARYOTES
ISBN 0 12 513780 x

303

I.   TRANSCRIPTION PRODUCTS OF *T. GLABRATA* mtDNA

Mitochondrial RNAs were glyoxalated, run on agarose gels (McMaster and Carmichael, 1977) transferred to nitro-cellulose (Thomas, 1980) and hybridized to labelled DNA from petite strains retaining defined regions of wild-type *S. cerevisiae* mtDNA.   We have identified by this method transcripts for cytochrome oxidase subunits I, II and III, cytochrome b, and ATPase subunits 6 and 9 (Fig. 1). Precursors to cytochrome oxidase subunit I and II were also detected.   ATPase subunit 9 probe hybridizes to two RNA species, the larger species is due to sequence homology between ATPase subunit 9 and cytochrome oxidase subunit I and in fact homology between these two genes in *S. cerevisiae* has been detected by sequence comparison. Inclusion of excess unlabelled petite mtDNA retaining the cytochrome oxidase I sequence eliminated labelling of the larger RNA (not illustrated).

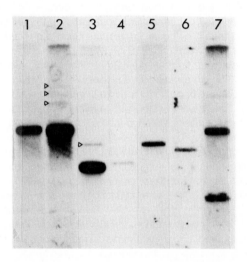

FIGURE 1.   Nitrocellulose blots of RNA separated on agarose gels were hybridised to DNA probes containing the sequences for cytochrome oxidase subunit I (1 and 2, short and long exposures respectively); subunit II (3); subunit III (4); cytochrome b (5); ATPase subunit 6 (6); and subunit 9 (7). Open triangles show the precursors.

Northern blot experiments presented above, and S1 endonuclease protection experiments (data not shown) did not reveal the presence of joint transcripts even for adjacent genes, suggesting independent "start-transcription" sites for each of the RNAs.

II. IDENTIFICATION OF PRIMARY TRANSCRIPTS BY
     GUANYLYL TRANSFERASE (CAPPING ENZYME)

Guanylyl transferase catalysed addition of GMP to the 5' polyphosphate termini of nacent RNA molecules, has been used

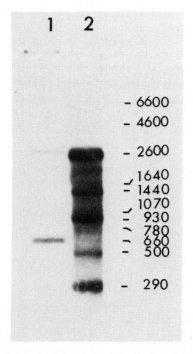

FIGURE 2. Electrophoretic pattern of in vitro capped RNA. (1) cytoplasmic RNA from petite mutant lacking mitochondrial DNA and (2) RNA from mitochondria enriched fractions from wild-type cells. The values on the side of the figure represent sizes (in bases) of the transcripts.

to identify primary transcripts in *S. cerevisiae*
mitochondria (Levens *et al.*, 1981). When this technique was
applied to mtRNA isolated from glucose repressed *T. glabrata*
we found at least 11 discrete sized RNAs which were
"cappable" (Fig. 2). Of these, 9 were of mitochondrial
origin while two or possibly three were derived from the
cytoplasm as they are present in a petite mutant lacking
mtDNA. Thus, the presence of several "cappable" RNA species
strongly favour multiple initiation sites for RNA
synthesis. Subsequent studies involving sequence analysis
of isolated capped transcripts should help to establish this
view.

REFERENCES

Clark-Walker, G.D., McArthur, C.R. and Daley, D.J.
    (1981). *Current Genetics 4*, 7.
Clark-Walker, G.D. and Sriprakash, K.S. (1981).
    J. Mol. Biol. 151, 367.
Levens, D., Ticho, B., Ackerman, E. and Rabinowitz,
    M. (1981).   *J. Biol. Chem. 256*, 5226.
Li, M. and Tzagoloff, A. (1979).   *Cell 18*, 47.
McMaster, G.K. and Carmichael, G.G. (1977).
    Proc. Natl. Acad. Sci., U.S.A., *74*, 4835.
Thomas, P.S. (1980).   *Proc. Natl. Acad. Sci.*,
    U.S.A. 77, 5201.

Identification of Sequence Homologies between Maxicircle DNA of *Leishmania tarentolae* and Specific Yeast Mitochondrial Structural Genes

Terence W. Spithill,[1] Agda M. Simpson and Larry Simpson

Biology Department and Molecular Biology Institute
University of California
Los Angeles, California

INTRODUCTION

The kinetoplast DNA of the hemoflagellate protozoan Leishmania tarentolae is composed of two molecular species, minicircles (~0.9 kilobases, kb) and maxicircles (30 kb), catenated together to form a giant network of mitochondrial DNA (mtDNA) (Borst and Hoeijmakers, 1979). The maxicircle DNA appears to represent the equivalent of the informational mtDNA found in other cells. This DNA is transcribed into two abundant RNA species of size 9S (520 b) and 12S (1,020 b) (Simpson and Simpson, 1978) and similar RNAs exhibiting a high degree of sequence conservation have been observed in several kinetoplastid species (Simpson and Simpson, 1980; Simpson et al., 1980) suggesting that these RNAs represent the presumptive ribosomal RNAs in kineto-plastida (Borst and Hoeijmakers, 1979; Simpson et al., 1980). In addition, several less abundant polyadenylated transcripts of maxicircle DNA, ranging in size from 0.1 to 1.8 kb, have been described for L.tarentolae (Simpson et al., 1980; Simpson et al., 1982b) and Trypanosoma brucei

[1]Present address: The Walter and Eliza Hall Institute of Medical Research, Parkville, Australia.

MANIPULATION AND EXPRESSION
OF GENES IN EUKARYOTES
ISBN 0 12 513780 x

(Hoeijmakers et al., 1981) suggesting that maxicircle DNA
may also specify information for several mitochondrial
structural genes.

An intriguing feature of mitochondrial genomes studied
to date has been the retention in mtDNAs of a limited number
of structural genes whose amino acid and to a lesser extent
nucleotide sequences have been predominantly conserved
although the physical arrangement of these structural genes
within the genome has varied (Tzagoloff et al., 1979;
Anderson et al., 1981; Clark-Walker and Sriprakash, 1981).
The conservation of nucleotide sequences in mitochondrial
genes has allowed the rapid mapping of genes in heterologous
mitochondrial genomes using yeast mtDNA sequences as hybridization probes (Macino et al., 1980).

In an attempt to further define maxicircle DNA structure
and function, we have similarly used yeast mitochondrial
structural gene fragments as probes to locate homologous
sequences in the maxicircle DNA of L.tarentolae. Although
these results provide tentative gene localizations, they
represent the first indication that maxicircle DNA encodes
the structural genes for cytochrome b, ATPase subunit 6 and
cytochrome oxidase subunits I and II (Simpson et al.,
1982a,c).

RESULTS AND DISCUSSION

Maxicircle DNA from L.tarentolae, linearized at the
unique EcoRI site (RIMaxi DNA), was isolated from kinetoplast DNA networks in CsCl gradients (Simpson, 1979). An
EcoRI - BamHI fragment of maxicircle DNA cloned in pBR 322
was isolated as described (Masuda et al., 1979). The purified RIMaxi DNA and cloned insert DNA were digested with
several restriction endonucleases, the DNA fragments were
size fractionated in agarose gels and blotted onto nitrocellulose. As hybridization probes, we obtained from
A. Tzagoloff a series of yeast petite mutants which retained
various mitochondrial structural genes, isolated the petite
mtDNAs and purified specific restriction fragments bearing
DNA sequences contained entirely within the various
structural genes (see Simpson et al., 1982a). These gene
fragments were labelled by nick translation and hybridized
to the maxicircle blots at low stringency (5% Formamide, 1M
NaCl, 37°C, 48h). The blots were washed in 6 x SSC at 41°C

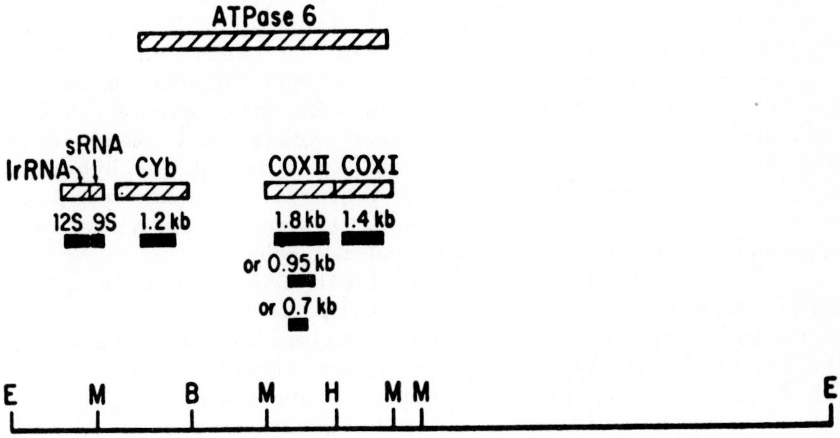

ATPase 6

FIGURE 1. Organization of structural genes and tentative RNA transcripts of the genes on the maxicircle DNA of L.tarentolae. The maxicircle (30 kb) is linearized at the EcoRI (E) site and the Msp1 (M), Bam H1 (B) and Hha I (H) sites are shown. The hatched bars represent the structural gene localizations and the solid bars represent the positions of RNA transcripts (size in kb) which have been mapped and tentatively assigned to the CYb, COX I and COX II genes (Simpson et al., 1982b). The 9S and 12S RNAs are transcripts of the putative small and large ribosomal RNA genes.

for 2h, exposed to film and then rewashed in 6 x SSC at increasing temperature to determine the thermal stability of the hybrids.

Hybridization to maxicircle DNA was observed with probes representing fragments of the yeast genes for cytochrome b (exon b1) (CYb), ATPase subunit 6, and cytochrome oxidase subunits I (exon A5) and II (COX I, COX II). The probe for ATPase subunit 9 failed to hybridize. As the stringency of the post hybridization washes was increased, it was possible to localize the regions of homology to one or two adjacent maxicircle fragments although the hybridization to the ATPase 6 probe extended over several fragments (Fig. 1).

The thermal stabilities of the hybrids were assessed in order to gain information on the relative degree of mismatch between the various yeast probes and the homologous maxicircle sequences. The order of melting and the approximate melting temperature was: CYb (80°), ATPase 6 (70°), COX I and II (60°). With an average GC content for a mitochondrial gene of 30% (Nobrega and Tzagoloff, 1980), the Tm for such a gene under our hybridization conditions is 90°. Assuming a 1° decrease in Tm per 1.5% base sequence mismatch (Laird et al., 1969), there is approximately a 15-45% sequence mismatch between those yeast and maxicircle sequences sufficiently homologous to form stable hybrids. These results suggest a greater degree of sequence homology between the cytochrome b (exon b1) genes of yeast and Leishmania than any other mitochondrial gene.

The identification of maxicircle DNA sequences which form thermally stable hybrids with defined yeast mitochondrial structural gene sequences implies that the maxicircle of L.tarentolae encodes these structural genes but DNA sequence analysis is needed to prove this. The localizations of the putative structural genes on the maxicircle DNA restriction map are shown in Fig. 1 together with the positions of several maxicircle transcripts which may represent the transcripts of these genes (Simpson et al., 1982b). The CYb (exon b1) gene is mapped within the maxicircle fragment cloned in pBR 322 and this region is currently being sequenced to confirm the presence of this gene. The COX I (exon A5) and COX II genes are positioned on adjacent DNA fragments although the COX I gene may overlap the Hha I site. The ATPase 6 gene cannot be precisely mapped to date since the yeast probe hybridized to several maxicircle fragments in a thermally stable fashion. In each case, candidate transcripts of appropriate size have been mapped to the positions of the genes in the maxicircle.

The question of the presence in maxicircle DNA of genes for ATPase subunit 9 and cytochrome oxidase subunit III is uncertain. The failure to observe hybrids with the ATPase 9 probe suggests either that this gene is absent from maxicircle DNA, as in humans, Neurospora and Aspergillus (Tzagoloff et al., 1979; Macino et al., 1980; Anderson et al., 1981) or that this maxicircle gene has evolved to such an extent that stable hybrids cannot be obtained. With the COX III probe, we observed extensive hybridization to several non-adjacent maxicircle fragments but these hybrids melted at a low temperature (48°) implying extensive mismatch. These results suggest that in Leishmania genes for some mitochondrial proteins may either be found in the nucleus (ATPase 9) or may evolve more rapidly than others (COX III) and this rate of sequence divergence may limit the application of the heterologous gene mapping procedure in evolutionary diverse organisms. Despite these limitations, this procedure has provided the first evidence for the

physical presence and arrangement in maxicircle DNA of mitochondrial structural genes homologous to those found in
yeast.

REFERENCES

Anderson, S., Bankier, A., Barrell, B., de Bruijn, M.,
    Coulson, A., Drouin, J., Eperon, I., Nierlich, D.,
    Roe, B., Sanger, F., Schreier, P., Smith, A.,
    Staden, R., and Young, I. (1981). Nature 290, 457.
Borst, P., and Hoeijmakers, J. (1979). Plasmid 2, 20.
Clark-Walker, G.D., and Sriprakash, K.S. (1981).
    J. Mol. Biol. 151, 367.
Hoeijmakers, J., Snijders, A., Janssen, J., and Borst, P.
    (1981). Plasmid 5, 329.
Laird, C.D., McConaughy, B.L., and McCarthy, B.J. (1969).
    Nature 224, 149.
Macino, G., Scazzocchio, C., Waring, R., Berks, M., and
    Davies, R. (1980). Nature 288, 404.
Masuda, H., Simpson, L., Rosenblatt, H., and Simpson, A.M.
    (1979). Gene 6, 51.
Nobrega, F., and Tzagoloff, A. (1980). J. Biol. Chem.
    255, 9828.
Simpson, L. (1979). Proc. Natl. Acad. Sci. USA 76, 1585.
Simpson, L., and Simpson, A.M. (1978). Cell 14, 169.
Simpson, L., and Simpson, A.M. (1980).
    Mol. Biochem. Parasitol. 2, 93.
Simpson, L., Simpson, A., Kidane, G., Livingston, L.,
    and Spithill, T. (1980). Am. J. Trop. Med. Hyg.
    29 (Suppl), 1053.
Simpson, L., Spithill, T.W., and Simpson, A.M. (1982a).
    Mol. Biochem. Parasitol. (in press).
Simpson, A.M., Simpson, L., and Livingston, L. (1982b).
    Mol. Biochem. Parasitol. (in press).
Simpson, L., Simpson, A.M., Spithill, T., and
    Livingston, L. (1982c). In "Mitochondrial Genes",
    Cold Spring Harbor Symposium (P. Borst, G. Attardi,
    and P. Slonimski, eds.), p. 435. Cold Spring Harbor
    Laboratory, New York.
Tzagoloff, A., Macino, G., and Sebald, W. (1979).
    Ann. Rev. Biochem. 48, 419.

50

The Expression of the Mitochondrial Genes for the Subunits of the ATP-Synthetase and Cytochrome *c* Oxidase Complexes

A. M. Kroon, Etienne Agsteribbe, Jenny C. de Jonge,
Hans de Vries, M. Holtrop, Alex A. Moen, John Samallo
and Peter van 't Sant

Laboratory of Physiological chemistry
State University Medical School
Groningen, The Netherlands

I. INTRODUCTION

The biogenesis of mitochondria within the eukaryotic cell is a complicated cell-biological process, which has fascinated many scientists in the last 25 years (1,2), One of the reasons for this interest is the presence of indispensable genetic information within the mitochondria. This mitochondrial DNA (mtDNA) has now been characterized in great detail for various organisms. The length and, therefore, the genetic potential of mtDNAs is quite variable among various organisms: 75 000 basepairs (bp) for mtDNA of *Saccharomyces cerevisiae*, 60 000 bp for mtDNA of *Neurospora crassa* and only 16 000 bp for mammalian mtDNA. The basesequence is known for large parts of the mtDNA of *Saccharomyces*, of *Neurospora* and of some other ascomycetes. For bovine, human, murine and rat mtDNA the complete sequences have been determined. None the less, the genetic functions even of the mammalian mtDNAs are not yet fully understood, because a number of reading frames have not yet been assigned to polypeptides of which the activity or function is known. As far as the information contained in

MANIPULATION AND EXPRESSION
OF GENES IN EUKARYOTES
ISBN 0 12 513780 x

mtDNA is identified it is clear that all mtDNAs have genes
for ribosomal and transfer RNAs and that all mtDNAs code for
subunits of the complexes III (cytochrome bc_1 complex),
IV (cytochrome c oxidase) and V (ATP-synthetase). These complexes are involved in oxidative phosphorylation and located
in the inner mitochondrial membrane. Moreover, yeast and
Neurospora contain a mt-gene for a polypeptide necessary for
the assembly or function of mitochondrial ribosomes, designated var 1 (3) and S5 (4) respectively. In the yeast
Saccharomyces cerevisiae a few reading frames, present within
the highly mosaic mitochondrial genes for apocytochrome b and
for subunit 1 of cytochrome c oxidase have been identified as
coding entities for socalled maturases, nucleolytic enzymes,
involved in the processing of mitochondrial transcripts. The
various aspects of the expression of the genes involved in
the biogenesis of mitochondria in yeast will be covered in
this volume by others. In this paper we want to present a
brief survey of our recent data on the structure and expression of the mitochondrial genes for the subunits of the complexes IV and V in the two organisms under investigation in
our laboratory, the rat and the mould *Neurospora crassa*.

II. THE ATP-SYNTHETASE COMPLEX

The ATP-synthase complex or complex V consists of about
10-12 polypeptide subunits. The complex catalyzes the synthesis of ATP coupled to respiration. Its activity is based
on the ability to translocate protons through the mitochondrial membrane. The complex can also hydrolyze ATP to ADP and
inorganic phosphate. This latter activity can be displayed by
the part of the enzyme complex that is protruding into the
matrix space, even if it is not any longer incorporated in
the complex. This part, called F_1, consists of 5 subunits
($\alpha-\varepsilon$) which are all coded in nuclear DNA. The membrane part
is responsible for protontranslocation and contains the other
5-7 subunits. It is designated F_o, because it harbours the
subnunits which are responsible for the oligomycin-sensitivity
of complex V. The F_O part of complex V has a dual genetic
origin: it is partly coded and expressed via the nucleocytoplasmic system and partly by the mitochondrial system. This
conception is based on *in vivo* experiments in which mitochondrial proteinsynthesis was inhibited by either chloramphenicol or another drug specifically interfering with mitochondrial biosynthetic processes. Using this approach it was observed that the ATPsynthase activity was strongly impaired,

whereas the ATPhydrolase activity was hardly affected. The latter activity became, however insensitive to oligomycin (5,6,7). The number and nature of the subunits coded by the mtDNA is not the same for mitochondria of different origin. In yeasts a 19 kD[1] subunit and the 8 kD DCCD-binding[2] protein are mitochondrial translation products. For *Neurospora* it was originally thought that a 19 kD and a 11 kD polypeptide were of mitochondrial origin. The 11 kD subunit is not anymore considered to be part of the ATP-synthase complex. In *Xenopus laevis* three subunits of 24, 21 and 17.5 kD are synthesized inside the mitochondria (8). For the mammalian complex V the situation will be discussed in the next paragraph.

A. *The mammalian complex V*

We have shown previously that *in vivo* inhibition of mito-chondrial proteinsynthesis leads to the formation of an oligomycin-insensitive, membrane-bound ATPase complex in mitochondria of regenerating rat-liver (9). This oligomycin-insensitive complex could be isolated by the same procedures as normal ATPase complex. *In vivo* labeling experiments re-vealed that 3 out of 12 subunits, namely those numbered 6, 7 and 10, were under-represented in the preparations obtained from mitochondria of drug-treated animals. Polypeptides with apparent molecular weights similar to the subunits 6 and 7 were synthesized by isolated mitochondria. Subunit 7 was even incorporated into the ATPase complex under these *in vitro* conditions (10). It was further shown, that the DCCD-binding protein was not a mitochondrial translation product in rat (11). From these studies the picture emerged that in mammalian mitochondria possibly 2 subunits of the complex, *viz.* the subunits 6 and 7 with apparent molecular weight of 25 and 22 kD respectively, are the products of mitochondrial trans-lation activity. It was furthermore concluded that the latter activity is indispensible for the correct assembly of the 10 kD subunit 10.

When the complete sequence of human mitochondrial DNA was published, one of the reading frames was assigned to subunit 6 of the ATPsynthase complex (12). This assignment was based on the homology of the basesequence with that of the gene for the 19 kD subunit of the yeast complex. Since there were clearly 2 candidates for this gene we sequenced the corres-

---

[1] kD = kiloDalton
[2] DCCD = N,N'-dicyclohexyl carbodiimide

5' A T G A A C G A A A A T C T A T T T G C C T C T T T C 3'

NH$_2$-asn - glu - asn - leu - phe - ala - ser - phe -

FIGURE 1. The 5' nucleotide sequence of the mt-gene ascribed to an ATPase subunit N-terminal aminoacid sequence of the 25 kD subunit of the ATPsynthase complex of the Wistar rat.

ponding gene of rat-mtDNA (13). We further determined the N-terminal aminoacid sequence of the subunit 6 and 7 of the rat-ATPsynthase complex. The main data are shown in Fig. 1. The basesequence can be perfectly aligned with the aminoacid sequence of the 25 kD subunit, incidentally also subunit 6 as in the yeast complex. The results further show that the primary mitochondrial translation product is N-terminally processed because the peptide lacks the formyl-methionine. It can be noted that at baseposition 19 a C instead of a G has been found in the mtDNA of Sprague-Dawley rats (14). This should lead to the substitution of the alanine residue by a proline.

The situation for subunit 7 is more complicated. At the initial steps of aminoacid analysis pairs of aminoacids were found, indicating that band 7 contains two different polypeptides. This possibility was already raised previously, because band 7 is supposed to contain the oligomycin sensitivity conferring protein (OSCP), which at least in yeast is a product of nucleocytoplasmic origin. The data, gathered so far do not allow yet any conclusion about the genetic origin of either of the two polypeptides.

B. *The* Neurospora *Complex V*

Two subunits of the *Neurospora* complex V have been reported to be mitochondrial translation products: a 19 kD and a 11 kD polypeptide (15). The 11 kD subunit is, however, not indispensable for normal ATP-synthase activity. In a stopper mutant isolated and studied in our laboratory this 11 kD subunit is absent, whereas the mitochondria contain a normal oligomycin-sensitive F_1F_0-ATPase (16). The true nature of the 11 kD polypeptide observed is not known. In a mutant deficient in the cytochrome b and aa_3 the 11 kD polypeptide band is missing. Furthermore, on electrophoresis of isolated labeled cytochrome b a labeled product of about 11 kD was detected. We consider it likely, therefore, that the 11 kD polypeptide is a breakdown product of (apo)cytochrome b which is selectively copurified with the ATP-synthase complex.

From the above data the inevitable conclusion seems that only one subunit of the complex V of *Neurospora* is of mitochondrial origin. However, the situation may be more complicated, for *Neurospora* mtDNA contains a gene for another ATP-synthase subunit, *viz* for a proteolipid-type, DCCD-binding protein (17). The expression of this gene has not been shown upto now. Under the conditions tested so far the DCCD-binding protein present in the ATP-synthase complex is the product of a nuclear gene, synthesized in the cytoplasm as a precursor and transported into the mitochondria. None the less it is tempting to argue that the mt-gene will be active under certain, yet unknown conditions. Otherwise it is very difficult to envisage how the intact gene, in which all essential characteristics of a DCCD-binding protein have been conserved, has survived the coexistence with the nuclear gene. The latter point is especially relevant in view of the relatively low degree of amino acid homology between the nuclear and the mitochondrial gene products. If one assumes a common origin for the two genes, the divergence should have taken place early in the evolution of metazoa. It is further worthwhile noting that the upstream flanking region of the gene shows structural characteristics which are very similar to those of some other mitochondrial genes (see below).

C. *The control of subunit synthesis and assembly*

It has already been mentioned that the F_1-part of complex V consists of subunits of nucleocytoplasmic origin. F_1 is present in petite yeast cells which are devoid of mitochondrial protein synthesis (5). Also in animal mitochondria deficient in polypeptides of mitochondrial biosynthetic origin, F_1 activity is not coördinately decreasing together with the F_1F_0-activity. *I.e.* high ATPase activity can be measured *in vitro*. This activity is not reflecting a physiological function, however. It has been shown that at least animal mitochondria remain coupled in spite of the relative surplus of F_1 as compared to functional F_1F_0 (7). Apparently, the F_1 is complexed to rudimentary F_0 only consisting of the subunits of nucleocytoplasmic origin. In fact these odd complexes can be purified by the same procedures as the whole F_1F_0-complexes (9). The relative excess of F_1 under conditions of impaired mitochondrial protein synthesis is not a general phenomenon. This can be seen in Table I, which shows that the inhibitory effects of the antibiotics chloramphenicol (CAP), thiamphenicol (TAP) and oxytetracycline (OTC) on ATPase (=F_1) and cytochrome *c* oxidase

TABLE 1.

Effects of treatment with mitochondriotropic antibiotics on ATPase and cytochrome c oxidase activities in mitochondria from different tissues in the rat. The data given are expressed as percent of the activity in the tissues of animals treated in the same way with 0.15 M NaCl instead of the drugs.

Tissue (treated)	ATPase	Cytochrome *c* oxidase	Cyt. *c* ox. / ATPase
Intestinal epithelium			
(TAP; 2.5 days)	66	35	0.5
(OTC; 0.7 days)	100	75	0.8
Regenerating liver			
(CAP; 2 days)	103	65	0.6
(CAP; 3 days)	110	70	0.6
(CAP; 4 days)	92	67	0.7
Skeletal muscle			
(OTC; 17 days)	55	69	1.2

activities are not the same for all tissues or for all conditions. Skeletal muscle offers an example of a coördinate decrease of both activities. There is no obvious relation to the degree of reduction of the cytochrome *c* oxidase activity. We have considered the possibility that the F_1 of the treated animals was lost to other cell-fractions during isolation of the muscle mitochondria. However, with the aid of an anti-F_1 antibody this could be ruled out. Remains the question if there is an accumulation of unassembled F_1-subunits, either outside or inside the mitochondria. This question cannot yet be answered for the animal system. However, experiments with the stopper-mutant mentioned above, are indicative that such a situation may exist in *Neurospora crassa* (16). In this mutant the ATP-synthase as well as the ATP-hydrolase activity were strongly diminished as compared to the parent strain. Both acitivities showed comparable sensitivity to oligomycin in the wildtype and mutant strains. There were no clear differences in the subunit composition. However, the mutant mitochondria contained larger amounts of the unassembled α, β and γ subunits. This suggests that the F_1-ATPase is hardly assembled in the absence of a normal F_o-part of the complex. Of the organisms studied so far, this feature is only observed for *Neurospora*. Apparently synthesis and assembly can be regulated at various levels. The mechanisms of regulation remain poorly understood as yet.

III. THE CYTOCHROME C OXIDASE COMPLEX

The data presented refer to the three large subunits of the *Neurospora* enzyme complex, which have been recognized as mitochondrial translation products for a long time already.

A. *The translation products*

We have shown previously that at least 3 precursor polypeptides with apparent molecular weights of 45, 36 and 25 kD can be detected among the mitochondrial translation products (18). The 45 kD polypeptide is the precursor of subunit 1 of cytochrome c oxidase. The accumulation of such a protein in the mitochondrial mutant [mi-3] had been shown before (19). Our experiments show that the precursor is the primary translation product of subunit 1 under physiological conditions and does not exist thank to the mutagenic event in [mi-3]. By various experimental approaches, including immunological characterization, the relation of the 45 kD polypeptide to subunit 1 could be established unequivocally. The relation of the 36 kD polypeptide to subunit 2 was only infered from the absence of a formylmethionine residue in the mature subunit. For a similar reason the 25 kD precursor polypeptide can be related either to subunit 3 of cytochrome c oxidase or to the 19 kD ATPase subunit. The precursors accumulate at low temperature and in cells pretreated with cycloheximide. Under these conditions they are, as opposite to be mature subunits, loosely membrane-bound.

A further feature is the existence of the mature subunit 1 in two forms differing in apparent molecular weight: a 41 and a 50 kD form. The 50 kD polypeptide was one of the hitherto unidentified mitochondrial translation products. In cyanide-insensitive mutants, the 50 kD protein is absent or only present in low concentrations. In the mutant [mi-3] the 50 kD protein is completely absent and subunit 1 of cytochrome c oxidase is present in low concentration when the cells are grown and labeled at 25°C. However, both polypeptides are synthesized in considerable amounts if the experiments are performed at 30°C (20). This indicates that the proteins are related. Immunoprecipitation experiments showed that the 50 kD protein is associated with the cytochrome c oxidase complex in wild type. Moreover, peptide mapping reveals that this protein and subunit 1 of cytochrome c oxidase give a similar proteolytic pattern, supporting the relation further (Fig. 2). This is further confirmed by using a combination of peptide-mapping and immunodetection. A papain fragment from the 50 kD

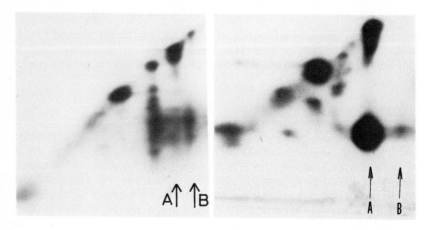

FIGURE   2. *Peptide mapping (left) and immunodetection of peptides (right) of mitochondrial translation products (20). 500 µg papain was used for degradation of ^{35}S-labeled mt-translation products. A fluorogram of a dried SDS/PAA gel (left). 200 µg mitochondrial protein was separated on a 15% SDS/PAAG slab. The whole lane was used for peptide mapping in the second dimension with 1 mg papain. The protein was next blotted onto nitrocellulose. The blot was treated with anti-subunit 1 antiserum and ^{125}I-labeled protein A. An autoradiogram is shown (right). The arrows indicate the positions of subunit 1 of cytochrome c oxidase (A) and the 50 kD polypeptide (B).*

protein is recognized by an antiserum against subunit 1 of cytochrome *c* oxidase (Fig. 2), despite the fact that the antibody does not recognize the integral 50 kD polypeptide (not shown). So, the 50 kD protein is a modification of the 41 kD subunit 1 of cytochrome *c* oxidase. Pulse-chase labeling experiments demonstrate that the modification is a posttranslational process. 10-20% of subunit 1 of cytochrome *c* oxidase is converted to the 50 kD polypeptide.

B. *The mitochondrial genes*

We have sequenced the gene for subunit 1 of cytochrome *c* oxidase completely and that of subunit 2 partially. The gene for subunit 3 has been completely sequenced by others (21). These analyses reveal that the gene for subunit 1 contains no introns in the first 1470 basepairs. Intron I1 up to I6 of the comparable yeast gene are definitely absent in *Neurospora.*

As a consequence the "Box"effect, known from yeast and based
on splicing of intron 4 of subunit 1 by a product of intron 4
of the cytochrome *b* gene is absent as well. A 45 basepair sequence near the carboxylterminus may be an intron. This remains
to be shown by aminoacid analysis. The 45 basepairs present a
continuous reading frame in phase with the rest of the gene
and without stopcodons. The amino acid sequence shows no homology to comparable genes of other organisms, also not with the
sequence of intron 7 of the yeast gene. There is, however,
some 50% homology between the carboxyl termini of subunit 1
of *Neurospora*, yeast, man and rat. The gene of *Neurospora* is
about 75 basepairs longer. One may wonder if the real carboxyl
terminus has at all been identified. In this respect it is
worthwhile noting that a transcript has been detected in different laboratories, including ours, with the surprising length
of about 6 500 bases.

Comparison of the N-terminal parts of the genes for the
subunits 1 and 2 with the aminoacid sequences shows N-terminal
extensions in the reading frames of 6   and 36 basepairs respectively. Since there does not exist a clear linear relation
between the molecular weights and the mobility in SDS/PAA gels
for hydrophobic proteins, we consider these extensions compatible with the differences between the apparent molecular
weights of the precursor polypeptides, discussed above, and
the mature subunits. Subunit 3 has no precursor (21).

IV. CONCLUDING REMARKS

From the data presented here and in other contributions to
this volume the diversity in the structure of similar genes
in mitochondria of various organisms is quite striking. The
regulation of the expression of the mitochondrial genes and
the processes balancing the contributions of the two cellular
genetic systems are yet poorly understood. It has been noticed
that a number of genes on *Neurospora* mtDNA is flanked by GC-
rich palindromic sequences (21) which are believed to act as
signals for RNA-processing (22). We have found similar structures at about the same distance from the N-termini as in the
subunit 3 gene (21) for subunit 1 of cytochrome *c* oxidase and
the proteolipid-type gene for an ATPase subunit. The latter
sequence is somewhat longer and contains a small loop. The
results are shown in Fig. 3. Further analysis is needed to
establish similar structures down stream the genes. A palindromic structure has not been detected for the gene of subunit 2 so far.

```
                                                       A T A
                                                       G   G
                                                       G - C
                                                       G - C
                                                       A - T
      C - G              T - A                         T - A
      A - T              G - C                         T - A
      T - A              A - T                         A - T
      C - G              G - G                         C - G
      C - G              G - C                         C - G
      A - T              T - A                         G - C
      C - G              C - G                         A - T
      T - A              C - G                         T - A
-158  A - T         -163 C - G                   -171  G - C
      TCA   CCA          CCCA    GGGG                  ATT   TTGA

    Subunit 1 of        Subunit 3 of cyt.           Proteolipid
    cyt. *c* ox.          *c* ox. (21)               (ATPase)
```

FIGURE 3. Palindromic structures near various genes of Neurospora *mtDNA.*

As to the codon usage there is a strong tendency also in *Neurospora* to exclude C and G from the wobble position in the reading frames. This holds for the identified as well as for the unidentified frames. The latter observation may indicate that these frames are indeed expressed. The identification of the products of these genes is the major effort to make in the near future. Elucidation of their function may give the key to a further understanding of the regulation of mitochondrial membrane assembly.

V. ACKNOWLEDGEMENTS

Support was received from the Dutch Organization for the Advancement of Pure Research, ZWO. The authors thank Karin van Wijk for the preparation of the manuscript.

VI. REFERENCES

1.   Kroon, A.M. and Saccone, C., Eds. "The Organization and Expression of the Mitochondrial Genome", Elsevier/North-Holland Biomedical Press, Amsterdam (1980).
2.   Slonimski, P., Borst, P. and Attardi, G., Eds. "Mitochondrial Genes, Cold Spring Harbor Laboratory", New York (1982).

3. Terpstra, P. and Butow, R.A., *J. Biol. Chem.* 254, 12662-12669 (1979).
4. LaPolla, R.J. and Lambowitz, A.M., *J. Biol. Chem.* 256, 7064-7067 (1981).
5. Schatz, G. and Mason, T.L., *Ann. Rev. Biochem.* 43, 51-87 (1974)
6. Gijzel, W.P., Strating, M. and Kroon, A.M., *Cell Differentiation* 1, 191-198 (1972).
7. De Jong, L., Holtrop, M. and Kroon, A.M., *Biochim. Biophys. Acta* 501, 405-414 (1978).
8. Koch, G., *J. Biol. Chem.* 251, 6097-6107 (1976).
9. De Jong, L., Holtrop, M. and Kroon, A.M., *Biochim. Biophys. Acta* 548, 48-62 (1979).
10. De Jong, L., Holtrop, M. and Kroon, A.M., *Biochim. Biophys. Acta* 606, 331-337 (1980).
11. De Jong, L., Holtrop, M. and Kroon, A.M., *Biochim. Biophys. Acta* 608, 32-38 (1980).
12. Anderson, S., Bankier, A.T., Barrell, B.G., De Bruijn, M.H.L., Coulson, A.R., Drouin, J., Eperon, I.C., Nierlich, D.P., Roe, B.A., Sanger, F., Schreier, P.H., Smith, A.J.H., Staden, R. and Young, J.G., *Nature* 290, 457-465 (1981).
13. Moen, A.A., Holtrop, M., Pepe, G., Gadeleta, G., Saccone, C. and Kroon, A.M., submitted.
14. Groszkopf, R. and Feldman, H., *Curr. Genet.* 4, 151-159, (1982).
15. Jackl, G. and Sebald, W., *Eur. J. Biochem.* 54, 97-106, (1975).
16. De Vries, H., De Jonge, J.C., Van 't Sant, P., Agsteribbe, E. and Arnberg, A., *Curr. Genet.* 3, 205-211 (1981).
17. Van den Boogaart, P., Samallo, J. and Agsteribbe, E., *Nature* 298, 187-189 (1982).
18. Van 't Sant, P., Mak, J.F.C. and Kroon, A.M., *Eur. J. Biochem.* 121, 21-26 (1981).
19. Werner, S. and Bertrand, H., *Eur. J. Biochem.* 99, 463-470 (1979).
20. Van 't Sant, P., "PhD. Thesis", University of Groningen (1982).
21. Browning, K.S. and RajBhandary, U.L., *J. Biol. Chem.* 257, 5253-5256 (1982).
22. Yin, S., Heckman, J.E. and RajBhandary, U.L., *Cell* 26, 325-332 (1981).

51

Studies on the Evolutionary History of the Mammalian Mitochondrial Genome

C. Saccone, C. De Benedetto, G. Gadaleta, C. Lanave, G. Pepe,
E. Sbisà

Istituto di Chimica Biologica, Università di Bari
e Centro Studio Mitocondri e Metabolismo
Energetico CNR, Bari, Italy

P. Cantatore, R. Gallerani, C. Quagliariello

Dipartimento di Biologia Cellulare, Università
della Calabria, Cosenza, Italy

M. Holtrop and A. M. Kroon

Laboratory of Physiological Chemistry,
University of Groningen, The Netherlands

The mammalian mitochondrial (mt) DNA is a circular molecule of about 16,000 bp containing information for ribosomal (r) transfer (t) and messenger (m) RNAs. The gene organization has been clarified by determining the nucleotide sequence of the entire molecule. Human (Anderson et al.,1981), bovine (Anderson et al.,1982) and mouse (Bibb et al.,1981)mtDNAs have been completely sequenced; the sequence of rat mtDNA, performed by various groups (Saccone et al.,1981; Cantatore et al.,1982; Grosskopf and Feldmann, 1981; Kobayashi et al., 1981) is almost complete. The availability of nucleotide sequences from various mammals enables us to undertake comparative studies at nucleotide level which would provide knowledge on the process of molecular evolution of the mt genome. We can compare the homologous genes and the entire genomes. The latter approach has the advantage to offer a wider outlook of the evolutionary events and would provide more insight into the nature of the functional constraints operating at the level of the molecule as a whole, independently of the expressed products. In this paper we report a comparative study

MANIPULATION AND EXPRESSION
OF GENES IN EUKARYOTES
ISBN 0 12 513780 x

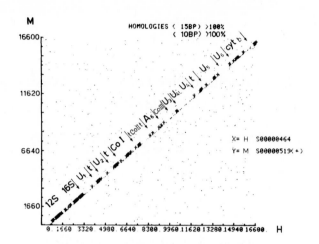

Fig. 1. Graphic representation of the homology between mouse and human mtDNA. The symbols indicate blocks of 15 nucleotides (+) or of 10 nucleotides (dots), having 100% homology.

performed on the entire mt genomes as well as on the various classes of mt genes from rat, mouse, bovine and human cells.

COMPARISON OF THE ENTIRE MT GENOMES

A comparative study has been carried out by using our graphic computer representation already described (Saccone et al., 1981). For the data analysis we have devised a new programme, written in FORTRAN-VAX. This programme, by a particular option, searches first for larger blocks (50 nucleotides max), then in progressive order for smaller ones, having 100% homology, in order to find the best alignment between two sequences. The number of matches and their frequency are reported and eventually presented by graphics. Fig. 1 shows the comparison of the mouse and human mtDNAs. The most conserved and hypervariable regions are clearly identified by searching for homology with strings of different length (15 nucleotides and 10 nucleotides respectively). The first part of the mtDNA molecule is highly conserved: long homologous stretches are found in the rRNA region. Only some of the tRNA genes appear to be highly conserved as already shown in a previous paper (Cantatore et al.,1982). Among protein coding genes, the CoI and cytochrome *b* genes are the less divergent. The genes coding for URFs possess only a few short regions of high homology, the remaining

parts being extremely variable. In Table I the four mammalian mt genomes have been compared also with respect to the
four base content. It appears very clearly that the G+C
content of mtDNA in mammals is not low, going from 36% in
mouse to 44% in human. The distribution of G and C between
the two strands is however highly asymmetric, the light (L)
strand being very poor in G. The L strand of animal mtDNA
has the same sequence as the majority of mt products which
are transcribed on the heavy (H) strand. In fact, even if
both strands are symmetrically transcribed, most of the L
strand transcripts are degraded with the exclusion of the
genes for eight tRNAs and the genes for possibly two proteins
as yet still unidentified (Attardi, 1981). As it will be
shown later, all the mt products transcribed on the H-strand
have a base composition similar to that reported in Table I.
In addition comparison of homologous genes reveals a pattern
of nucleotide substitution during evolution which suggests
a strong selective pressure depending on the bias in nucleotide usage in the two mtDNA strands.

THE rRNA and tRNA GENES

Although the overall homologies of the rRNA genes are not
remarkably high, as shown in Table II, the region of the mt
DNA containing the genes for 16S and 12S rRNA can be considered the most conserved among mammals. When the entire
genomes are compared, only in this region there are long
stretches highly preserved (up to 50 adjacent nucleotides
with 100% homology in the mouse:human comparison). This can
be inferred also from Fig. 1. At the level of primary

TABLE I. *General features of the mt genomes in mammals[a]*

Total length (bp)	Base composition (%)				Base homology (%)			
	A	C	T	G	H	B	M	R
H 16569	31	31	25	13	100	54.4	52.5	nd
B 16338	33	26	27	14	53.6	100	57.3	nd
M 16295	35	24	29	12	51.7	57.2	100	nd
R nd	34	26	25	15	nd	nd	nd	100

[a]*For the mtDNA (L strand) from mouse, bovine and human cells
the overall base composition is reported; for the rat molecule the data are concerning all the genes sequenced by us or
available in the literature. The base homologies have been
obtained by using the programme described in the text.
H= human; B= bovine; M= mouse; R= rat; nd= not determined.*

TABLE II. *Percentage of homology of mt rRNAs from mammals*[a]

| Homology | 12S rRNA | | | | 16S rRNA | | | |
(%)	R	M	B	H	R	M	B	H
R	100	91	78	76	100	76	70	70
M	91	100	81	74	77	100	71	71
B	78	81	100	73	70	71	100	76
H	76	74	73	100	70	71	76	100

[a]*R = rat;   M = mouse;   B = bovine;   H = human.*

structure the smaller rRNA species is more conserved than the
larger one.   Studies on the secondary structure of the rRNAs
now in progress will clarify if the degree of conservation is
maintained also at this level.   The base content of the rRNA
genes reflects that of the L strand of mtDNA.   The same
holds for the tRNAs whose information is on the same strand
(Table III.)   With regard to the nature of the substitutions
among the genes of different animals the following conclusions
can be drawn.   In rRNA genes there is a clear predominance
of C-T changes over G-A changes and also the A-Y (pyrimidine)
changes are much higher in number than the G-Y changes.   For
tRNAs we have previously shown that the D-loop, T-loop and
extra arm are the most variable parts of the molecule
(Cantatore et al.,1982).   A strong bias in avoiding transitions and transversions involving G in these regions is
observed.   This seems to suggest that the use of G is
strictly confined to those positions which play an essential
role in the function of tRNA.

THE PROTEIN CODING GENES

The codon usage in the genes coding for mitochondrial
proteins confirms and extends the above reported observations.
A large number of codons is only scarcely used, especially
those bearing the G in the 3' end position in the codon.   The
codon usage does not markedly differ between identified and

TABLE III.   *Base composition of rRNA and tRNA genes*[a]

Genes	A	C	T	G
rRNAs	36.6 ± 1.5	22.1 ± 2.4	24.0 ± 1.7	17.3 ± 0.8
tRNAs (L)	36.9 ± 1.2	18.6 ± 1.2	29.3 ± 1.0	15.2 ± 0.4
tRNAs (H)	26.6 ± 2.8	14.0 ± 0.2	33.8 ± 0.5	25.1 ± 2.3

[a]*Average values (± s.d.) from rat, mouse, bovine and human
genes.*

TABLE IV. *Homologies and types of substitutions between rat and other mammalian genes[a]*

	Homology (%)		Transitions (%)				Transversions (%)									Total s (%)	Total r (%)
	Nucleotides	Amino acids	C-T s	C-T r	G-A s	G-A r	G-C s	G-C r	G-T s	G-T r	A-C s	A-C r	A-T s	A-T r		
CoIII																
R/M	82	96	54	1	3	1	1	1	1	0	17	2	17	2	93	7
R/B	78	85	33	7	4	6	3	1	1	1	17	9	12	6	70	30
R/H	78	87	33	8	9	5	1	1	1	1	26	6	4	5	74	26
URF3																
R/M	76	83	33	16	9	4	0	3	0	0	13	1	12	9	67	33
R/B	67	69	32	11	6	7	2	1	0	4	12	11	6	8	58	42
R/H	65	65	27	11	5	4	0	5	2	3	9	13	8	13	51	49

[a] R = rat; M = mouse; B = bovine; H = human; s = silent substitutions; r = replacements

not identified products.   In both cases a strong bias in avoiding the use of G from the wobble position is observed during evolution.   Table IV shows that either in the case of the CoIII gene or in the case of the URF3 gene the transitions generally exceed the transversions.   There is a predominance of C-T changes over G-A changes as well as a predominance of A-Y over G-Y changes.   This means that the mitochondrial genes are rather constrained in the silent substitutions.   This constriction, observable particularly at the third codon position, might explain, according to some of us (Saccone and Lanave, submitted), also the deviation of the mt genetic code from the universal one.   The change of a terminator codon UGA into a codon for tryptophane could probably be due to the necessity to find an alternative for the UGG codon whose occurrence in the L strand of mtDNA should be extremely rare.   The same holds for the initiator codon AUG which is, in some cases, substituted by the more abundant AUA.   However, although constrained by nucleotide usage, the rate of silent substitutions in mt genes is extremely high.   In Table V we report the sequence divergence of several mt genes from rat compared to that of nuclear genes.   The K values, the corrected number of nucleotide substitutions per base, have been calculated according to Kimura (1981).   It can be observed that in mt genes the Ks values are roughly constant.   From the data reported in Table V we can calculate the rate of synonymous substitutions for mt genes.   Using the formula $Vs=Ks/2T$ and considering $T=8 \times 10^7$ years for all the genes reported in Table V, we found a rate of $4.38 \pm 0.14 \times 10^{-9}$ substitutions/site/year, which is about twice that of nuclear genes.   This value is much lower than that calculated by the divergence in the endonuclease restriction pattern of mtDNA (Brown et al.,1979). However, taking into account the above mentioned strategy of the mt genetic code this value is probably underestimated. In fact, when the K3 for the CoIII gene was calculated with the two frequency class (2FC)model of Kimura, which considers the unequal distribution of the four bases at the third codon position, we obtained a K3 value of 1.39, which is significantly higher than that reported in Table V.   An underestimate of the rate of silent substitutions in mtDNA could also depend on the saturation of the silent sites during the evolution. Further calculation for a larger number of mt genes and between more closely related species will offer a more valuable estimate of the evolutionary rate of the mtDNA.

CONCLUSIONS

A simple graphic computer program allows the identification of the most conserved and hypervariable regions of the mt genome in mammals.   The analysis of the base content,

TABLE V. Evolutionary distance values (± s.e.) of mt genes for identified and unidentified products compared with nuclear genes (Kimura, 1981)[a].

Comparison	Evolutionary distance per nucleotide site (model 3ST)			
	K1	K2	K3	Ks
H/R presomatotropin	0.26 ± 0.04	0.18 ± 0.03	0.53 ± 0.07	0.44 ± 0.07
H/R I preproinsulin	0.04 ± 0.03	0.00	0.46 ± 0.12	0.38 ± 0.12
H/R CoII	0.25 ± 0.04	0.13 ± 0.03	0.93 ± 0.14	0.72 ± 0.14
H/R CoIII	0.11 ± 0.02	0.07 ± 0.02	0.93 ± 0.18	0.77 ± 0.18
H/R ATPase	0.23 ± 0.04	0.10 ± 0.02	0.85 ± 0.12	0.67 ± 0.12
H/R URF3	0.43 ± 0.08	0.14 ± 0.04	0.91 ± 0.18	0.71 ± 0.17
H/R URF4	0.39 ± 0.04	0.13 ± 0.02	0.99 ± 0.10	0.66 ± 0.10

[a] H = human; R = rat.

pattern of nucleotide substitutions and codon usage clearly shows that the mammalian mt genome is under selective pressure being constrained by a peculiar nucleotide usage in the two strands. In particular we have found that the strand which contains information for the majority of mt products has a very low G content. The bias in avoiding the use of G, particularly in the third codon base, might explain also the deviation of the mt genetic code from the universal one. Measurement of the evolutionary rate of synonymous substitutions in the protein coding sequences demonstrates that it is at least twice as high as the corresponding rate of nuclear genes confirming a rapid evolution of mtDNA.

REFERENCES

Anderson, S., Bankier, A.T., Barrel, B.G., de Bruijn, M.N.L., Coulson, A.R., Drouin, J., Eperon, I.C., Nierlich, D.P., Roe, B.A., Sanger, F., Schreier, P.H., Smith, A.J.H., Staden, R., and Young, I.G. (1981). *Nature* 290, 457.
Anderson, S., de Bruijn, M.H.L., Coulson, A.R., Eperon, I.C., Sanger, F., and Young, I.G. (1982). *J. Mol. Biol.* 156, 683.

Attardi, G. (1981). *Trends Biochem. Sci. 6,* 100.

Bibb, M.J., Van Etten, R.A., Wright, C.T., Walberg, M.W., and Clayton, D.A. (1981). *Cell 26,* 167.

Brown, W.M., George, M.J., and Wilson, A.C. (1979). *Proc. Natl. Acad. Sci. USA 76,* 1967.

Cantatore, P., De Benedetto, C., Gadaleta, G., Gallerani, R., Kroon, A.M., Holtrop, M., Lanave, C., Pepe, G., Quagliariello, C., Saccone, C., and Sbisà, E. (1982). *Nucleic Acids Res. 10,* 3279.

Grosskopf, R., and Feldmann, H. (1981). *Curr. Genet. 4,* 151.

Kimura, M. (1981). *Proc. Natl. Acad. Sci. USA 78,* 454.

Kobayashi, M., Seki, T., Yaginuma, K., and Koike, K. (1981). *Gene 16,* 297.

Saccone, C., Cantatore, P., Gadaleta, G., Gallerani, R., Pepe, G., and Kroon, A.M. (1981). *Nucleic Acids Res. 16,* 4139.

Part VI
Viruses

Introduction

Bacterial viruses were crucial to the elucidation of the basic principles of prokaryote molecular biology, and animal and plant viruses have served a similar role in our understanding of the molecular biology of eukaryotes. For example, a study of gene expression of an animal virus first demonstrated the presence of the intervening sequences in the coding region of a eukaryote gene.

Probably the most significant application of animal viruses to eukaryote molecular biology has been their development as vectors in gene transformation in mammalian cultured cells. The SV40 system is particularly well characterized, and the analysis of faulty RNA splicing in α-thalassaemia (Chapter 52) provides an elegant demonstration of the power of this viral vector system. Other chapters describe the application of SV40 to studies on the role of RNA secondary structure in splicing (Chapter 53) and on recombination in mammalian cells (Chapter 54).

The study of animal viruses themselves remains important. Detailed analysis of the genome of the influenza virus is rapidly increasing our understanding of the evolution of virus strains and of the onset of epidemics (Chapter 56). We are also beginning to probe the mechanisms of viral-induced cell transformation; one example is given for cultured rat cells (Chapter 57). This is an exciting field—recent findings have shown that some of the transforming genes in mammalian cancer cells may be identical to genes found in the genomes of some cancer viruses.

We are just beginning to understand the control of viral and host gene expression, but we certainly know less about the interactions of plant cells and their viruses. However, the analysis of the genome of cucumber mosaic virus (Chapter 60) shows that application of recombinant DNA methodology will result in a rapid advance in knowledge. This, in turn, is likely to lead to some exciting developments in the diagnosis and prevention of viral diseases in agricultural plants.

Part VI
Viruses

Introduction

52
Use of an SV40 Vector to Demonstrate Abnormal Splicing in α-Thalassemia

Barbara K. Felber, Dean H. Hamer

Laboratory of Biochemistry
National Cancer Institute
National Institutes of Health
Bethesda, Maryland

Stuart H. Orkin

Division of Hematology and Oncology
Children's Hospital Medical Center
and The Sidney Farber Cancer Institute
Department of Pediatrics
Harvard Medical School
Boston, Massachusetts

I. INTRODUCTION

SV40, a small DNA tumor virus, can be used as a vector to introduce new genetic information into cultured cells. We have shown previously that the mouse α-globin gene signals for transcription initiation, polyadenylation, splicing and translation are appropriately utilized in African green monkey kidney cells infected with encapsidated SV40 recombinants (Hamer et al., 1980). Subsequently, similar results have been obtained for the rabbit β-globin (Banerji et al., 1981), human β-globin (Moschonas et al., 1981; Busslinger et al., 1981) and human α-globin (Mellon et al., 1981) genes introduced into cultured monkey or human cells by the more convenient method of direct DNA transfection with plasmid SV40 recombinants. The faithful recognition of globin

MANIPULATION AND EXPRESSION
OF GENES IN EUKARYOTES
ISBN 0 12 513780 x

regulatory sequences cloned in SV40 suggested that this
might provide a useful system for studying the thalassemias –
hereditary defects in human globin chain synthesis.

In this article, we describe our work on the α2-globin
gene from an Italian α-thalassemia patient (Orkin et al.,
1979). Whereas in normal individuals the duplicated α gene
complex directs the synthesis of stable α1 and α2-globin
mRNAs in the ratio 2:3 (Orkin and Goff, 1981), this patient
has a profound deficiency of α2-globin mRNA. Restriction
mapping showed that the α2-globin gene is completely deleted
from one chromosome but present on the other chromosome in
this patient. However, sequence analysis of the remaining
gene revealed a pentanucleotide deletion in the first intron
(Orkin et al., 1981). The deletion lies immediately 3' to
the first exon and destroys the G-T dinucleotide that is
common to the 5' ends of all known introns and thought to
act as part of the recognition sequence for processing
enzymes (Breathnach et al., 1978; Lewin, 1980). To test the
consequences of this sequence alteration, we have inserted
the thalassemic gene into an SV40 plasmid vector and studied
its expression in transfected monkey cells. Futher, in
order to exclude possible artefacts of the SV40 system, we
have also analyzed the expression of the thalassemic gene
directly in the bone marrow of the patient. These complementary approaches allow us to demonstrate that the thalassemic
phenotype of the patient results from an abnormality in RNA
splicing. A detailed report of this work has appeared
recently (Felber et al., 1982).

II. RESULTS

A. *Expression of the Normal and Thalassemic Globin Genes in Transfected Monkey Cells*

We used a plasmid SV40 vector to compare the expression
of the normal and thalassemic α2-globin genes in transfected
monkey cells. As shown in Figure 1, the viral portion of
this vector retains the origin of SV40 DNA replication, an
intact early gene region, and the extreme 5' and 3' portions
of the late gene region. The inserted globin gene fragments
contain the complete normal or thalassemic structural
sequences together with about 590 bp of 5' and 100 bp of 3'
flanking sequences and are joined to the vector in the same
orientation as SV40 early transcription. The recombinant
molecules were introduced by the DEAE-dextran method into
monkey kidney cells, the permissive host for SV40, and

Deletion in $\alpha_2{}^t$ Globin Gene

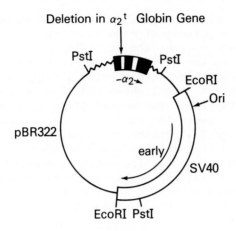

FIGURE 1. SV40 plasmids carrying the normal α2-globin gene
and the thalassemic α2t globin gene.

total cell polyA$^+$ RNA was extracted after 3 days. As
controls, parallel cultures were transfected either with no
DNA or with wild-type SV40 DNA. Analysis of these samples
by gel transfer hybridization to a labelled globin probe
showed that the normal and thalassemic genes were expressed
into stable, approximately full-length globin RNA with the
same efficiency (results not shown).

B. *The First Intron of the Thalassemic RNA is Spliced from
    a New Donor Site to the Normal Acceptor Site*

The structure of the thalassemic RNA was established by
S1 nuclease mapping using a series of end-labeled probes.
Figure 2 (left panel) shows the mapping of the 5' end of
intron 1 with a probe 3' end-labeled in exon 1. As predicted, RNA from cells transfected with the normal gene
protected a predominant cluster of bands lying approximately
111 bases from the end label (lane n). Alignment with a
sequence ladder of the probe showed that the two major bands
occur immediately 5' to the junction between exon 1 and
intron 1. In contrast, RNA from cells transfected with the
thalassemic RNA (lane t) protected fragments mapping within
exon 1 at approximately 62 bases from the end label. Comparison with the sequence ladder showed that the major band
lies between the G and T residues of a donor-like sequence.
Thus the pentanucleotide deletion in the thalassemic gene
eliminates splicing from the normal donor site at the junction

of exon 1 and intron 1 but activates a new donor lying within
the structural sequences of exon 1.

To determine whether the first intron of the thalassemic
RNA is spliced to the normal 3' acceptor site, we repeated
the S1 nuclease mapping with a probe 5' end-labeled in exon
2. Figure 3 (first panel) shows that RNA from monkey cells
transfected with either the normal (lane 4) or thalassemic
gene (lane 5) protects the same 180 base fragment as authentic
globin mRNA (lane 2). Furthermore, we did not detect any
larger protected fragments as would be predicted if intron 1
was not spliced out or was incompletely removed. These
results indicate that the thalassemic RNA is efficiently
spliced from the alternative donor site in exon 1 to the
usual acceptor site at the junction of intron 1 and exon 2.

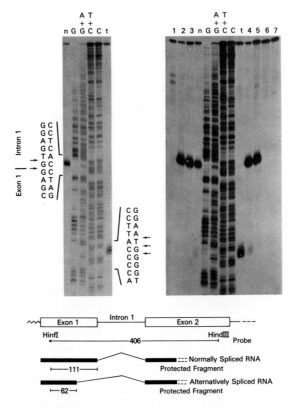

FIGURE 2. Nuclease S1 Mapping of the Alternative Donor Site.
The samples analyzed in the different lanes are described in
the text. The right panel shows a 15-times longer exposure
of the same gel shown in the left panel.

C. The Rest of the Structure of the Thalassemic RNA is Normal

The structure of the thalassemic RNA was further analyzed by hybridization to probes designed to detect the 5' and 3' termini of the RNA and the donor and acceptor sites for the second intron. Figure 3 shows that, with each probe, the thalassemic RNA gave the same protected fragment as authentic α-globin mRNA. Thus the abnormal processing of intron 1 does not alter transcription initiation or termination nor does it prohibit the removal of intron 2.

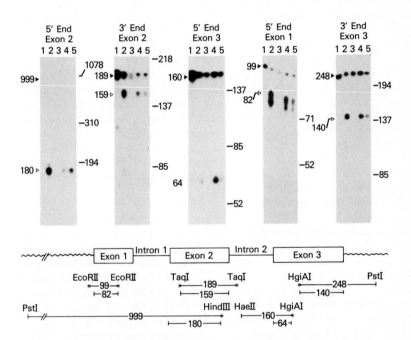

FIGURE 3. Nuclease S1 Mapping of Normal and Thalassemic α2-globin Gene Transcripts: 5' end of exon 2, intron 2 splice junction and 5' and 3' ends. The numbers to the left of each panel indicate the lengths of the probes (solid arrows) and protected fragments (open arrows) and the numbers to the right refer to the end-labeled restriction fragments used as markers. Each panel contains (1) no added RNA; (2) human peripheral blood RNA; (3) RNA from monkey cells mock-transfected with no DNA; (4) RNA from monkey cells transfected with the normal α2-globin gene; (5) RNA from monkey cells transfected with the thalassemic α2-globin gene.

D. *Alternative Splicing in the Thalassemia Patient*

We next wished to determine if alternative splicing also occurs in the patient from whom the thalassemic gene was isolated. For this purpose RNA from the patient's bone marrow, the site of erythroid cell production, was S1 mapped using the probe 3' end-labeled in exon 1. Figure 2 (right panel, lane 4) shows that this RNA protected a predominant cluster of bands of the length corresponding to normally spliced RNA. This was expected because the patient retains a normal αl-globin gene (Orkin et al., 1979) and because the sequences of the αl and α2-globin genes are identical over the region covered by the hybridization probe (Orkin and Goff, 1981). In addition, however, we detected a significant level of fragments indicative of splicing from the alternative donor site in the middle of exon 1. These fragments, which exactly comigrated with the fragments protected by RNA from monkey cells transfected with the cloned thalassemic gene (lane t), were not seen in reactions containing RNA from mock (lane 7) or wild-type SV40 (lane 6) transfected cells but were detected at a very low level in a control reaction containing no added RNA (lane 1). RNAs from normal peripheral blood (lane 2), normal bone marrow (lane 3) and β^+-thalassemic bone marrow (lane 5) protected only background levels of the alternative splice fragments.

These observations were confirmed by using a cDNA probe 5' end-labeled in exon 2. As shown in Figure 4, hybridization of normally spliced RNA to this probe should protect a 275 base fragment extending to the A-T joint between the cDNA and plasmid sequences. In contrast, RNA spliced from the alternative donor is expected to protect a 180 base fragment ending at the junction between exon 1 and exon 2. These predictions were confirmed by hybridizing the probe to RNA from monkey cells transfected with the normal (lane n) or thalassemic (lane t) gene and aligning the protected fragments with the sequence ladder (left panel). When bone marrow RNA from the thalassemic patient was hybridized to this probe, protected fragments indicative of alternative splicing could readily be visualized (right panel, lane 6). Peripheral blood RNA from the patient showed a significant but substantially lower quantity of alternatively spliced RNA (lane 5, barely visible in this reproduction of the autoradiogram). This agrees with the previous observation of a deficiency of circulating α2-globin RNA in the patient (Orkin et al., 1981) and suggests that the alternatively spliced RNA is unstable. No alternative splice fragments were protected by RNA from normal bone marrow (lane 3), normal peripheral blood (lane 4) or β^+-thalassemic bone marrow (lane 7), although all of these samples contained large quantities of normally spliced transcripts.

E.   Use of the Alternative Splice Donor in the Normal Globin Gene in Transfected Monkey Cells

As described above, most of the RNA made in monkey cells transfected with the normal α2-globin gene is spliced from the normal first donor site. However, long exposures of the autoradiograms revealed the presence of low levels of RNA spliced at the alternative site in exon 1. Figure 2 (right panel, lane n) shows a long exposure of the S1 mapping experi-

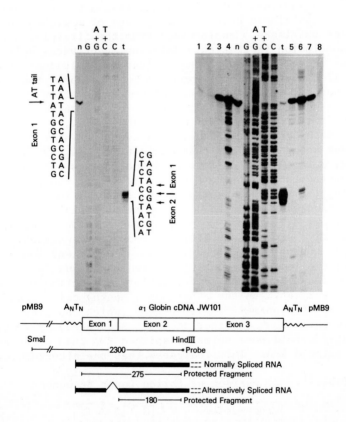

FIGURE 4.   Detection of Alternatively Spliced RNA with a cDNA Probe.   The samples analyzed in the different lanes are described in the text.   The right panel in a 15 times longer exposure of the same gel shown in the left panel.

ment using the probe 3' end-labeled in exon 1. Alternative
splice fragments are clearly visible at a level exceeding
the background encountered with this probe. Clearer evidence
was obtained using the cDNA probe (Figure 4, right panel,
lane n). A low level of alternative splice fragments was
again observed, and in this experiment no background bands
were found at this position in the gel. The amount of RNA
spliced from the alternative donor is less than 1% the amount
spliced from the normal site in transfected monkey cells.

III. DISCUSSION

We have analyzed the in vivo expression of an α-thalassemia
gene that contains a complete complement of α-globin coding
sequences but lacks 5 nucleotides from the first intron. Our
major conclusion is that this deletion inactivates splicing
from the normal donor site and activates an alternative donor
in the middle of the first exon. Table 1 compares the sequences
of the normal, alternative and deleted donor sites to the
consensus donor sequence characteristic of a variety of
eukaryotic splice junctions (Breathnach et al., 1978; Lewin,
1980). Both the normal and alternative sites differ from the
consensus sequence by a single nucleotide. In contrast, the
inactive mutated site differs by 4 nucleotides and lacks the
G-T dinucleotide common to the 5' ends of all known introns.
Screening the sequence of exon 1 and intron 1 for additional
G-T dinucleotides revealed eight other potential donor sites.
One of these differs by two nucleotides and the remainder by
three to five nucleotides from the consensus sequence. The
fact that none of these functions as an efficient donor
supports the idea that homology to the consensus donor
sequence is an important criterion for recognition by the
splicing enzymes. On the other hand, monkey cells transfected with the normal α2-globin gene, which contains
both the normal and alternative donor sequences, synthesize
at least 100-fold more normally spliced RNA than alternatively spliced RNA. This indicates that the alternative

Table 1. Comparison of Splice Donor Sequences

Consensus Donor	5' A	G	G	T	A	A	G 3'
Normal Donor	A	G	G	T	g	A	G
Alternative Donor	g	G	G	T	A	A	G
Mutated Donor	A	G	G	c	t	c	c

site is recognized less efficiently than the normal site and
that homology to the consensus donor sequence is not the sole
factor influencing recognition by the processing enzymes.
One possible explanation for this result is that splicing
occurs by a 3' to 5' scanning mechanism which recognizes the
first available donor more readily than the second (Sharp,
1981). Alternatively, the secondary or tertiary structure of
the precursor RNA may be such that the normal site is more
accessible to the processing enzymes than the alternative
site.

Alternatively spliced RNA was also detected in bone marrow from the thalassemic patient although not in bone marrow
from a normal individual or from an anemic β^+-thalassemic
patient. This demonstrates that alternative splicing of the
mutated $\alpha2$-globin gene is not simply an artefact of the
heterologous SV40-monkey cell system. However, while the
patient contains one $\alpha2$-thalessemic gene and one normal $\alpha1$-
globin gene, the ratio of alternatively to normally spliced
RNA is far less than 1. It seem unlikely that this is due to
decreased transcription of the thalassemic gene because it is
expressed as efficiently as the normal gene both in an in
vitro transcription system (Orkin et al., 1981) and in transfected monkey cells. A more plausible explanation is that
the alternatively spliced RNA is unstable. Support for this
later hypotheses comes from the observation that there is
less alternatively spliced RNA in the mature reticulocytes
from the patient's peripheral blood than in the relatively
immature cells in his bone marrow. The reason for the dis-
crepancy between the stability of the thalassemic transcripts
in monkey cells and the human is unclear, but may reflect an
erythroid mechanism that recognizes and degrades nonfunctional RNA species.

The alternatively spliced RNA encoded by the thalassemic
gene lacks 48 to 50 bases of structural sequences frome exon
1. If the splice follows the G-T "rule" (Breathnach et al.,
1980; Lewin, 1978) the resulting RNA will contain a stop
codon in exon 2. Splicing at the next nucleotide also generates a stop codon in exon 2 whereas splicing in frame would
give a globin polypeptide lacking 16 amino acids (lys[16] to
arg[31]) from exon 1. In any event, it is clear that the truncated RNA from this gene is incapable of encoding a normal
globin polypeptide. It is curious, then, that we observed a
small amount of alternatively spliced RNA in monkey cells
transfected with the normal $\alpha2$-globin gene. Because alternatively spliced RNA could not be detected in normal or β-
thalassemic bone marrows, presumably because of its instability, it seems unlikely that such RNA plays a significant

physiological role. A more plausible explanation is simply
that the splicing mechanism is not completely accurate in
distinguishing between closely related donor sites. Such
inaccuracy has been postulated as an important factor in the
evolution of discontinuous genes (Gilbert, 1978; Doolittle,
1978; Darnell, 1978).

We have studied the expression of the α-thalassemia gene
both in SV40 transfected monkey cells and in the patient. The
main advantages of the SV40 system are that (i) it is possible to formally prove that abnormal expression is due to a
sequence change within the cloned gene rather than at a distant location; (ii) the expression of the mutated gene can be
studied in the absence of its normal counterpart; and (iii)
relatively large amounts of RNA can easily be prepared. However, a potential drawback of this system is that the expression of the cloned gene into stable globin mRNA in transfected monkey cells may not accurately reflect the situation in
erythroid cells. Our observation that the thalassemic gene
directs the synthesis of stable RNA with the same efficiency
as the normal gene in monkey cells, but not in bone marrow,
highlights this difficulty. For this reason it was essential
to corraborate the results obtained with the SV40 vector by
analysis of patient material. Clearly a combination of both
approaches is necessary to correlate specific alterations in
cloned genes with inherited human diseases.

ACKNOWLEDGEMENTS

    We thank Gail Taff for excellent preparation of the man-
uscript. B.K.F was supported by a fellowship from Schweize-
rischer Nationalfonds. S.H.O. was supported by grants from
NIH, the National Foundation and the NIH Career Development
Award.

REFERENCES

Banerji, J., Rusconi, S. and Schaffner, W. (1981). Cell 27,
    299.
Breathnach, R., Benoist, C., O'Hare, K., Gannon, F. and
    Chambon, P. (1978). Proc. Natl. Acad. Sci. USA 75, 4853.
Busslinger, M., Moschonas, N. and Flavell, R.A. (1981). Cell
    27, 289.
Darnell, J.E. Jr. (1978). Science 202, 1257.

Doolittle, W.F. (1978). Nature 272, 581.

Felber, B.K., Orkin, S.H. and Hamer, D.H. (1982) Cell, 29, 895.

Gilbert, W. (1978). Nature 271, 501.

Hamer, D.H., Smith, K.D., Boyer, S.H. and Leder, P. (1979). Cell 17, 725.

Hamer, D.H., Kaehler, M. and Leder, P. (1980). Cell 21, 697.

Lewin, B. (1980). Cell 22, 324.

Mellon, P., Parker, V., Gluzman, Y. and Maniatis, T. (1981). Cell 27, 279.

Moschonas, N., deBoer, E., Grosveld, F.G., Dahl, H.H.M., Wright, S., Shewmaker, C.K. and Flavell, R.A. (1981). Nucl. Acids Res. 9, 4391.

Orkin, S.H., Old, J., Lazarus, H., Altay, C., Gurgey, A., Weatherall, D.J. and Nathan, D.G. (1979). Cell 17, 33.

Orkin, S.H. and Goff, S.C. (1981). Cell 24, 345.

Orkin, S.H., Goff, S.C. and Hechtman, R.L. (1981). Proc. Natl. Acad. Sci. USA 78, 5041.

Sharp, P.A. (1981). Cell 23, 643.

53

Complementary Relationship between the SV40 Pre-Promotor Region and the Early Gene Splice Junction Exon Sequences

Leslie Burnett

Department of Clinical Biochemistry
Royal Prince Alfred Hospital
Camperdown, N.S.W.
Australia

I. INTRODUCTION

Several groups have described and characterised mutants of the Simian Virus 40 (SV40) which carry deletions in the region of the early gene promotor. These studies have revealed that TATA box is not essential for promotor function, but that a second region remotely 5' and upstream of the TATA box, is essential for promotor function (1-3). This pre-promotor region enhances the expression of the downstream early genes, and is active in either orientation. It may also be active when it is present in other positions of the genome not upstream to the early gene and also may enhance the expression of non-SV40 genes placed in the genomic position of the SV40 early genes (4).

I have previously described a computer assisted genome analysis of SV40, which revealed (at a low level of resolution) that there were sequences in the region close to the pre-promotor region which were highly complementary to the SV40 early gene splice junction exons. This study was

This work was supported by a grant from the National Health and Medical Research Council of Australia.

was undertaken to determine, at a high degree of resolution, the precise location of the complementary sequences in the pre-promotor region.

II.   COMPUTER SEARCH STRATEGY

The SV40 nucleotide sequence was analysed using a computer program to determine whether the genome contained sequences capable of forming complementary interactions with the SV40 big-T antigen splice junction (6). The 3400 most statistically significant regions of the genome complementary to sequences adjacent to, and occurring on the same coding strand as, the splice junction were extracted. The least stable of these structures were excluded from further analysis, while the remaining complementary sequences were analysed by a computer program to ensure there were no sequences elsewhere in the genome capable of forming interactions which would have been energetically more stable than those interactions formed with the splice junctions. Interactions with 3000 potential interacting sequences were evaluated.

III.   RESULTS AND DISCUSSION

This study has revealed that there are two sequences in the pre-promotor region which are highly complementary to the upstream and downstream exons of the big-T splice junction and the downstream exon of little-t splice junction, but not complementary to the upstream exon of the little-t splice junction (Fig. 1 and 2). The location of the complementary sequences in the pre-promotor region lie partly within the 72-base major repeat regions, and are characterised by extreme GC-richness; the two complementary regions are separated by a short segment which is not complementary to any of the SV40 splice junctions.

The predicted thermodynamic stability of the complementary interactions between the pre-promotor region and the early gene splice junction exon sequences is extremely high (ΔG = -44 to -67 Kcal/mol), and the structures are of comparable stability to transfer RNA molecules. The statistical significance of the occurrence of complementarity of this degree is also high (p <0.001).

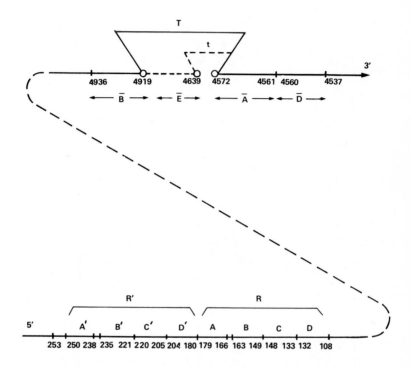

FIGURE 1.   Location of the pre-promotor sequences complementary to the SV40 early gene splice junctions.

The location of the complementary sequences correspond exactly to the location of the pre-promotor enhancing sequences.   Deletions on either side of the pre-promotor complementary region have little effect on the promotor, but deletions extending more than 1 nucleotide into the complementary regions appear to abolish promotor activity (2).

IV.   REFERENCES

1.   Gluzman, Y., Sambrook,JF. and Frisque, RJ. (1980)
         Proc. Natn. Acad. Sci. USA 77:   3898-3902
2.   Benoist, C. and Chambon, P. (1981) *Nature 290:*   304-310
3.   Mathis, DJ. and Chambon, P. (1981) *Nature 290:*   310-315
4.   Banerji, J., Rusconi, S. and Schaffner, W. (1981)
         Cell 27:   299-308
5.   Burnett, L., Buchman, AR., McMahon, JE., Tinoco, I. Jr.
         and Berg, P. (1980) *Proc. Aust. Biochem. Soc. 13:*   84
6.   Burnett, L. (1982) *J. Theoret. Biol.*   (in press)

350   *Leslie Burnett*

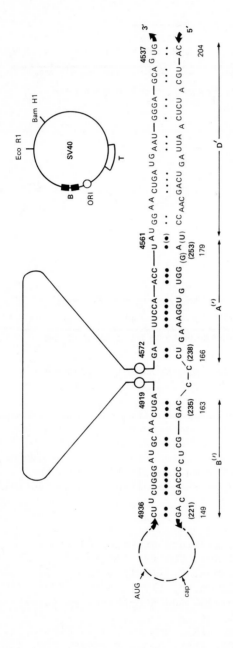

FIGURE 2. Proposed complementary interactions between the exons of the SV40 big-t antigen splice junction and the pre-promotor sequences. (The structure has been arbitrarily represented as an intramolecular interaction)

54
Mammalian Cell Functions Mediating Recombination of Genetic Elements

Peter Upcroft

Recombinant DNA Unit
Queensland Institute of Medical Research
Brisbane, Queensland

Although recombination, or what we now know as the exchange of genetic information from one segment of DNA to another, was proposed in 1931 when Belling (1931) formulated the first hypothetical mechanism and paired chromosomes were first observed in 1918 (Pascher, 1918), very little is understood at the molecular level. At the DNA level recombination has been regarded as an homologous exchange of DNA strands (Holliday, 1964) with various other models arising to fit the genetic data. However, the mechanisms of recombination of bacteriophage Mu, IS sequences and transposons in *E.coli* have emphasised non-homologous exchange between genetic elements. Gene control elements capable of inducing chromosome breakage and rearrangement and transposition from one chromosome to another were described originally in maize by McClintock (1956). Transposition of elements of the *copia* and dispersed, repeated gene families has been described for *D.melanogaster*, thus demonstrating the mobility of some genetic elements in eukaryotes.

I have initiated a series of experiments to define at a molecular level the various functions that mammalian cells encode to mediate recombination of elements of genetic information. These studies are the basis for a systematic approach to the understanding of DNA strand exchange in mammalian cells and have used the well characterised SV40 genome as a miniature chromosome in the initial studies. Some of the implications for stable genome rearrangements are in the control

MANIPULATION AND EXPRESSION
OF GENES IN EUKARYOTES
ISBN 0 12 513780 x

of immunoglobulin diversity, intron and exon shuffling, development and differentiation, cell transformation, gene
amplification and the evolution of multi-gene families.

I. RECOMBINATION BETWEEN DNA SEGMENTS THAT HAVE FLUSH OR SHORT COHESIVE TERMINI IN MAMMALIAN CELLS

The cohesive nature of the termini generated by Eco RI
cleavage of the SV40 genome was demonstrated when these unit
length linear (fIII) genomes were transfected into African
green monkey kidney (AGMK) cells (Mertz and Davis, 1972).
Plaques were generated at a frequency of 10% of those produced by supercoiled (fI) SV40 DNA. Deletion mutants have been
generated by resection of the SV40 genome with the lambda 5'-
exonuclease by a length of 15-50 nucleotides after cleavage
with different restriction endonucleases (Carbon et al.,
1975) and by random cleavage with HaeIII (Cole et al., 1977).
These early experiments thus demonstrated that SV40 DNA transfected into AGMK cells could be subject to joining reactions
apparently facilitated by weak homology of single-stranded
DNA.
    To demonstrate that the joining reactions may not be due
simply to random weak homology Upcroft et al. (1980b) showed
that fIII SV40 DNA generated by cleavage with HpaI, which
yields flush termini, produced plaques at the same frequency
as Eco RI cleaved genomes. When the SV40 genome was cleaved
with two restriction endonucleases that recognise unique but
different sites in the genome and the products were transfected into AGMK cells, these segments with cohesive termini re-
sealed to produce plaques at a frequency of 0.4% of those
produced by fI DNA (Wilson, 1977). Furthermore, cleavage of fI
SV40 DNA into two specific segments by the flush-terminus
generating HpaI, followed by agarose gel purification and
transfection into AGMK cells, demonstrated plaque production
at a frequency of 0.08% of fI SV40 DNA. All four HpaI sites
in these plaque isolates were retained (Upcroft et al.,
1980b), thus demonstrating faithful closure of the flush termini. SV40 transfected AGMK cells therefore supported joining
of DNA segments that have weak single-stranded homology,
short homologous single-stranded termini (4bp) and flush
termini.

II. RECOMBINATION BETWEEN DNA SEGMENTS THAT HAVE DOUBLE-STRANDED HOMOLOGY AND EXPRESS NO BIOLOGICAL FUNCTION

The previous studies led to the question whether SV40 genome segments could be utilised as a system to analyse 'generalised recombination' in mammalian cells (Upcroft et al., 1980a). This choice was based on the extensive biophysical and genetic data compiled for the SV40 genome, its limited encoding capacity, behaviour as a miniature chromosome and its use in model studies of chromosome structure.

Segments of the SV40 genome were generated by restriction endonuclease cleavage such that the terminal portions of each segment overlapped and shared some homology with a terminus of the other segment (Upcroft et al., 1980a). The amount of homology varied from a few bases to half the genome in length. In all cases recombination between the segments was detected after transfection into AGMK cells as plaques at a frequency of the order of 0.1% of the number of plaques generated by fI SV40 DNA. Background levels of contamination were one hundred times lower, thus establishing an efficient recombination system with the particular advantage of being able to manipulate the genetic structure of the molecules to be recombined at the molecular level. The frequencies of recombinant plaque formation were the same as observed for *in vitro* constructed SV40 fIII concatamers (Wake and Wilson, 1979).

The hypothetical requirements for replication and unusual DNA structures, such as palindromes, in recombination were excluded in a series of experiments utilising suitably cleaved and genetically marked segments of SV40 DNA with homologous overlapping termini (Upcroft et al., 1980a). Further experiments demonstrated that transcription and translation of viral genes were not necessary (Upcroft et al., 1980b). Since these DNA segments expressed no biological function and are therefore neutral elements of genetic information, their recombination must have been executed by host cell encoded enzymatic machinery. These recombination functions then are applicable to any element of genetic information.

Although the basic assay system is efficient for the above types of analyses, I have utilised other approaches when applicable, *e.g.* the use of microinjection (Capecchi, 1980) to examine single cell capacity for recombination, phosphate coprecipitation (Graham and van der Eb, 1973) to examine recombination functions related to chromosomal integration and the appearance of recombination functions related to drug induced sister chromatid exchange (P.Upcroft,

unpublished data). The assay system has evolved into the use
of SV40 as a vector to carry other suitable genes and exon-
intron arrangements of interest to recombination studies. I
have used as the model system the plasmid pBR322/SV-O+T/TK
(Capecchi, 1980), which has inserted into a deleted portion
of the late gene region of SV40, pBR322 and the TK gene of
herpes simplex, the latter two segments being an effective
intron for the late gene region of SV40 across which recombination can occur (P.Upcroft, ms. submitted). These types of
constructs have the further advantage of use under both short
term (lytic) and long term (non-lytic) conditions that approach those of chromosomes (Upcroft et al., 1978).

ACKNOWLEDGMENTS

    I wish to thank B.Carter for technical assistance and C.
Kidson for support.

REFERENCES

Belling, J. (1931). *Univ. Calif. Publ. Bot. 16*, 311.
Capecchi, M. (1980). Cell *22*, 479.
Carbon, J., Shenk, T.E. and Berg, P. (1975). *Proc. Natl.
    Acad. Sci. 72*, 1392.
Cole, C.N., Landers, T., Goff, S.P., Manteuil-Brutlag, S. and
    Berg, P. (1977). *J. Virol. 24*, 277.
Graham, F.L. and van der Eb, A.J. (1973). *J. Virol. 52*, 456.
Holliday, R. (1964). *Genet. Res. 5*, 282.
McClintock, B. (1956). *Cold Spring Harbor Symp. Quant. Biol.
    21*, 197.
Mertz, J.E. and Davis, R.W. (1972). *Proc. Natl. Acad. Sci.
    69*, 3370.
Pascher, A. (1918). *Ber. Dtsch. Bot. Ges. 36*, 136.
Upcroft, P., Carter, B. and Kidson, C. (1980a). *Nuc. Acids
    Res. 8*, 2725.
Upcroft, P., Carter, B. and Kidson, C. (1980b). *Nuc. Acids
    Res. 8*, 5835.
Upcroft, P., Skolnik, H., Upcroft, J.A., Solomon, D., Khoury,
    G., Hamer, D.H. and Fareed, G.C. (1978). *Proc. Natl.
    Acad. Sci. 75*, 2117.
Wake, C.T. and Wilson, J.H. (1979). *Proc. Natl. Acad. Sci.
    76*, 2876.
Wilson, J.H. (1977). *Proc. Natl. Acad. Sci. 74*, 3503.

55

Attempts of Permanent Introduction of SV40 Genome into Mouse Gametes and Early Embryos

Jan Abramczuk[1,2]

The Wistar Institute
Philadelphia, Pennsylvania

Purified virions and DNA were isolated from primary cultures of African Green Monkey Kidney cells infected at low multiplicities of infection (log MOI = -4 to -3) with plaque-purified, large plaque mutant RH 911 of Simian virus 40. DNA of ts A58 mutant was also used in one experiment.

To introduce the viral genome into gametes: a) females were injected intravenously with virions (log infective units = 9) one week before ovulation, or at the ovulation time, and mated to normal males; b) sperm was infected with virions (log MOI = 3), and used for *in vitro* fertilization. Early embryos were exposed to SV40 by: a) infecting two-cell stage embryos with virions (log MOI = 3) or DNA (0.5 mg/ml, calcium method); b) injecting virions (log inf units = 3) or DNA (5 - 10 pg) into blastocyst cavity; c) injecting virions (log inf units = 7) or DNA (4 µg) into amniotic cavity.

DNA was extracted from internal organs of mice obtained from the gametes or early embryos exposed to SV40 (one sample from each immature, and four different samples from each adult animal) and tested for the presence of viral genome by hybridization with 32-P-labelled, nick-translated SV40 DNA. The principal technique was the analysis of reassociation kinetics (sensitivity 0.03 - 0.27 SV40 DNA copy/diploid genome, except for some brain DNA samples). In addition,

[1]Present address: Australian National University, Canberra
[2]Supported by NIH grant ROI ESO 1866

MANIPULATION AND EXPRESSION
OF GENES IN EUKARYOTES
ISBN 0 12 513780 x

TABLE. *Number of Animals Tested for the Presence*
of SV40 Genome

Developmental stage and method of exposure to SV40	Tested at age of:	
	<1 month	1-15 months
Oocyte, virions injected into female	1	8
Sperm, virions in vitro	8	0
2-cell embryo, DNA in vitro	8	0
2-cell embryo, virions in vitro	0	10
Blastocyst, DNA in vitro	20	29
Blastocyst, ts A58 DNA in vitro	4	0
Blastocyst, virions in vitro	1	10
Fetus, DNA in vivo	4	4
Fetus, virions in vivo	8	4

"Southern" blotting was used for testing some brain DNA
samples (sensitivity 0.01 - 0.02 SV40 DNA copy/genome). Total
of 124 animals were tested (TABLE). Special emphasis was
placed on confirming the existing report (Jaenisch & Mintz,
1974) of persistence of SV40 DNA in large proportions of mice
injected with the viral DNA while in a blastocyst stage, by
testing 64 offspring developed from blastocysts exposed to
SV40.

Only two animals - both derived from blastocysts injected
with DNA - were found to carry SV40 DNA sequences in their
tissues. In one mouse, killed at age of 6 months, viral DNA
was present in the liver (0.07 copy/diploid genome) and brain
(1.80 copy/genome), but not in pancreas + lungs, nor in
kidneys + spleen. In another animal, killed while newborn,
SV40 DNA was found in the quantity equivalent to 1.20 copy/
genome in the pool of internal organs. Digestion with
restriction endonucleases showed that the viral genome
existed in this animal in a circular, non-integrated form.

These results demonstrate that mouse embryos are
relatively resistent to permanent insertion of SV40 DNA
propagated under conditions minimizing genome alterations.

REFERENCE

Jaenisch, R. and Mintz, B. (1974). *Proc. Nat. Acad. Sci.*
US, *71*, 1250.

56
Antigenic Change and Haemagglutinin Gene Evolution among Influenza Viruses

M. J. Sleigh and G. W. Both

CSIRO Molecular and Cellular Biology Unit, Sydney

I. ANTIGENIC SHIFT AND DRIFT AND THE ORIGIN OF INFLUENZA OUTBREAKS

Studies with influenza viruses since their first isolation in 1933, as well as retrospective studies of the antibodies carried by people exposed to the virus before this time, have suggested that humans are infected by a limited number of viral subtypes which periodically make an appearance in the population (reviewed in Webster and Laver, 1975). The key element in distinguishing one subtype from another is the haemagglutinin (HA), the viral surface protein against which neutralising antibodies are directed (Laver and Kilbourne, 1966).

Haemagglutinins from different viral subtypes have equivalent functions and share sufficient structural features to suggest that they evolved from a common origin. However, the amino acid homology between different HA types can vary considerably e.g. 69% between H1 and H2 but only 37% between H1 and H3 (Winter *et al.*, 1981). A subject for continuing debate is the process by which the different subtypes evolved initially, and the reservoir from which they emerge, in some cases in a form almost identical to that of isolates from many years earlier (Kendal *et al.*, 1978).

Hong Kong subtype influenza, which carries a haemagglutinin of type 3 (H3), reappeared in the human population in 1968 after an absence of 70 years (Masurel and Marine, 1973; Fedson *et al.*, 1972). Between 1968 and 1971 there was little change in the antigenic characteristics of H3 viruses in

MANIPULATION AND EXPRESSION
OF GENES IN EUKARYOTES
ISBN 0 12 513780 x

circulation and the infection rate began to slow as immunity in the population increased. However in 1972 and following years there were renewed outbreaks as antigenic variants of the initial prototype strain emerged (reviewed in Stuart-Harris, 1981). It is now apparent that this antigenic drift was due to evolution of the virus, with amino acid changes accumulating in the viral HA under the selective pressure imposed by the need to escape host immunity.

II. ANTIGENIC DRIFT ANALYSED BY HA GENE NUCLEOTIDE SEQUENCING

Our present understanding of the nature of antigenic drift in the Hong Kong subtype has come from serological studies (for example, Fazekas de St. Groth, 1978) and from analysis of the HA by comparative peptide mapping and protein sequencing (Laver *et al.*, 1980). A more detailed picture has been provided by nucleotide sequencing of the viral gene coding for the HA of different H3 isolates (Min Jou *et al.*, 1980; Verhoeyen *et al.*, 1980; Both and Sleigh, 1981; Sleigh and Both, 1981).

The genetic information coding for influenza virus haemagglutinin is contained in the fourth largest of eight single stranded RNA segments making up the viral genome. The cloning of a DNA copy of this gene and its structure and relationship to the mature HA protein, have been described in detail elsewhere (Both and Sleigh, 1980). We have extended this analysis to a comparison of nucleotide sequences of HA genes taken from sixteen strains of the Hong Kong (H3) subtype isolated between 1968 and 1980 (Sleigh *et al.*, 1981; Both and Sleigh, 1981; Both, Sleigh, Cox and Kendal, manuscript in preparation).

The data collected in these and related studies demonstrate that evolution is progressive and global. Virus isolates from many different countries have acquired the same HA gene mutations, and these mutations are generally retained and supplemented in successive isolates. Thus there appears to be a dominant evolutionary line for the subtype, connecting putative progenitor strains with the most recent isolates. Some instances of branching evolutionary lines have been seen. The most successful of these diverged from the dominant line in 1974 and produced the epidemic strain Vic/3/75, but the line apparently disappeared after further evolution to 1976 strains.

From epidemiological data, isolates can be divided into those causing either major or minor outbreaks. Successive

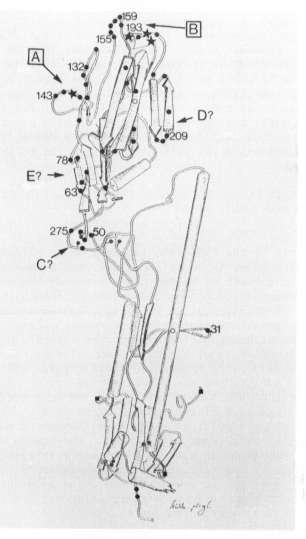

FIGURE 1. *The three dimensional structure of influenza haemagglutinin as described for a 1968 H3 strain by Wiley et al., (1981). Groups of surface amino acids making up putative antigenic sites are marked A-E. Shown on the figure are locations of all amino acid changes seen during antigenic drift from 1968 to 1980 using data for the eighteen strains listed in the legend to Fig. 2. HA2 sequences are known for only five of these strains. HA2 changes (■); HA1 changes (●); amino acids showing sequential changes (★).*

epidemic strains differ considerably in antigenic properties as well as in HA gene sequence, consistent with the ability to escape neutralisation by antibodies directed against the previous epidemic strain. Minor strains include those too similar to previous strains to escape neutralisation, but also some terminal strains. The latter appear to carry favourable antigenic changes but failed to cause significant outbreaks or to evolve to provide further variants. Possibly the terminal strains failed to compete with other circulating strains because some of the changes they had acquired disrupted HA function as well as affecting its antigenicity.

III. HIGHLY LOCALISED CHANGES EFFECT ANTIGENIC CHANGES IN
     H3 HAEMAGGLUTININS

In all of the 1968-1980 isolates whose HA gene sequences were determined, amino acid changes were concentrated in HA1, the large subunit of HA and the section of the molecule containing the functional and antibody binding sites (Jackson *et al.*, 1979; Wiley *et al.*, 1981). Changes are also found largely in surface regions of the molecule (Fig. 1). The locations of antigenic regions on the molecule have not been precisely determined, but have been inferred from data on the sites of amino acid changes in field strains and in antigenic variants isolated in the laboratory (Wiley *et al.*, 1981). The importance of the A and B sites and the area between them, (Fig. 1) has been clearly demonstrated from studies on the effects of particular HA amino acid changes on monoclonal antibody binding (Underwood, 1982). The status of sites C-E is less clear and is discussed elsewhere (Sleigh and Both, 1981; Both *et al.*, manuscript in preparation).

A closer look at events in the amino acid loops making up the A and B regions reveals that during antigenic drift there is a remarkable concentration of amino acid changes in particularly prominent residues within these sites (Figs. 2 and 3). Even more remarkable is that some surface residues undergo no change at all (e.g. residue 192 in the B site – Fig. 2) while nearby residues, in which changes might affect antibody binding by altering loop conformation (e.g. residues 140-142 in the A site) also remain constant (Fig. 3).

It seems likely that the residues that are changing are those actually involved in antigen-antibody interactions: the interaction is prevented by the insertion of a residue with a bulkier side chain or with an altered charge (Both *et al.*, manuscript in preparation) which might prevent the hydro-

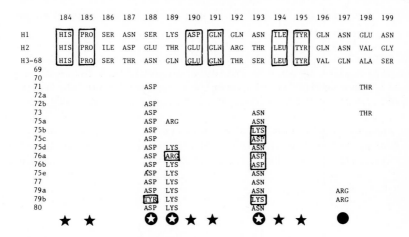

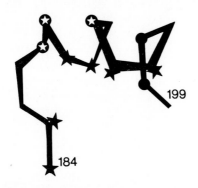

FIGURE 2. *Amino acid sequence changes, within and between influenza subtypes, in residues 184-190 of the B antigenic region of HA. (★) "Anchor" residues unchanged in all influenza subtypes. (●) Residues that have changed once in H3 subtype development. (✪) Residues that have undergone sequential changes. The strains shown are H1 subtype: PR/8/34 (Winter et al., 1981); H2 subtype: Japan/ 307/57 (Gething et al., 1980); H3 subtype-68: NT/60/68: H3-69: Eng/878/69; H3-70: Qu/7/70; H3-72b: Mem/102/72 (Sleigh et al., 1981); H3-72a: Eng/42/72; H3-73: Port Chalmers/1/73; H3-75a: Singapore/4/75; H3-75b: Mayo Clinic/4/75; H3-75c: Tokyo/ 1/75; H3-75e: Eng/864/ 75; H3-76a: Allegheny County/ 29/76; H3-76b: Vic/112/76; H3-79b: Bangkok/2/79; H3-80: Shanghai/31/ 80 (Both et al., in preparation); H3-75d: Vic/3/ 75 (Min Jou et al., 1980): H3-77: Texas/1/77 (Moss and Brownlee, in prepration); H3-79a: BK/1/79 (Both and Sleigh, 1981).*

phobic interaction involved in antibody attachment (Fazekas de St. Groth, 1978).

The limitations of amino acid changes during antigenic drift to such specific locations, with a second round of changes beginning in key residues as evolution proceeds (Both *et al.*, manuscript in preparation) suggests that stringent selective pressures in favour of altered antigenicity are operating. In addition, the need to maintain virus viability may further limit the changes permitted, since terminal variants contain antigenically-favourable changes but do not survive to pass on the changes they bear to further variants.

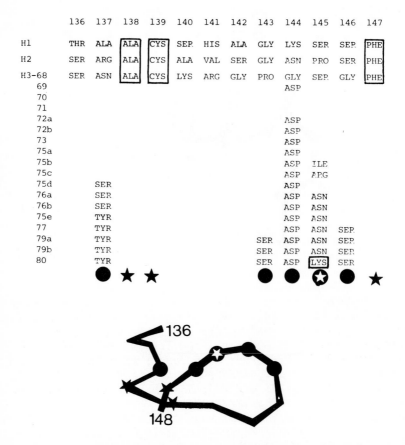

FIGURE 3. Amino acid sequence changes within and between influenza subtypes in residues 136-147 of the A antigenic region of HA. All details are as for Fig. 2.

IV. ARE THE ANTIGENIC REGIONS THE SAME IN HAs FROM DIFFERENT SUBTYPES?

Although the three-dimensional structure shown for HA was determined for a 1968 isolate (Fig. 1) it seems likely that the overall structure of all HA types will be the same.   In addition to conservation of cysteine and other key residues (Gething *et al.*, 1980; Both and Sleigh, 1980), anchor sites for the loops of antigenic amino acids are also preserved (Wiley *et al.*, 1981; Figs. 2 and 3).   Whether the residues involved in antibody binding are the same for all HA types will become apparent as antigenic variants from subtypes other than H3 are analysed.   A change at residue 160 in a monoclonal variant of an H1 strain has already been reported (Winter *et al.*, 1981).   It is interesting that some residues that remain constant during H3 subtype evolution are highly variable when different subtypes are compared (e.g. residues 140-142 in the A region, 187-192 and 196 in the B region - Figs. 2 and 3).   Perhaps changes in these residues contribute to a change in surface loop configuration, so that a new arrangement of epitopes is achieved.

It is not yet clear whether changes producing such conformational shifts would survive the stringent selective pressures which appear to be operating during H3 subtype evolution in humans.   An alternative is that the different subtypes arose during evolution in a species such as birds, where the rapid turnover in population would remove herd immunity as a selective element (Webster and Laver, 1975). The amino acid sequence of the HA of the avian influenza strain A/Duck/ Ukraine/1/63 is so similar to that of early H3 isolates from 1968 that it was proposed that the human and duck strains had a common and rather recent precursor (Fang *et al.*, 1981; Ward and Dopheide, 1981).   Significantly, only a small proportion of the differences were in potential antigenic regions (Fang *et al.*, 1981), supporting the idea that evolution had occurred under selective pressures different from those producing antigenic drift in human strains. Such evolution might ultimately supply the diverse HA types which periodically emerge in man.

REFERENCES

Both, G.W., and Sleigh, M.J. (1980). *Nucl. Acids Res. 8*, 2561.
Both, G.W., and Sleigh, M.J. (1981). *J. Virol. 39*, 663.
Fang, R.X., Min Jou, W., Huylebroeck, D., Devos, R., and Fiers, W. (1981). *Cell 25*, 315.
Fazekas de St. Groth, S. (1978). *Topics in Infectious Diseases 3*, 25.
Fedson, D.S, Huber, M.H., Kasel, J.A., and Webster, R.G. (1972). *Proc. Soc. Exptl. Biol. Med. 139*, 825.
Gething, M.-J., Bye, J., Skehel, J., and Waterfield, M. (1980). *Nature 287*, 301.
Jackson, D.C., Dopheide, T.A., Russell, R.J., White, D.O., and Ward, C.W. (1979). *Virology 93*, 458.
Kendal, A.P., Noble, G.R., Skehel, J.J., and Dowdle, W.R. (1978). *Virology 89*, 632.
Laver, W.G., Air, G.M., Ward, C.W., and Dopheide, T.A. (1980). *Nature (London) 283*, 454.
Laver, W.G., and Kilbourne, E.D. (1966). *Virology 30*, 493.
Masurel, N., and Marine, W.M. (1973). *Amer. J. Epidem. 97*, 44.
Min Jou, W., Verhoeyen, M., Devos, R., Saman, E., Fang, R., Huylebroeck, D., and Fiers, W. (1979). *Cell 19*, 683.
Sleigh, M.J., and Both, G.W. (1981). *In* "Genetic Variation Among Influenza Viruses" (D.P. Nayak, ed.) p. 341. Academic Press, New York.
Sleigh, M.J., Both, G.W., Underwood, P.A., and Bender, V.J. (1981). *J. Virol. 37*, 845.
Verhoeyen, M., Fang, R., Min Jou, W., Devos, R., Huylebroeck, D., Saman, E., and Fiers, W. (1980). *Nature (London) 286*, 771.
Underwood, P.A. (1982). *J. Gen. Virol.* In press.
Ward, C.W., and Dopheide, T. (1981). *In* "Genetic Variation Among Influenza Viruses" (D.P. Nayak, ed.), p. 323. Academic Press, New York.
Webster, R.G., and Laver, W.G. (1975). *In* "The Influenza Viruses and Influenza" (E.D. Kilbourne, ed.), p. 269. Academic Press, New York.
Wiley, D.C., Wilson, I.A., and Skehel, J.J. (1981). *Nature 289*, 373.
Winter, G., Fields, S., and Brownlee, G.G. (1981). *Nature 292*, 72.

57

Adenovirus Transforming Gene(s) in Early Region E1A Alter the Program of Gene Expression during the Cell Growth Cycle

Alan J. D. Bellett, Antony W. Braithwaite,[1] Li Peng
and Brian F. Cheetham[2]

John Curtin School of Medical Research,
Australian National University,
Canberra

Transformation of rat cells by adenovirus type 5 (Ad5) requires expression of only two viral early regions, E1A and E1B (Tooze, 1980). E1A also controls expression of other viral early regions (Shenk et al., 1979; Ricciardi et al., 1981). Region E2 affects the frequency of transformation indirectly. Ad5 acts at a control point late in G1 to induce cell DNA replication and thymidine kinase without many of the events in the normal G1 program, and alters cell cycle progression in growing cells (Braithwaite et al., 1981; Cheetham and Bellett, 1982; Murray et al., 1982). Because this may be related to transformation, we tested the ability of 9 mutants in E1 (Fig. 1) and 3 in E2 to induce cell cycle alterations in rat cells.

Wild type (w.t.) Ad5 caused a decrease in cells with G1 DNA content, and increases in cells with S phase, G2+M and aneuploid (>4n) DNA contents (Fig. 2). The E1A mutants did not cause these changes (Fig. 2), chromosome aberrations, cell

[1]Present address: Research School of Biological Sciences,
Australian National University, Canberra.
[2]Present address: University of California, Santa Barbara,
California.

MANIPULATION AND EXPRESSION
OF GENES IN EUKARYOTES
ISBN 0 12 513780 x

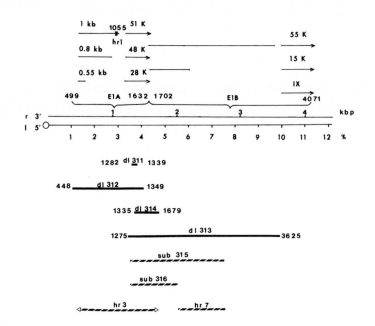

FIGURE 1. *The transforming region (E1) of Ad5, showing (top) the mRNAs, polypeptides and the hr1 mutation (Tooze, 1980; Ricciardi et al., 1981), and (bottom) locations in base pairs of deletion (dl) mutations (Shenk et al., 1979; N. Jones, personal communication) and approximate positions of substitution (sub) and host range (hr) mutations.*

DNA replication or increased thymidine kinase (TDK) activity, and did not synthesise E2 products (Table 1). Thus E1A expression is required for cell cycle alterations and early viral protein synthesis in rat cells. Mutants hr 7 (E1B) and dl 313 (E1B phenotype) (Fig. 2, Table 1), ts125 (E2A), ts36 and ts37 (E2B) were positive in all tests. Thus expression of these regions is not required to induce cell cycle alterations. Moreover, ts125 induced peaks of cells with 8n, 16n and 32n DNA contents, probably because the mutation stabilises E1A mRNA (Babich and Nevins, 1981).

Mutants sub315 and sub316 (Fig. 1) synthesised E2 products in rat cells, but caused no cell cycle alterations (Fig. 2, Table 1). We conclude that alteration of cell cycle progression is a function of transformation region E1A, independent of its control of other viral early regions.

TABLE 1.   Cell Cycle Effects in Rodent Cells Infected by Mutant Adenoviruses

Virus	Region mutated	Early[a] proteins	Cell DNA[b] induction	Cell TDK[b] induction	Cycle progression[c] alterations	Chromosome aberrations
w.t.	None	+	+	+	+	+
dl312	EIA	−	−	−	−	−
dl314	EIA	−	ND[d]	ND	−	ND
hr1	EIA	−	−	−	−	−
hr3	EIA	+	ND	ND	−	ND
dl311	EIA	+/−	ND	ND	−	ND
Sub315	EIA,EIB	+/−	ND	ND	−	ND
Sub316	EIA,EIB	+	ND	ND	−	ND
dl313	EIA,EIB	+	+	+	+/−	+
hr7	EIB	+	+	ND	+	+
ts125	E2A	+	+	+	+	+
ts36	E2B	+	+	+	+	+
ts37	E2B	+	+	+	ND	+

[a]Synthesis of early proteins outside the region mutated.
[b]Induction in cells arrested in G1 by serum starvation.
[c]Data obtained by flow cytometry.
[d]No data.

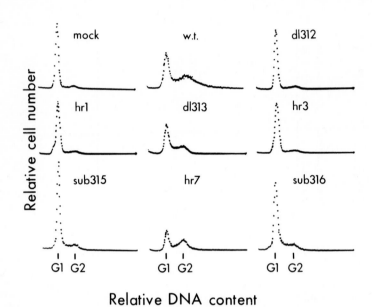

Relative DNA content

Figure 2.   Flow cytometry of rat cells, infected as indicated.   DNA content profiles were obtained using a FACS IV (Braithwaite et al., 1981; Murray et al., 1982).

REFERENCES

Babich, A., and Nevins, J.R. (1981). *Cell 26,* 37.
Braithwaite, A.W., Murray, J.D., and Bellett, A.J.D. (1981). *J. Virol. 39,* 331.
Cheetham, B.F., and Bellett, A.J.D. (1982). *J. Cell. Physiol. 110,* 114.
Murray, J.D., Bellett, A.J.D., Braithwaite, A.W., Waldron, L.K., and Taylor, I.W. (1982). *J. Cell. Physiol. 110,* in press.
Ricciardi, R.P., Jones, R.L., Cepko, C.L., Sharp, P.A., and Roberts, B.E. (1981). *Proc. Natl. Acad. Sci. U.S.A., 78,* 6121.
Shenk, T., Jones, N., Colby, W., and Fowlkes, D. (1979). *Cold Spring Harbor Symp. Quant. Biol. 44,* 367.
Tooze, J. (ed.) (1980). "DNA Tumor Viruses". Cold Spring Harbor, New York.

58
Origins of Two Human Milk Viral-like Reverse Transcriptases[1]

Yan-Hwa Wu Lee [2]

Shu-Chen Fang

Department of Biochemistry
Natl. Yang-Ming Medical College
Taipei, Taiwan

Heung-Tat Ng
Tse-Jia Lieu

Department of Obsterics and Gynecology
Department of Surgery
Veterans General Hospital
Taipei, Taiwan

Two distinct forms of viral-like reverse transcriptase were found in Chinese women milk particulates (Lee *et al.*, 1982b). The two enzymes differed in sizes, elution position from poly(rC)-sepharose and preference for synthetic template poly(rCm)·oligo(dG). In attempts to investigate the viral origins of these two viral-like reverse transcriptases, ultracentrifugation was employed for separating the different viral-like particles of human milk (Lee *et al.*, 1982a). Our results indicated that the endogenous RNA-dependent DNA polymerase activity of human milk could be separated into four den-

[1] This work was supported by NSC grant NSC70-0412-B010-18 Republic of China.
[2] Y.-H. W. Lee received a research chair award from the Tjing-Ling Basic Medical Foundation.

MANIPULATION AND EXPRESSION
OF GENES IN EUKARYOTES
ISBN 0 12 513780 x

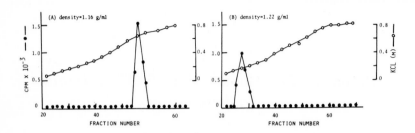

FIGURE 1.   Poly(rC)-sepharose chromatography of viral-like reverse transcriptases of human milk particles with density (A) 1.16 g/ml (B) 1.22 g/ml in CsCl gradient.

sity regions in CsCl and sucrose gradients. The elution behaviors of viral-like reverse transcriptases from different density particles of human milk were next analyzed on poly(rC)-sepharose column. Our results indicated that the two viral-like reverse transcriptases of human milk were originated from different viral-like particles with separable buoyant densities in CsCl gradient. The larger enzyme (5.8S), which was eluted from poly(rC)-sepharose column at 0.62-0.65 M KCl, was found in the particles with density of 1.16 g/ml (Fig. 1A); whereas the smaller enzyme (5S), which was eluted from poly(rC)-sepharose at 0.17-0.20 M KCl, was associated with particles with density of 1.22 g/ml (Fig. 1B). The question that whether these human milk particles, banding at 1.16 or 1.22 g/ml respectively, exhibit any structural similarity to the morphologies of mouse mammary tumor virus, murine leukemia-sarcoma virus, and Mason-Phizer monkey virus will be further studied.

REFERENCES

Lee, Y.-H. W., Fang, S.-C., Ng, H.-T., and Liu, T.-J. (1982a). *Fed. Proc. Fed. Am. Soc. Exp. Bio. 41,* 1276.
Lee, Y.-H. W., Ng. H.-T., and Lieu, T.-J. (1982b). *IRCS Medical Science 10,* 546.

A General Strategy for Cloning Double Stranded RNA: Preliminary Characterization of Cloned Simian-11 Rotavirus Genes

G. W. Both

CSIRO Molecular and Cellular Biology Unit, Sydney
Australia

A. R. Bellamy

J. E. Street

Department of Cell Biology
University of Auckland
Auckland, New Zealand

Viruses of the *Reoviridae* family possess genomes which consist of a number (usually 10-12) of segments of double stranded (ds) RNA. Several viruses having medical and veterinary importance are members of this group. For example rotaviruses, which are principal aetiological agents of diarrheoal disease in young children and domestic animals are members of this family and possess a genome consisting of eleven segments of dsRNA.

Until very recently, most rotaviruses have proved to be difficult to grow in tissue culture. Frequently, the viruses could only be isolated from faecal samples in small amounts. Thus a procedure for obtaining and cloning dsDNA copies derived from viral dsRNA gene segments would provide a means to obtain basic information about gene structure, function and variation among rotaviruses, as has been done for influenza viruses.

We have developed a generalized cloning strategy for dsRNA viruses based on work with the cultivable Simian-11 rotavirus. The protocol should have general utility for members of the *Reoviridae* family. Total genomic dsRNA was

MANIPULATION AND EXPRESSION
OF GENES IN EUKARYOTES
ISBN 0 12 513780 x

extracted from purified SA11 virus and polyadenylated with *E. coli* terminal riboadenylate transferase either as the ds form or, after heat denaturation, as ssRNA. Adenylated RNA was denatured and transcribed into cDNA using AMV reverse transcriptase and oligo dT(12-18) as primer. The RNA template was destroyed with pancreatic ribonuclease and the total mixture of ss cDNAs annealed to produce ds cDNA segments whose size distribution was similar to that of the original mixture of ds RNA segments. The ds cDNAs were filled in with reverse transcriptase to complete any partial copies and make them blunt ended, tailed with dCMP residues using terminal transferase and inserted into the PstI site of dGMP-tailed pBR322. Recombinant plasmids were used to transform *E. coli* RR1 and those colonies of the correct phenotype were analysed to determine the size of the DNA insert. Large inserts were labelled by nick translation and assigned to particular RNA genome segments by Northern hybridization i.e. ds RNA gene segments were resolved by polyacrylamide gel electrophoresis, electrophoretically transferred to DBM paper and probed with labelled insert DNA.

We have so far obtained clones representing DNA copies of genes 1, 8, 9, 10 and 11 from SA11 virus. Not all recombinant plasmids have yet been analysed. The cloned copy of gene 1 is a partial one (~1000 base pairs). Terminal sequence analysis of the others has revealed that full length gene copies for genes 8 and 11 have been obtained so far. The gene 10 copy is missing only seven nucleotides from one end, the gene 9 copy is missing 70 nucleotides from one end. The sequence of the missing nucleotides was determined by copying SA11 rotavirus mRNA in the presence of dideoxynucleoside triphosphates using reverse transcriptase and a DNA primer fragment derived from the cloned gene. This procedure also allowed the orientation of the cloned genes to be deduced since the cDNA sequence copied from the mRNA was of the negative sense.

Characterization of the terminal sequences for four of the genes cloned indicates that there is a conserved octamer at the 3' end of the plus strand and a conserved hexamer at the 5' end. The cloned genes are being sequenced to determine the gene structures and used for hybrid selection of rotavirus mRNAs to assist in completing coding assignments for the SA11 viral proteins.

60
Gene Content and Expression of the Four RNAs of Cucumber Mosaic Virus

Robert H. Symons, Dalip S. Gill, Karl H. J. Gordon
and Allan R. Gould

Adelaide University Centre for Gene Technology
Department of Biochemistry
University of Adelaide
Adelaide, South Australia

I. INTRODUCTION

Eukaryotic viral genomes replicate in host cells through the successful subversion of cellular processes. The study of such viral infections can tell us a great deal, not only about the molecular details of viral genome structure, genetic organization and function, but also about the molecular biology of the uninfected cell.

The model system we have chosen to study viral replication and interaction with host processes is the infection of cucumber seedlings with cucumber mosaic virus (CMV). The CMV genome consists of three single-strand (+) RNAs (RNAs 1, 2 and 3) which must be present simultaneously for successful infection (Peden and Symons, 1973). A fourth sub-genomic RNA (RNA 4) is always found in purified virions. The translation and replication of this multipartite viral genome may be regulated by novel protein-protein, protein-RNA or RNA-RNA interactions. Thus, the nucleotide sequences and structural features of the genomal RNAs need to be determined to provide a basis for our understanding of their different functions as messenger RNAs and replicative templates and of the generation of sub-

MANIPULATION AND EXPRESSION
OF GENES IN EUKARYOTES
ISBN 0 12 513780 x

genomic mRNAs or the use of overlapping reading frames to extend the viral coding potential.

II.   STRUCTURE AND GENE CONTENT OF THE CMV GENOME

Our current knowledge of the coding capacity and primary structure of the four RNAs of the Q strain of CMV is summarized in Fig. 1 and 2. *In vitro* translation of these

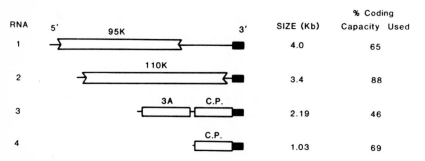

FIGURE 1. *Size and gene content of Q-CMV RNAs. The exact location of the in vitro translation products of RNAs 1 and 2 of M_r 95,000 and 110,000, respectively, are not known. Further details on the in vivo translation products of RNA 3, the 3A protein, and of RNA 4, the coat protein (C.P.), are given in Fig. 2. The black box at the end of each RNA is the conserved region of about 265 residues. All RNAs contain a 5'-m^7G cap.*

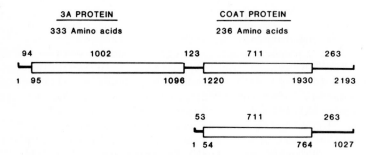

FIGURE 2. *Structure and gene content of Q-CMV RNAs 3 and 4. The lengths in residues of the 5'- and 3'-untranslated regions and the intercistronic region of RNA 3 are given above the RNAs. Residue numbers from the 5'-end of each RNA are given below the RNA.*

RNAs has given a single polypeptide product for each of the
RNAs with molecular weights of 95,000, 110,000, 36,000 and
26,200 for RNAs 1 to 4 respectively (Schwinghamer and
Symons, 1975, 1977; Gould and Symons, 1982; Gordon *et al.*,
1982). The sequence determination of the 1027 residues of
RNA 4 has confirmed that this RNA is monocistronic coding
for the coat protein of 236 amino acids (M_r 26,200; Gould
and Symons, 1982). In the case of RNA 3, the sequence
determination of the 2,193 residues has confirmed that it is
di-cistronic with the 5'-terminal cistron coding for the 333
residues of the 3A protein (M_r 36,700) found on *in vitro*
translation (Gould and Symons, 1982). The 3'-terminal
cistron of RNA 3 was confirmed as that for the coat protein
as deduced from biological (Habili and Francki, 1974) and
hybridization data (Gould and Symons, 1977). The coat
protein of RNA 3 is therefore silent on *in vitro* translation
and presumably also *in vivo*. RNA 4 is taken as the
subgenomic mRNA for the CMV coat protein *in vivo*, although
unequivocal evidence is lacking. However, RNA 3 must serve
as a source of RNA 4 *in vivo* but the mechanism for the
generation of RNA 4 from RNA 3 is unknown.

The sequence studies on RNAs 1 and 2 are not yet
complete. However, it is known from hybridization analysis
that there is no significant sequence homology between RNA
1, 2 and 3 except for a remarkably conserved region of about
270 residues at the 3'-terminus of all RNAs (Gould and
Symons, 1977; Symons, 1979). A similar terminal sequence
homology is found for the RNAs of the related cucomovirus,
tomato aspermy virus (TAV; Wilson and Symons, 1981) and for
the four RNAs of brome mosaic virus (Ahlquist *et al.*,
1981). Less extensive homology was found for the four RNAs
of alfalfa mosaic virus (A1MV) (Gunn and Symons, 1980a).

There is an intriguing difference in the proportion of
RNA 1 and RNA 2 required to code for their *in vitro*
translation products of M_r 95,000 and 110,000, respectively
(Fig. 1). For RNA 2, 88% of the sequence is required but
only 65% for RNA 1; this suggests that a second cistron is
present at the 3'-end of RNA 1. Given that the presumed 5'-
and 3'-untranslated regions of RNA 1 plus the putative
intercistronic region total 400 residues, there are 1,000
residues unaccounted for in RNA 1 after allowing for the
2,600 residues required for the M_r 95,000 protein; this is
sufficient to code for another protein of M_r 35,000. The
sequence determination of RNA 1 will indicate if such
calculations are justified.

Another intriguing dilemma is that the full length
in vitro translation products of RNAs 1, 2 and 3 have never

been found *in vivo*. On the other hand, the coat protein of M_r 26,200 is readily detected in cowpea protoplasts infected with CMV after labelling with radioactive amino acids followed by gel electrophoresis of protein extracts (Gonda and Symons, 1979). It can even be detected in extracts of cucumber seedlings by one-dimensional gel electrophoresis followed by staining (unpublished). Numerous attempts to find the translation products of RNAs 1, 2 and 3 in CMV-infected cowpea protoplasts or cucumber seedlings by both one- and two-dimensional gel electrophoresis of protein extracts using various detection methods have been unsuccessful (Gonda and Symons, 1979; Gordon *et al.*, 1982; and unpublished data). Possible explanations for these failures are that these proteins are only formed in very small amounts and are thus difficult to detect or that we have yet to develop procedures to extract them from plant tissue in a soluble form. We must also consider the possibility that the primary translation products of RNAs 1, 2 or 3 are rapidly processed to smaller functional products.

III. REPLICATION OF CMV RNAS

A study of the enzyme system involved in the replication of CMV RNAs should hopefully elucidate the roles of viral gene products and of any host proteins involved. CMV infection of cucumber seedlings induces high levels of RNA dependent RNA polymerase activity (RNA replicase) in both soluble and particulate fractions; there is no activity detectable in healthy plants (Kumarasamy and Symons, 1979; Gill *et al.*, 1981) although low levels of RNA dependent RNA polymerase activity have been found in extracts of other healthy plants (e.g., Romaine and Zaitlin, 1978; Dorssers *et al.*, 1982). Both the soluble and particulate (after solubilization) forms of the enzyme have been extensively purified to produce very similar preparations (Kumarasamy and Symons, 1979; Gill *et al.*, 1981). The specific activities of the highly purified forms, using poly(C) as template, were about 900 units/mg protein (1 unit equals 1 nmol GMP incorporated into RNA per minute at 37°C), comparable with 800 units/mg for the 95% purified bacteriophage Qβ RNA replicase (Blumenthal, 1979).
Comparison of the highly purified soluble and particulate forms of the enzyme by SDS-polyacrylamide gel electrophoresis showed that they have the same polypeptide composition. Three polypeptides of M_r 110,000, 100,000 and

35,000 were not present in similarly purified control
extracts from healthy plants. More extensive
characterization of these enzyme preparations by affinity
chromatography showed that the RNA dependent RNA polymerase
activity copurified with the M_r 100,000 polypeptide (in some
preparations virtually no other polypeptides could be
detected) which indicated that it is the catalytic component
(Gordon *et al.*, 1982). The close similarity in size between
these three polypeptides and those of the *in vitro*
translation products of Q-CMV RNAs 2, 1 and 3 respectively,
led us to compare them in greater detail using two
experimental approaches.

   In vitro translation of genomal RNAs from two other
strains of CMV (P and T, Habili and Francki, 1974; Rao
et al., 1982) showed that the translation products of T-CMV
RNAs 1 and 3 were slightly larger on SDS-polyacrylamide gel
electrophoresis than the corresponding products of Q-CMV
RNAs while that of P-CMV RNA 2 was slightly smaller than
that of Q-CMV RNA 2. In contrast to these variations,
highly purified RNA replicase from particulate fractions of
cucumber plants infected with these three strains of CMV
showed no corresponding variation in the sizes of their
component polypeptides upon SDS-polyacrylamide gel
electrophoresis. This indicated that the M_r 110,000,
100,000 and 35,000 polypeptides in the highly purified
enzyme were not viral coded. This conclusion was supported
by peptide mapping studies using V8 protease from
Staphylococcus aureus, trypsin or cyanogen bromide cleavage
(Gordon *et al.*, 1982).

   The highly purified CMV RNA replicase showed little
template specificity. Thus, it copied, in addition to CMV
RNA, alfalfa mosaic virus RNA, tobacco mosaic virus RNA and
even yeast ribosomal RNA but at a reduced level relative to
the viral RNAs (Kumarasamy and Symons, 1979). More
recently, it has been shown to copy avocado sunblotch viroid
and the virusoid of subterranean clover mottle virus
(unpublished). Because of this lack of specificity and of
our long held view that the *in vivo* replication of CMV RNAs
occurs in the particulate fraction of infected plants (May
et al., 1970), we investigated the properties of the RNA
replicase in the well washed particulate fracton which was
the starting material for the solubilization and
purification of the particulate enzyme (Gill *et al.*,
1981). This fraction contains readily detectable amounts of
the four CMV RNAs and the CMV coat protein, both being
detected by staining after gel electrophoresis of the
appropriate extracts of the particulate fraction. When this

well washed fraction was incubated in the presence
of $\alpha-{}^{32}$P-NTPs, label was incorporated into the four full
length CMV RNAs. Because this incorporation was only
slightly inhibited by 10 μg/ml heparin, compared with 50%
inhibition of the solubilized enzyme by 50 ng/ml heparin
(unpublished), it was concluded that the particulate RNA
replicase was completing pre-existing chains rather than
re-initiating new ones. The particulate fraction did not
respond to exogenous RNAs.

In view of the above results, our current model is for
the replication of CMV RNAs to occur in a particulate
replication complex in which the host-coded M_r 100,000
protein, the synthesis of which is induced by CMV infection,
is the catalytic sub-unit. The replication complex also
contains the three viral coded proteins of M_r 110,000,
95,000 and 35,000 which are considered to regulate the
assembly and specificity of the replication complex. When
the RNA dependent RNA polymerase activity is solubilized by
treatment of the particulate fraction with $MgSO_4$ (Gill
et al., 1981), only host proteins, including the M_r 100,000
protein, are solubilized.

IV.   RELATIONSHIP OF RNA STRUCTURE TO REPLICATION

The 3′-termini of the four CMV RNAs are of interest
because of their presumed participation in the initiation of
RNA replication and their ability to be aminoacylated with
tyrosine by appropriate aminoacyl-tRNA synthetases (Kohl and
Hall, 1974). The approximately 125 residues nearest the
3′-terminus can form a tRNA-like structure (Symons, 1979) as
can also the RNAs of another cucomovirus, TAV (Wilson and
Symons, 1981) and of the bromoviruses (Gunn and Symons,
1980b; Ahlquist *et al.*, 1981). The cucumoviruses and
bromoviruses do not require coat protein for the initiation
of replication in contrast to another group of tripartite
plant viruses of which AlMV is the best characterized (van
Vloten-Doting and Jaspars, 1977). There is a strong binding
of coat protein to the 3′-terminal region of the AlMV RNAs
(Stoker *et al.*, 1980) which cannot be folded into a tRNA-
like structure (Gunn and Symons, 1980a). Hence, it seems
that the initation of replication in tripartite plant
viruses requires either a 3′-terminal tRNA-like structure
(as in the cucumo- and bromoviruses) or strong binding of
coat protein to the 3′-terminus as for AlMV (van
Vloten-Doting and Jaspars, 1977).

The functional significance of the ability of CMV RNAs to have a 3'-terminal tRNA-like structure and to be aminoacylatable by tyrosine *in vitro* is both intriguing and unresolved. Whether or not the CMV RNAs are aminoacylated *in vivo* is unknown. It is feasible that the initiation of replication is preceded by aminoacylation which then allows the RNA replicase to bind and hence RNA synthesis of the minus RNA strand to start. It is worth noting that the bacteriophage Qβ RNA replicase contains the host coded proteins Ts and Tu, the protein elongation factors, in addition to the ribosomal protein S1 and the viral coded catalytic subunit (Blumenthal and Carmichael, 1979). Although Qβ RNA is not aminoacylated (Hall, 1979), it is fascinating that the protein Tu, as a Tu-GTP complex, can bind aminoacyl-tRNA and transfer it to the ribosome. Perhaps one of the components of the replication complex in CMV-infected plants is an eukaryotic elongation factor involved in binding to tyrosyl-CMV RNAs with GTP being the first incorporated nucleotide.

In an attempt to define the sequence and structural features of the cucumovirus RNAs important in the initiation of replication, we have sequenced the 3'-terminal regions of the RNAs from four strains of CMV and of two strains of TAV (Symons, 1979; Gunn and Symons, 1980b; Wilson and Symons, 1981; Barker, Wilson and Symons, unpublished). About half (61) of the 125 residues forming the proposed tRNA-like structures are conserved amoung the four CMV RNA strains. These conserved residues are heavily concentrated in that part of the structure which corresponds to the aminoacyl and TΨ C stems of conventional tRNAs.

Qβ RNA replicase has, however, been found to bind with internal sites on Qβ RNA and only weakly with the 3'-terminus (Meyer *et al.*, 1981). Similarly, recognition sites other than the 3'-terminus may be involved in initiation of CMV RNA replication, especially since the subgenomic messenger RNA 4 also has a 3'-terminal tRNA-like structure and is not thought to be replicated independently (Gonda and Symons, 1979).

REFERENCES

Ahlquist, P., Dasgupta, R., and Kaesberg, P. (1981). *Cell* *23*, 183-189.
Blumenthal, T. (1979). *Methods Enzymol. 60*, 628-638.

380   *Robert H. Symons et al.*

Blumenthal, T., and Carmichael, G.G. (1979). *Ann. Rev. Biochem. 48*, 525-548.

Dorssers, L., Zabel, P., van der Meer, J., and van Kammen, A. (1982). *Virology 116*, 236-249.

Gill, D.S., Kumarasamy, R., and Symons, R.H. (1981). *Virology 113*, 1-8.

Gonda, T.J., and Symons, R.H. (1979). *J. Gen. Virol. 45*, 723-736.

Gordon, K.H.J., Gill, D.S., and Symons, R.H. (1982). *Virology*, in press.

Gould, A.R., and Symons, R.H. (1977). *Nucleic Acids Res. 4*, 3787-3802.

Gould, A.R., and Symons, R.H. (1982). *Eur. J. Biochem.*, in press.

Gunn, M.R., and Symons, R.H. (1980a). *FEBS Lett. 109*, 145-150.

Gunn, M.R., and Symons, R.H. (1980b). *FEBS Lett. 115*, 77-82.

Habili, N., and Francki, R.I.B. (1974). *Virology 61*, 443-449.

Hall, T.C. (1979). *Int. Rev. Cytol. 60*, 1-26.

Kumarasamy, R., and Symons, R.H. (1979). *Virology 96*, 622-632.

May, J.T., Gilliland, J.M., and Symons, R.H. (1970). *Virology 41*, 653-664.

Meyer, F., Weber, H., and Weissmann, C. (1981). *J. Mol. Biol. 153*, 631-660.

Peden, K.W.C., and Symons, R.H. (1973). *Virology 53*, 487-492.

Rao, A.L.N., Hatta, T., and Francki, R.I.B. (1982). *Virology 116*, 318-325.

Romaine, C.P., and Zaitlin, M. (1978). *Virology 86*, 241-253.

Schwinghamer, M.W., and Symons, R.H. (1975). *Virology 63*, 252-262.

Schwinghamer, M.W., and Symons, R.H. (1977). *Virology 79*, 88-108.

Stoker, K., Koper-Zwarthoff, E.C., Bol, J.F., and Jaspars, E.M.J. (1980). *FEBS Lett. 121*, 123-126.

Symons, R.H. (1979). *Nucleic Acids Res. 7*, 825-837.

Van Vloten-Doting, L., and Jaspars, E.M.J. (1977). *Comprehensive Virol. 11*, 1-53.

Wilson, P.A., and Symons, R.H. (1981). *Virology 112*, 342-345.

Index

shared, 77, 129
SV40, 347–350
Prothrombin, 13–16
Protein(s)
functional domains, 8
phosphorylated, 26
post-translational modification, 28,
316, 319
precursor(s), 3–9, 61–63, 316, 319,
321
three-dimensional structure,
359–363
Protoplast, plant, 242–244
Puff, DNA, 113–119
Pyrimidine dimers, 46

R

Rat
ATP-synthetase complex, 315–318
mitochondrial DNA, 325–332
phenylalanine hydroxylase, 27–28
Rearrangement
DNA, 29–30, 33–36, 41–44, 225
in genomes, 351
plasmid, 123–130
Reassociation kinetics, 255, 355
Recombination
immunoglobulin genes, 41
mammalian cells, 351–354
Relaxin, 21
Renin, 11–12
Repair, DNA, 45–54
Repeated sequence(s)
human Alu family, 14–15
mitochondrial DNA, 273–274,
279–289
SV40 genome, 348–350
transcript, 91–94
Reverse transcriptase, 369–370
Rhynchosciara americana, 113–119
Ribosomal protein, mitochondrial,
270, 275
Ribosomal RNA genes
lily, 239–240
mammalian mitochondria, 327–328
spinach chloroplast, 256
Xenopus 5S RNA, 83–89
Ribulose bisphosphate carboxylase,
229–236
RNA
capping at 5'end, 305–306

cloning of viral, 371–372
double stranded, 371–372
kinetoplast, 309
mitochondrial, 141–150, 303–306,
310
nuclear, 91–94
precursor cleavage, 37–38
from repeated sequence, 91–94
structure of viral, 373–380
3' termini, 33–38
translation of viral, 373–380
viral gene content, 374
RNA replicase, 376–378
Rotavirus, Simian-11, 371–372

S

Saccharomyces cerevisiae
ATPase, F_1 sector, 151–158
*CBP*1, *CBP*2 genes, 141–150
*FAS*1, *FAS*2 genes, 131–140
fatty acid synthetase, 131–140
fusion with *Kluyveromyces lactis*,
159–166
mitochondrial genes, 257–268,
269–277, 280, 308–311
nuclear genes in cytochrome *b*
expression, 141–150
petite mutants, 257–268, 279–289,
291–292
reversion of promoter deletion,
123–130
transformation, 113–119, 136–140,
146, 152, 293–302
Scale, chicken, 65–72
Secondary structure, nucleotide
sequences, 348–350
Seed proteins, 193–201, 203–211
Sequence comparison
DNA, 65–72, 206–208, 279–289,
325–332
protein, 157, 259–260
Sequence homology, mitochondrial
and maxicircle DNA, 310
Shrunken locus, 221
Simian-11 rotavirus, 371–372
Somatostatin, 95–98
Soybean, 193–201
Sperm, mouse, 355
Spinach, 247–254, 255–256
Submandibular gland, 11–12

3 4 5 6 7 8 9 0 1 2
A B C D E F G H I J